Quantitation and Mass Spectrometric Data of Drugs and Isotopically Labeled Analogs

Quantitation and Mass Spectrometric Data of Drugs and Isotopically Labeled Analogs

Ray H. Liu
Sheng-Meng Wang
Dennis V. Canfield

With the assistance of
Meng-Yan Wu and Bud-Gen Chen
Fooyin University

and

Robert J. Lewis and Roxane M. Ritter
U.S. FAA Civil Aerospace Medical Institute

CRC Press
Taylor & Francis Group
Boca Raton London New York

CRC Press is an imprint of the
Taylor & Francis Group, an **informa** business

CRC Press
Taylor & Francis Group
6000 Broken Sound Parkway NW, Suite 300
Boca Raton, FL 33487-2742

© 2010 by Taylor and Francis Group, LLC
CRC Press is an imprint of Taylor & Francis Group, an Informa business

No claim to original U.S. Government works

Printed in the United States of America on acid-free paper
10 9 8 7 6 5 4 3 2 1

International Standard Book Number: 978-1-4200-9497-8 (Hardback)

Visit the Taylor & Francis Web site at
http://www.taylorandfrancis.com

and the CRC Press Web site at
http://www.crcpress.com

Contents

PART ONE
ISOTOPICALLY LABELED ANALOG AS INTERNAL STANDARD
FOR DRUG QUANTITATION — METHODOLOGY

PART TWO
MASS SPECTRA OF COMMONLY ABUSED DRUGS AND
THEIR ISOTOPICALLY LABELED ANALOGS IN
VARIOUS DERIVATIZATION FORMS

PART THREE
CROSS-CONTRIBUTIONS OF ION INTENSITY BETWEEN
ANALYTES AND THEIR ISOTOPICALLY LABELED ANALOGS
IN VARIOUS DERIVATIZATION FORMS

Foreword

The whole is more than the sum of its parts. — Aristotle

The sum of all sums is eternity. — Lucretius

To say that mass spectrometric analysis of drugs in biological media is similar to archeology may be a bit of a stretch to some, but consider the parallels. The archeologist looks at fragments and sees patterns suggesting whole structures. A pottery shard becomes the intact vessel that in turn reveals cultural aspects of past generations. Likewise, when the forensic toxicologist is presented with a biological specimen, they perform an archeological "dig" for evidence of drug residues. Instead of a shovel or trowel, mass spectrometry becomes the tool for uncovering remains. Pattern analysis of the evidence, a technique used in virtually all fields of scientific endeavor, becomes essential in drug interpretation. Comparisons to standards of known purity are essential. Bodily processes frequently alter pharmaceutical products and illicit drugs to metabolites more suitable for elimination. The "remains" of biological analysis are the analytical report that identifies and provides quantitative information on what was present in the specimen.

The first and foremost goal of the analyst is to provide accurate and precise drug identifications and measurements. The power of chromatographic separation coupled with mass spectrometry allows this modern miracle to occur on drug residues that cannot be seen with the naked eye. The analytical report, thus, provides evidence of drug exposure based on what was present and identifiable and how much was present in the specimen. In many cases, the outcome of drug analysis is not a trivial issue and may be used in many circumstances such as guiding therapeutic outcome, accident, death and criminal investigations, and as a requirement in securing or continuing employment. Consequently, the analyst has to get it right! The results must be inconvertible. That is what this book is all about. One of the authors (RHL) and I have discussed the need for documentation of mass spectrometric data on drugs for many years. This work by the authors represents years of work compiling mass spectra of the many forms and derivatives of drugs and their metabolites and their isotopically labeled counterparts. This compilation should well serve those involved in drug analyses of biological specimens and those involved in interpretation of results.

Edward J. Cone, Ph.D.

Preface

The analysis of drugs and their metabolites in biological media are now expected to routinely achieve ±20% accuracy in the ng/mL concentration level. This is possible mainly because of the incorporation of the internal standard method, using isotopically labeled analogs of the analytes as the internal standards into the analytical protocols. The availability of various isotopically labeled analogs for a wide variety of drug analytes from commercial sources is also a helpful contributing factor.

Using isotopically labeled analogs of the analytes as the internal standards, the most important issue affecting the accuracy of the quantitation results and the achievable linear calibration range is the cross-contribution to the intensities of ions designating the analytes and their isotopically labeled analogs serving as the internal standards. Thus, the availability and the selection of quality ion-pairs designating the analytes and their isotopically labeled analogs (internal standards) are crucial matters. Quality ion-pairs come from careful selections of the isotopically labeled analogs to serve as the internal standards and the derivatization groups for the analyte/internal standard pairs that require chemical derivatization and amenable to chromatography-mass spectrometry methods of analysis.

With these understandings in mind, this book is prepared in three parts. Part One of this book includes two descriptive chapters illustrating crucial issues related to quantitative analysis using isotopically labeled analogs as the internal standards in the analytical protocols. Part Two of this book is a systematic compilation of full-scan mass spectra of drugs and their isotopically labeled analogs in various derivatization forms. Part Three of this book is a systematical compilation of cross-contribution data for ion-pairs, derived from various combinations of isotopically labeled analogs and chemical derivation groups that are potentially useful for designating the analytes and their internal standards. One hundred and three drugs along with 134 isotopically labeled analogs included in this study are grouped into 7 categories and accordingly presented in Parts Two and Three. Information included in these three parts should be of routine reference value to individuals and laboratories engaged in the analysis of drugs in biological media.

The preparation of this book was conceptualized during the summer of 1990 when one of the authors (RHL) was on an Intergovernmental Personnel Assignment serving as a visiting scientist at the U.S. Addiction Research Center's Laboratory of Chemistry and Drug Metabolism (Baltimore, Maryland), where Dr. Edward J. Cone then served as the Chief of the Laboratory. A major portion of the laboratory data was collected in 2004 under a contract (DTFAAC-04-C-00012) in the laboratory of Aeromedical Research Division, Civil Aerospace Medical Institute, U.S. Federal Aviation Administration (Oklahoma City, Oklahoma). Additional data collection, data preparation, and writing were completed at Fooyin University (Kaohsiung Hsien, Taiwan) with the support of a 3-year (2004–2007) grant from the Taiwanese National Science Council (NSC 93-2745-M-242-003-URD, NSC 94-2745-M-242-003-URD, NSC 95-2745-M-242-002-URD).

In addition to the financial supports mentioned above, the following colleagues have also made invaluable contributions to the completion of this book project: Chief Toxicologist Dr. Dong-Liang Lin of the Institute of Forensic Medicine (Taipei, Taiwan), Professor Dr. Wei-Tun Chang of Central Police University (Taoyuan Hsien, Taiwan), Principal Scientist Dr. Shiv Kumar of ISOTEC™

(Miamisburg, Ohio). We are also indebted to the skillful assistance provided by the following undergraduate students from Fooyin University: Meng-Jie Sie (2009), Yu-Shin Lan (2009), Chiung-Dan Chang (2007), Yi-Chun Chen (2007), and Chia-Ting Wang (2006).

Ray H. Liu, Ph.D.
Sheng-Meng Wang, Ph.D.
Dennis V. Canfield, Ph.D

About the Authors

Ray H. Liu received a law degree from Central Police University (then Central Police College, Taipei, Taiwan) and a Ph.D. degree in chemistry from Southern Illinois University (Carbondale, IL) in 1976. He is currently a professor in the Department of Medical Technology, Fooyin University (Kaohsiung Hsicn, Taiwan), and professor emeritus in the Department of Justice Sciences, University of Alabama at Birmingham.

Before pursuing his doctoral training in chemistry, Dr. Liu studied forensic science under the guidance of Professor Robert F. Borkenstein at Indiana University (Bloomington) and received internship training in Dr. Doug Lucas's laboratory (Centre of Forensic Sciences in Toronto, Canada). Dr. Liu has worked as an assistant professor at the University of Illinois at Chicago, as a chemist at the U.S. Environmental Protection Agency's Central Regional Laboratory (Chicago, IL), and as a center mass spectrometrist at the U.S. Department of Agriculture's Eastern Regional Research Center (Philadelphia, PA) and Southern Regional Research Center (New Orleans, LA). He was a faculty member at the University of Alabama at Birmingham for 20 years and retired in 2004 after serving for more than 10 years as the director of the University's Graduate Program in Forensic Science.

Dr. Liu's works have been mainly in the analytical aspects of drugs of abuse (criminalistics and toxicology), with a significant number of publications in the following subject matters: enantiomeric analysis, quantitation, correlation of immunoassay and GC-MS test results, specimen source differentiation, and development of analytical methodologies. He has authored (or co-authored) several books and book chapters; more than 100 articles in refereed journals; and approximately 150 presentations in scientific meetings. He is qualified by the New York State Department of Health to serve as a laboratory director in forensic toxicology and he has served as a technical director in a U.S. drug-testing laboratory that held major contracts with military, federal, local, and private institutions.

Dr. Liu has been an active member of the following professional organizations for more than (or close to) 30 years: the American Chemical Society, Sigma Xi—The Scientific Research Society, the American Academy of Forensic Sciences (fellow), and the American Society for Mass Spectrometry. He is also a member of the Society of Forensic Toxicologists and the American Society of Crime Laboratory Directors (academic affiliate). Dr. Liu consults with several governmental and nongovernmental agencies, including serving as a laboratory inspector for the U.S. and the Taiwanese workplace drug-testing laboratory certification programs. He is the editor-in-chief of *Forensic Science Review* (www.forensicsciencereview.com) and serves on the editorial boards of the following journals: *Journal of Forensic Sciences (1998–2008)*, *Journal of Analytical Toxicology*, *Journal of Food and Drug Analysis* (Taipei), *Forensic Toxicology* (Tokyo), *Forensic Science Journal* (Taoyuan, Taiwan), and *Fooyin Journal of Health Sciences* (Kaohsiung, Taiwan).

Sheng-Meng Wang received a B.S. degree in forensic science from Central Police University (Taoyuan, Taiwan) in 1988 and a Ph.D. degree in chemistry from National Tsing Hua University (Hsingchu, Taiwan) in 1997. Dr. Wang is currently professor of forensic science and director of scientific laboratories, Central Police University.

Dr. Wang has been a visiting associate professor at the Graduate Program in Forensic Science, University of Alabama at Birmingham, and conducted research at the U.S. Federal Aviation Administration's Civil Aerospace Medical Institute (Oklahoma City, OK). Dr. Wang has been working in various areas of forensic toxicology and his current research activities include: evaluation of various chemical derivatization approaches in the sample preparation process, application of solid-phase microextraction to the analysis of drugs in biological fluids, and the characterizations of drug depositions in various biological specimens.

Since 1988, Dr. Wang has been serving as a laboratory evaluator for the Drug Testing Laboratory Accreditation Program under the auspices of the (Taiwanese) National Bureau of Controlled Drugs. He has also been serving as the executive secretary for the Taiwan Academy of Forensic Science since 2006.

Dennis V. Canfield received a B.S. degree in biology from Lynchburg College (Lynchburg, VA) in 1971. He completed an M.S. degree in forensic science at John Jay College of Criminal Justice, City University of New York (New York, NY), in 1976. He earned a Ph.D. in forensic chemistry in 1988 at Northeastern University (Boston, MA). For the past 19 years, Dr. Canfield has been the manager of the Bioaeronautical Sciences Research Laboratory at the U.S. Federal Aviation Administration's Civil Aerospace Medical Institute (CAMI) in Oklahoma City, OK, conducting research into forensic toxicology, biochemistry, radiobiology, functional genomics, and bioinformatics.

Before joining CAMI, Dr. Canfield was a senior forensic chemist for the New Jersey State Police Crime Laboratory (Little Falls, NJ) for 5 years and worked as the director of forensic science at the University of Southern Mississippi (Hattiesburg, MS) for 10 years in a tenured associate professor position. Dr. Canfield has worked primarily in the areas of drug identification and toxicology, starting in 1971 at the New Jersey State Police Crime Laboratory, and has continued to the present. He has published numerous peer-reviewed articles on drug identification and toxicology and testified on numerous occasions in federal, state, and local courts as an expert in forensic science. Dr. Canfield has participated as an editor and author in *Selected Powder Diffraction Data for Forensic Materials*, and *Carbon Monoxide and Human Lethality: Fire and Non-Fire Studies*.

Dr. Canfield is a fellow in the American Academy of Forensic Sciences, a member of the Society of Forensic Toxicologists, Sigma XI Research Society, and the Executive Board of the National Safety Council's Committee on Alcohol and Other Drugs.

PART ONE

ISOTOPICALLY LABELED ANALOG AS INTERNAL STANDARD FOR DRUG QUANTITATION — METHODOLOGY

Chapter 1

Quantitation of Drug in Biological Specimen
— Isotopically Labeled Analog of the Analyte as Internal Standard —

INTRODUCTION

The detection of drugs and their metabolites (collectively referred to as *drugs* hereafter) in biological tissues and fluids (collectively referred to as *biological media* hereafter) has always been an important component in clinical diagnostic analysis, forensic testing, pharmacological research, and drug discovery study. With advances in *analytical instrumentation* and a greater understanding of metabolism, we can now analyze drugs at a much lower concentration that was previously undetectable. Recent emphasis on monitoring illegal drug use in the workplace calls for massive testing of urine specimens, which has inspired the development and significant advances in *specimen pretreatment* technologies.

Newer instrumentation, such as GC-MS/MS or LC-MS/MS, capable of providing greater *specificity* and *signal-to-noise* ratio, are advantageous for identifying unknown metabolites at low concentrations. On the other hand, robust GC-MS methods are routinely used under therapeutic drug monitoring, emergency room drug screening, and workplace drug testing settings, in which the drugs of interest have previously been well characterized and often present at higher levels.

Analytical instrumentation and specimen pretreatment technologies are "*hardware*" aspects of the analytical sciences; the development and implementation of complementary "*software*" components help reach the full potential made possible by hardware advances. For example, the development of the "internal standard" method [1,2], especially the adaptation of isotopically labeled analogs (ILAs) as the internal standards (ISs) [3,4], has greatly improved the accuracy in the quantitation of drugs in biological media. Developments related to the use of ILAs as the ISs for accurate quantitation are based on GC-MS technology and readily adapted into GC-MS/MS, LC-MS, and LC-MS/MS applications. While many GC-MS/MS, LC-MS, and LC-MS/MS studies utilize ILAs as ISs, they do not generate better quantitative results than GC-MS and we know none that was devoted to better understanding the methodology itself.

Significance of Accurate Quantitation

Recent government regulations in workplace drug testing activities include monitoring quantitative data [5]; thus, making quantitation an important aspect of quality control practices in the analysis of drugs in biological media. Furthermore, specific "cutoff" value has been adapted as one of the essential criteria for defining whether a specific test specimen is "positive" or "negative" for a targeted drug. Accurate quantitation has now become an essential part of the routine testing protocol; it has, in addition to being a scientific pursuit, evolved into a legal issue.

In many non-routine analytical settings, emphasis may be placed on the *detection of a drug at very low concentrations* and *interpretation of quantitative data with small inter-specimen drug concentration differences*. Furthermore, sample preparation approaches often result in a final aliquot with hundred- or thousand-folds concentration in drugs' content; raw analytical result derived from the measurement step are then multiplied by a factor (of two or three orders of magnitude), thereby grossly magnifying any inaccuracies embedded in the raw data. Thus, proper interpretation and utilization of analytical findings rely heavily on the accuracy of the raw analytical data. This is especially critical in circumstances where drugs are present at a very low concentration level and interpretations are based on small inter-specimen differences, e.g., in hair-related studies where the objectives are on:

a. Differentiating drugs derived from external contamination from incorporation through active ingestion [6];
b. Determining racial bias due to the drug incorporation process or drug recovery in the sample pretreatment step [7]; or
c. Assessing variation in susceptibility to environmental contaminations due to differences in race origin [8] or hair treatments [9].

Preferred Calibration Method

Accurate quantitation requires a proper calibration (standardization) procedure to fully account for artifacts derived from variations in specimen matrix, specimen preparation, and instrumental conditions. Three most commonly used calibration techniques are the *analytical* or *working curve*, *standard additions*, and *internal standard* methods [1,2].

Mass spectrometric methods have proven to be one of the most sensitive and specific methods for drug assay. In particular, selected ion monitoring (SIM) approach has been used for several decades to achieve better accuracy and precision in ion intensity measurements. This approach is still an integral part of the quantitation protocol involving various forms of mass spectrometric methods in where an internal standard method is used. A typical protocol involves monitoring several selected corresponding ions (referred to as "ion-pairs" hereafter) designating the targeted drug and the ILA adapted as the IS. One or several calibration standards, containing known amounts of the drug, are processed in parallel with test specimens throughout the entire analytical protocol. All calibration standards and test specimens are spiked with the same amount of the IS. Quantitation is achieved by comparing a selected drug-to-IS ion-pair intensity ratio observed in the *test specimen* with the same ratio observed in the *calibration standard(s)*.

With practically identical chemical property and mass spectrometric fragmentation characteristics, an ILA is a preferred IS because it offers the following advantages:

a. Errors derived from (i) incomplete recovery of the drug in the sample preparation process or (ii) varying gas chromatographic and mass spectrometric conditions are compensated for; and
b. The presence of interfering materials (or mechanisms) affecting the detection (or quantitation) of the drug will result in the absence of the IS in the final chromatogram [10] or altered response and ion intensity ratios [11]; thus, alerting the analyst to conduct further investigation.

I. INTERNAL STANDARD AND QUANTITATION IONS

Under low resolution measurement conditions, the intensities of ions designating the drug and the IS are representative of these compounds' concentrations only if the following conditions are met:

a. The ILA is *isotopically pure* (an extrinsic factor); and
b. An adequate number of the *labeling isotopes are positioned at appropriate locations* in the molecular framework, so that, after the fragmentation process, ions meeting the following requirements are present (an intrinsic factor): (i) with high-mass and significant intensities; (ii) retaining at least three labeling isotopes; and (iii) without (or with insignificant) *cross-contribution* or CC (*see* Section II in Chapter 2 for full description on this phenomenon) between the ions designating the drug and the IS.

A. Inadequate Isotopic Purity — An Extrinsic Factor

If the ILA is not manufactured with sufficient *isotopic purity*, the addition of the IS, especially when a high concentration of the ILA is used, will result in the observation of a significant amount of the drug in a truly negative specimen. For a truly positive specimen, the resulting quantitative data will include systematic errors.

This problem has been well illustrated by a benzoylecgonine (BZ) study [12] in which a high concentration of ILA IS (1,500 ng/mL BZ-d_3) was adapted.

At the time of the study, a high concentration of IS (BZ-d_3) was commonly used by laboratories engaged in testing workplace specimens. Since the concentration of BZ encountered in positive samples are typically high (>5,000 ng/mL), adapting a high IS concentration can minimize the following problems:

a. To reduce the intensities of the ions designating the analyte (BZ), solvent volume used to reconstitute the extract may be so large that the resulting IS become too dilute to generate adequate ion intensity for reliable quantitative determination; and

b. The contribution of the isotopic ions, derived from the naturally abundant ^{13}C-atoms in the analyte, to the intensity of ions designating the IS may become very significant when the concentration of the latter is disproportionally low.

This study [12] examined two lots of 0.1 mg/mL BZ-d_3 in methanol. With the addition of 4.5 μg BZ-d_3 IS into 3 mL of urine samples (corresponding to 1,500 ng/mL), followed by solid-phase extraction, derivatization, and concentration down to 100 μL for GC/MS analysis, ions designating BZ (m/z 331, 272, and 210) were observed in truly negative test specimens. For a negative specimen, the concentration (X) of the observed BZ caused by the addition of these two lots of IS were 7.080 and 28.99 ng/mL as calculated by **Equation 1-1**.

$$X / (1,500 - X) = \text{(Ion intensity of } m/z \ 210) / \text{(Ion intensity of } m/z \ 213) \quad (1-1)$$

(where ions m/z 210 and 213 were used to designate BZ and the IS.) These concentrations correspond to 0.472% and 1.87% impurity of BZ in these two lots of BZ-d_3 IS provided by that specific manufacturer in 1988.

Isotopically impure ISs also introduced systematic errors embedded in the quantitative data derived from positive specimens. Data shown in **Table 1-1** demonstrate the systematic errors exhibiting the following characteristics:

a. No error is introduced if the concentration of the BZ in the test specimen is at the exact level of the BZ in the calibration standard.

b. A higher apparent result will be observed if the concentration of the BZ in the test specimen is lower than the concentration of the BZ in the calibration standard, and vice versa.

c. The degree of the above deviations increases as the isotopic impurity in the adapted IS increases.

Table 1-1. Quantitation error as a function of the isotopic impurity level of the internal standard and the difference between the analyte concentration in the calibration standard and the test sample

Isotopic impurity[a,b]	Apparent concentration[a,c]	True concentration[a,c]	% Error
7.080	83.82	80.70	+3.87
	141.1	140.7	+0.284
	146.0	145.7	+0.206
	147.1	147.0	+0.0680
	147.3	147.2	+0.0679
	148.1	148.0	+0.0676
	252.1	256.9	−1.87
27.99	84.19	72.28	+16.5
	154.1	154.8	−0.453
	155.5	156.6	−0.702
	156.8	158.0	−0.759
	161.7	163.9	−1.34
	273.5	296.6	−7.79
	288.9	314.6	−8.17
	292.5	319.1	−8.34
	293.9	320.7	−8.36
	298.3	326.0	−8.50

[a] Concentration in ng/mL. The concentration of the IS is 1,500 ng/mL. The analyte's (benzoylecgonine) concentration in the calibration standard is 150 ng/mL.
[b] 7.080 and 27.99 ng/mL of benzoylecgonine are included in the 1,500 ng/mL of benzoylecgonine-d_3 IS.
[c] See the original reference [12] for the calculation of the apparent and true analyte concentrations.

B. Cross-Contribution Derived from Ion Fragmentation Mechanism — An Intrinsic Factor

Under typical GC-MS analytical conditions, the drug and the IS are chromatographically inadequately resolved; thus, a proposed ILA IS must generate at least one (preferably two or three) ions relatively free from CC by the drug. There must also be at least one ion designating the drug that is relatively free from CC by the proposed ILA IS. (Current practice requires at least three "interference-free" ions derived from the drug allowing monitoring two ion-intensity ratios as an important criterion for drug confirmation.)

To make this possible, the *labeling isotopes in the ILA must be positioned at appropriate locations* in the molecular framework, allowing the fragmentation process to generate a sufficient number of high-mass ions (with significant intensities) that (a) retain the labeling isotopes; and (b) will not interfere with the intensity measurement of ions derived from the drug. Otherwise, the [M + n] ion (derived from the drug) may, because of the naturally occurring isotope abundance, make a significant contribution to the intensity of the ion designating the ILA that corresponds to the [M] ion of the drug. ("M" is the mass of the ion derived from the drug and selected for monitoring; "n" is the nominal mass difference of the ions designating the drug and the ILA serving as the IS.)

If deuterium, as in most currently available commercial products, is used as the labeling isotope, a difference in three mass units (n = 3) between the drug and the ILA is sufficient under normal circumstances. (If the concentration of the analyte is disproportionally higher than the concentration of the IS included in the assay process, the intensity of the [M + 3] ion originated from the analyte may become significant enough to require an additional analysis using a diluted aliquot.)

Secobarbital/$^{13}C_4$-secobarbital (SB/$^{13}C_4$-SB) data shown in **Figure 1-1** [13] illustrate how CC (of the intensities of ions designating SB and the IS) affects the accuracy in quantitation. In this example, CCs between the first pair of ions (m/z 196/200) are so insignificant (*see* CC data shown in the legend of the figure) that the linearity of the "SB/IS ion-pair intensity ratio" versus "SB concentration" plot (Figure 1-1-a) extends through a wide analyte concentration range. On the other hand, CCs between the two ions in the second ion-pair (m/z 181/185) are much more significant. In this latter case, significant errors can occur if the ion-pair intensity ratio generated from the test specimen is used directly to determine the analyte's concentration using a linear calibration model. The error can become very serious if the drug's concentration in the test specimen is significantly higher or lower (Figure 1-1-b) than the drug concentration adapted in the calibration standard (*see* further discussion in the next section — Fitting Calibration Data).

II. FITTING CALIBRATION DATA

Series of ions, [M – H_n], are typically seen in the EI fragmentation process [14,15]. The [M – H_n] processes, the presence of the naturally abundant ^{2}H-atoms in the drug and the ^{1}H-atoms in the ^{2}H-labeled IS, the isotopic effect of the [M – H_n] processes [16,17], and varying conditions in each sample (test specimen or standard) prohibit quantitations based on direct comparison of intensities of ions derived from the drug and the corresponding ion of the IS. The effects of these phenomena are minimized by comparing the drug/ILA IS ion-pair intensity ratio observed in the test sample against those observed in one or a set of calibration standards.

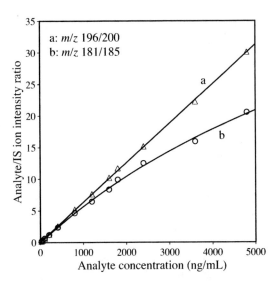

Figure 1-1. Fitting calibration data adapting linear (a) and hyperbolic (b) models using ion-pairs with different degrees of cross-contribution from the secobarbital/secobarbital-$^{13}C_4$ system. (a) m/z 196/200: 0.23% of the measured intensity of m/z 196 (designating secobarbital) is contributed by secobarbital-$^{13}C_4$; while 0.017% of the measured intensity of m/z 200 (designating secobarbital-$^{13}C_4$) is contributed by secobarbital. (b) m/z 181/185: 1.6% of the measured intensity of m/z 181 (designating secobarbital) is contributed by secobarbital-$^{13}C_4$, while 0.29% of the measured intensity of m/z 185 (designating secobarbital-$^{13}C_4$) is contributed by secobarbital [13].

A typical quantitative GC-MS protocol usually involves monitoring several selected ions from the drug and the ILA IS. Quantitation is achieved by comparing a selected drug-to-ILA ion-pair intensity ratio observed from the *test sample* against the same ratio observed from the *calibration standard*. The calibration standard contains the same amount of the IS (as those added to the test specimens) and a known amount of the drug, and is processed in parallel with the test specimens. The drug's concentration in the test specimen can be calculated using a one-point calibration approach as shown **Equation 1-2**.

The *one-point calibration* approach, in fact, is a *two-point linear calibration* method using only one empirical data point with the assumption that:

a. The drug-to-ILA ion-pair intensity ratio is zero when the drug's concentration in the test specimens or the standards is zero, i.e., the ILA IS will not contribute to the intensity of the ion monitored for the drug; and

b. The drug-to-ILA ion-pair intensity ratio will truly reflect the drug/ILA IS concentration ratio in the test specimens (and the standards), i.e., the drug will not contribute to the intensity of the ion monitored for the IS, and vice versa.

$$\frac{[\text{Int. of ion designating the analyte}) / (\text{Int. of ion designating IS})]_{\text{Test specimen}}}{[\text{Int. of ion designating the analyte}) / (\text{Int. of ion designating IS})]_{\text{Cal. standard}}} \times (\text{Analyte concentration})_{\text{Cal. standard}} \qquad (1\text{-}2)$$

In fact, these two assumptions are rarely valid for two reasons:

   a. The ILA IS often contains a small amount of isotopic impurity, i.e., the drug; and
   b. Ion fragmentation mechanisms often result in CC of the intensities of ions designating the drug and the IS.

Thus, drug concentrations derived from one-point calibration often include systematic errors as shown in Table 1-1 [12] and Figure 1-1-b (the m/z 181/185 ion-pair plot) [13]. The error is absent when the drug's concentration in the test specimen is the same as that in the calibration standard, but systematically increases as the drug's concentration in the test specimen is increasingly differently from that in the calibration standard.

Multiple-point linear calibration approaches, in which the observed drug-to-ILA IS ion-pair intensity ratios are plotted against the drug's concentration, are commonly used to extend quantitation to a wider concentration range. However, systematic errors still exist due to the inherently non-linear nature of the calibration curve. Basically, both of the abovementioned one-point and multiple-point approaches are based on linear models; thus, similar errors will be observed as long as the assumptions for the linear model are violated.

For calibration purposes, a "linear with zero intercept" relationship between the measured response and the drug's concentration is preferred. Thus, logarithmic-transformed ion-pair intensity ratio was proposed [18] to establish the standard curve, which was reportedly linear with an upper concentration of BZ up to 500,000 ng/mL.

Theoretical considerations [19] and various correction approaches [20–22] addressing the CC phenomenon have been reported. Our studies [23] have demonstrated that the hyperbolic model works well in cases where calibration data are derived from ion-pairs with significant CCs. This is clearly demonstrated in Figure 1-1-b (the m/z 181/185 ion-pair plot).

The effectiveness of the hyperbolic model is a result of its taking into account the CC phenomenon as shown in **Equation 1-3** [24].

$$
\begin{aligned}
y &= (xX + aA') / (xX' + aA); \\
&\quad (x + aA'/X) / (xX'/X + aA/X); \text{ or} \\
&\quad (x + C_1) / (C_2 x + C_3) \qquad (1\text{-}3)
\end{aligned}
$$

where
y = the observed ion-pair intensity ratio;
x = moles of the analyte in each standard;
X = intensity of the ion designating the analyte (generated by analyte);
a = moles of the IS in all standards;
A' = IS's cross-contribution to the the intensity of the ion designating the analyte;
X' = analyte's cross-contribution to the intensity of the ion designating the IS; and
A = intensity of the ion designating the IS (generated by IS).

Constant C_1 expresses the CC of the IS to the intensity of ion designating the drug; constant C_2 expresses the CC of the drug to the intensity of ion designating the IS; while constant C_3 reflects the moles of the IS used, the relative purity of the drug and the IS used in preparing the standard solutions, and the relative intensities of the ions designating the same amount of the drug and the IS.

Constant C_3 equals the concentration of the IS when both the drug and the IS are 100% pure (chemically and isotopically) with identical mass spectral responses (no isotopic effect). In the absence of CC between the drug and the IS, $C_1 = C_2 = 0$; thus, the relationship between the drug/IS ion intensity and concentration ratios reduces to a linear function (**Equation 1-4**).

$$y = Cx \qquad (1\text{-}4)$$

where $C = 1/C_3$.

III. ^{2}H- VERSUS ^{13}C-ANALOGS AS INTERNAL STANDARD

While evaluating the effectiveness of the $^{13}C_4$- and $^2H_5$-analogs of SB [25] and butalbital (BB) [26] in serving as the ISs, we have observed an interference phenomenon in cases where $^2H_5$-analogs of these two barbiturates were adapted. Specifically, *the intensity ratios of the ion-pairs designating these two drugs and their respective $^2H_5$-analogs increase as the volume of the solvent (ethyl acetate) used to reconstitute the extraction/derivatization residue increases.* This phenomenon was not observed when the respective $^{13}C_4$-analogs were used as the ISs in parallel experiments.

Since this interference phenomenon was observed only in the ^{2}H-labeled, but not in the ^{13}C-labeled systems, we do not believe the reported self-chemical ionization phenomenon [27,28] is the underlying cause. Since the ^{2}H-atoms in the SB-d$_5$ and BB-d$_5$ are placed at allylic positions, it was hypothesized that hydrogen/deuterium exchange has taken place at the ion source. This hypothesis was disproved [29] by the observation of the same phenomenon for drug/^{2}H-analog pairs with (such as SB, BB, and methohexital) and without (such as PB, and phenobarbital) this structural feature (**Figure 1-2, Table 1-2**).

Drug/^{13}C$_4$-analog systems differ from the corresponding drug/^{2}H$_5$-analog systems in displaying an identical retention time for the drugs and the ILA ISs. Thus, *retention time difference between the drug and the ^{2}H-analog IS* was hypothesized as the underlying factor causing the increase in the ion-pair intensity ratio observed for the drug/^{2}H-analog systems (but not for the drug/^{13}C-analog systems). To test this hypothesis, several series of experiments were performed, in which GC column temperature programming conditions were varied to modify the separation between the drug and the ^{2}H-analog IS [29]. The resulting drug-to-IS ion-pair intensity-ratio changes were characterized and evaluated.

SB/SB-^{13}C$_4$ system was again used as the control, the monitored ion-pair intensity ratio for this system remains constant as the reconstitution volume is increased and the temperature programming rate was changed from 30 to 15, and then to 5 °C/min. This was consistent with the hypothesis because the retention times for the drug and the ^{13}C-analog IS remain the same (no separation) regardless of the programming rate.

Data resulting from a series of parallel experiments for the SB/SB-d$_5$ system are shown in **Table 1-3**. Here, as the programming rate was reduced from 30 to 15, and then to 5 °C/min, the separation between the drug and the ^{2}H-analog IS increased, with the percentage of *m/z* 196 overlapped (by *m/z* 201) reducing from 89.5 to 77.7, and then to 70.2%. Under these three temperature-programming conditions and when the reconstitution volume was changed from 20 to 200 μL, the monitored ion-pair inten-

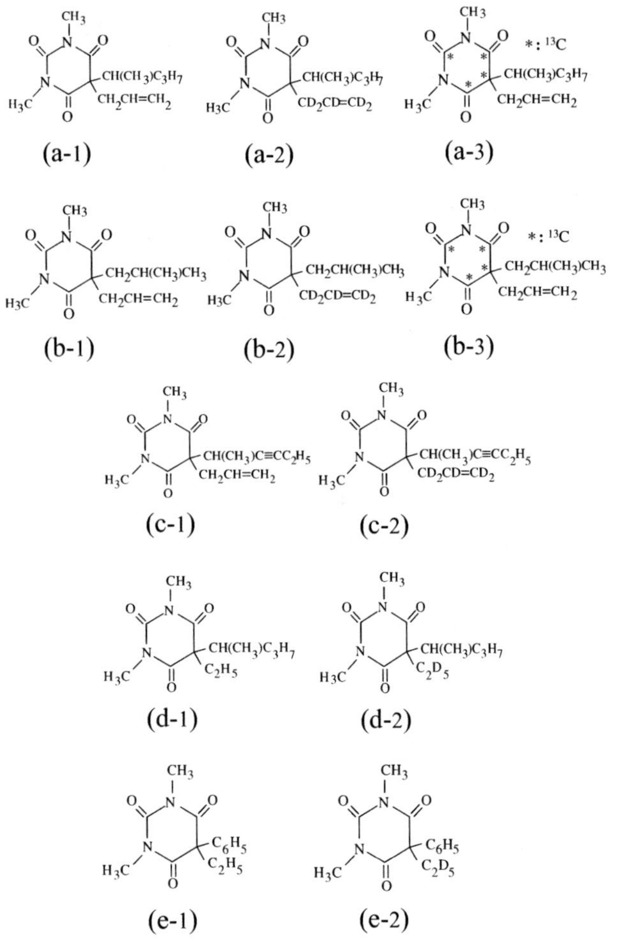

Figure 1-2. Structures of analytes/isotopically labeled analogs (all as methyl-derivatives): secobarbital/secobarbital-d$_5$/secobarbital-^{13}C$_4$ (a); butalbital/butalbital-d$_5$/butalbital-^{13}C$_4$ (b); methohexital/methohexital-d$_5$ (c); pentobarbital/pentobarbital-d$_5$ (d); phenobarbital/phenobarbital-d$_5$ (e).

sity ratio for the SB/SB-d$_5$ system changed 11.92%, 15.71%, and 18.35%, respectively.

Another series of experiments for the SB/SB-d$_5$ system were performed, in which the ^{2}H-analogous IS, rather than the drug, was the major component. In this latter case, as the programming rate was reduced from 30 to 15, and then to 5 °C/min, the separation between the drug and the IS similarly increases, with the percentage

Table 1-2. Analyte/isotopic analog ion-pair intensity ratio as a function of molecular abundance— Analyte: 2,500 ng/mL; isotopically labeled analog: 400 ng/mL

Reconstitute volume (μL)	Analyte/^{2}H$_5$-analog (*m/z*)					Analyte/^{13}C$_4$-analog (*m/z*)	
	Butalbital (196/201)	Secobarbital (196/201)	Methohexital (261/266)	Pentobarbital (184/189)	Phenobarbital (232/237)	Butalbital (196/200)	Secobarbital (196/200)
10	5.429	5.997	8.814	11.87	6.831	7.586	7.018
30	6.697	7.167	9.110	12.58	7.726	7.548	7.051
60	6.790	7.435	9.715	12.93	7.878	7.549	7.022
100	7.134	7.644	10.56	13.34	7.878	7.547	6.926
150	7.132	7.703	10.84	13.44	7.955	7.518	6.867

Table 1-3. Secobarbital/secobarbital-d$_5$ (SB/SB-d$_5$) ion-pair intensity ratio (*m/z* 196/201) as a function of molecular abundance under three temperature programming conditions resulting in different peak overlapping between SB and SB-d$_5$ — SB: 4,800 ng/mL; SB-d$_5$: 400 ng/mL

Rec. vol. (µL)	30 °C/min temperature ramp			15 °C/min temperature ramp			5 °C/min temperature ramp		
	Ion int. ratio	Ratio change (%)	Overlap[a] (%)	Ion int. ratio	Ratio change (%)	Overlap[a] (%)	Ion int. ratio	Ratio change (%)	Overlap[a] (%)
20	10.49		—[b]	10.82		—	11.50		—
30	10.69	1.91	97.3	11.32	4.62	80.5	11.89	3.39	70.0
40	11.03	5.15	—	11.67	7.86	—	12.35	7.39	—
60	10.99	4.77	83.4	11.90	9.98	78.0	12.50	8.70	71.8
80	11.41	8.77	—	11.89	9.89	—	12.87	11.91	—
120	11.39	8.58	89.6	12.35	14.14	79.1	12.99	12.96	73.2
160	11.62	10.77	—	12.38	14.42	—	13.36	16.17	—
200	11.74	11.92	87.6	12.52	15.71	73.2	13.61	18.35	66.0
Average			89.5			77.7			70.2

[a] Percentage of overlaps are calculated by dividing the area of *m/z* 196 that is overlapped with *m/z* 201 by the total peak area of *m/z* 196. Percentages of overlap with 30, 15, and 5 °C/min temperature ramps are approximately 89.5 ([97.3 + 83.4 + 89.6 + 87.6]/4), 77.7 ([80.5 + 78.0 + 79.1 + 73.2]/4), and 70.2 ([70.0 + 71.7 + 73.2 + 66.0]/4), respectively. Area calculations were done by rectangular summation method [30].

[b] Data not calculated.

of *m/z* 201 overlapped (by *m/z* 196) reducing from 100 to 94.3, and then to 62.4%. Under these three temperature programming conditions and when the reconstitution volume was changed from 20 to 200 µL, the monitored ion-pair intensity ratio for the SB/SB-d$_5$ system changed –7.65%, –14.2%, and –23.2%, respectively.

The phenomena observed from these three series of experiments are rationalized as follows:

- When two chromatographically closely-eluted compounds (such as drug/^{2}H-analog pairs) with their overlapping portions appearing at the ion source at the same time, the non-overlapping portions will have a higher ionization efficiency; thus, over-all ionization efficiency of the major component will be lower than that of the minor one.

- This difference in ionization efficiency between the major and the minor compounds becomes more significant when the total molecular population at the ion source is higher, i.e., with smaller reconstitution volume. This explains why, as the reconstitution volume is increased from 20 to 200 µL, the monitored ion-pair intensity ratios increase in Table 1-3 (SB as the major component), while decrease when SB-d$_5$ is the major component.

- As the drug and the ^{2}H-analog IS are *more closely eluted*, larger portions of these two compounds will appear at the ion source at the same time. Since these portions are proportionally affected by the decrease in their ionization efficiencies, the difference in the over-all ionization efficiency of these two compounds will decreases as they are more closely eluted. This explains why the rate of the changes (as the reconstitution volume is increased from 20 to 200 µL) in the monitored ion-

pair intensity ratio is much higher when the temperature-programming rate is decreased (drug and IS are better resolved).

The above reasonings are consistent with the observed peak overlapping data and ion-pair intensity ratio change characteristics shown in Table 1-3. They may also account for the reported interference on the quantitation of BZ caused by the coelution of fluconazole [11]. The authors attributed the observed "coeluting interference" to "saturation of the ionization chamber", but did not mention non-proportional variations in BZ/BZ-d$_5$ ionization efficiencies.

CONCLUDING REMARKS

Since the now well-known ion CC phenomenon exists in most (if not all) drug/ILA IS systems, calibration curves generated from these systems are likely non-linear. In systems where ion CC is absent, calibration curves generated from a drug/^{2}H-analog system is still inherently non-linear. This is due to the fact that the intensity ratio of an ion-pair (designating the drug/^{2}H-analog in a specific sample with the same concentration ratio) varies as their molecular abundances in the mass spectrometer ion source are changed. Variations can result from injecting different volume or injecting the same volume of a drug/^{2}H-analog mixture with different concentration but the same ratio. The non-linear characteristics of the calibration curve are also affected by the separation between the drug and its ^{2}H-analog IS, which is affected by the number and position of the ^{2}H-atoms placed in the molecular framework (an intrinsic factor) and the column temperature programming condition (an extrinsic factor). Thus, for most accurate quantitations, non-linear approaches [23] should be seriously considered for establishing the calibration curve.

REFERENCES

1. Willard HH, Merritt LL Jr, Dean JA, Settle A Jr: *Instrumental Methods of Analysis*, 7th ed; Wadsworth Publishing: Belmont, CA, p. 32; 1988.
2. Krull I, Swartz M: Quantitation in method validation; *LC•GC* 16:1984; 1998.
3. De Leenheer AP, Lefevere MF, Lambert WE, Colinet ES: Isotope-dilution mass spectrometry in clinical chemistry; *Advances in Clinical Chemistry*, Vol 24; Academic Press: London, UK; pp 111–161; 1985.
4. Garland WA, Barbalas MP: Applications to analytical chemistry: an evaluation of stable isotopes in mass spectral drug assays; *J Clin Pharmacol* 26:412; 1986.
5. U.S. Department of Health and Human Services (Substance Abuse and Mental Health Services Administration): Mandatory guidelines for federal workplace drug testing programs; *Fed Reg* 73:71858; 2008.
6. Blank DL, Kidwell DA: Decontamination procedures for drugs of abuse in hair: are they sufficient? *Forensic Sci Int* 70:13; 1995.
7. Mieczkowski T, Newel R: An evaluation of racial bias in hair assays for cocaine: black and white arrestees compared; *Forensic Sci Int* 63:85; 1993.
8. Sellers JK: *The Effects of Hair Treatment on Cocaine Contamination from External Exposure* — Master's thesis; Univ. of Alabama at Birmingham: Birmingham, AL, 1994.
9. Cirimele V, Kintz P, Mangin P: Drug concentration in human hair after bleaching; *J Anal Toxicol* 19:331; 1995.
10. Brunk SD: False negative GC-MS assay for carboxy THC due to ibuprofen interference; *J Anal Toxicol* 12:290; 1988.
11. Wu AH, Ostheimer D, Cremese M, Forte E, Hill D: Characterization of drug interferences caused by coelution of substances in GC-MS confirmation of targeted drugs in full-scan and selected ion monitoring modes; *Clin Chem* 40:216; 1994.
12. Liu RH, Baugh LD, Allen EE, Salud SC, Fentress JG, Ghadha H, Walia AS: Isotopic analogue as the internal standard for quantitative determination of benzoylecgonine: concerns with isotopic purity and concentration level; *J Forensic Sci* 34:986; 1989.
13. Liu RH, Lin T-L, Chang W-T, Liu C, Tsay W-I, Li J-H, Kuo T-L: Isotopically labeled analogues for drug quantitation; *Anal Chem* 74:618A; 2002.
14. Peterson DW, Hayes JM: In Hercules DM, Hieftje GM, Snyder LR, Evenson MA (Eds): *Contemporary Topics in Analytical and Clinical Chemistry*, Vol 3; Plenium Press: New York; pp. 217–252; 1978.
15. MacCoss MJ, Toth MJ, Matthews DE: Evaluation and optimization of ion-current ratio measurements by selected-ion-monitoring mass spectrometry; *Anal Chem* 73:2976; 2001.
16. Low IA, Liu RH, Barker SA, Fish F, Settine RL, Piotrowski EG, Damert WC, Liu J-Y: Selected ion monitoring mass spectrometry: parameters affecting quantitative determination; *Biomed Mass Spectrom* 12:633; 1985.
17. Benz W: Accuracy of isotopic label calculations for spectra with a (molecular ion – hydrogen) peak; *Anal Chem* 52:248; 1980.
18. Corburt MR, Koves EM: Gas chromatography/mass spectrometry for the determination of cocaine and benzoylecgonine over a wide concentration range (<0.005–5 mg/dL) in postmortem blood; *J Forensic Sci* 39:136; 1994.
19. Pickup JF, McPherson L: Theoretical considerations in stable isotope dilution mass spectrometry for organic analysis; *Anal Chem* 48:1885; 1976.
20. Bush ED, Trager WF: Analysis of linear approaches to quantitative stable isotope methodology in mass spectrometry; *Biomed Mass Spectrom* 8:211; 1981.
21. Thorne GC, Gaskell SJ, Payne PA: Approaches to the improvement of quantitative precision in selected ion monitoring: high resolution applications; *Biomed Mass Spectrom* 11:415; 1984.
22. Barbalas MP, Garland WA: A computer program for the deconvolution of mass spectral peak abundance data from experiments using stable isotopes; *J Pharm Sci* 80:922; 1991.
23. Whiting TC, Liu RH, Chang W-T, Bodapati MR: Isotopic analogs as internal standards for quantitative analyses of drugs/metabolites by GC/MS — Non-linear calibration approaches; *J Anal Toxicol* 25:179; 2001.
24. Thorne GC, Gaskell S, Payne PA: Approaches to the improvement of quantitative precision in selected ion monitoring: high resolution applications; *Biomed Mass Spectrom* 11:415; 1984.
25. Chang W-T, Lin D-L, Low I-A, Liu RH: ^{13}C$_4$-Secobarbital as the internal standard for the quantitative determination of secobarbital — A critical evaluation; *J Forensic Sci* 45:659; 2000.
26. Chang W-T, Liu RH: Mechanistic studies on the use of 2H- and ^{13}C-analogs as internal standards in selected ion monitoring GC-MS quantitative determination — Butalbital example; *J Anal Toxicol* 25:659; 2001.
27. Derrick PJ: Isotope effects in fragmentation; *Mass Spectrom Rev* 2:285; 1983.
28. Derrick PJ: Dynamics of unimolecular ionic decompositions: intermolecular kinetic isotope effects; In Beynon JH, McGlashan ML (Eds): *Current Topics in Mass Spectrometry and Chemical Kinetics*; Heyden: London, UK; p. 61; 1982.
29. Chang W-T, Smith J, Liu RH: Isotopic analogs as internal standards for quantitative GC/MS analysis — Molecular abundance and retention time difference as interference factors; *J Forensic Sci* 47:873; 2002.
30. Hinney RT, Thomas GB Jr: *Calculus*; Addison-Wesley: Reading, MA; Chap 5; 1990.

Chapter 2

Isotopically Labeled Analog of the Analyte as Internal Standard for Drug Quantitation
— Chemical Derivatization and Data Collection and Evaluation —

INTRODUCTION

With *internal standard method* as the preferred calibration approach and an isotopically labeled analog (ILA) of the analyte as the internal standard (IS) of choice [1], the cross-contribution (CC) phenomenon—contribution of IS to the intensities of the ions designating the analyte, and vice versa—is undoubtedly the most significant interference factor in a quantitative determination protocol [2]. Successful quantitation of a drug in biological matrices relies on the availability of ion-pairs that are free of (or with minimal) CC for designating the analyte and the IS.

An ideal complementary ILA of the analyte serving as the IS, that can produce ion-pairs with the desirable characteristics, may be synthesized with full understanding of the analyte's ion fragmentation pathways and skillful positioning of the labeling isotopic atoms in the analyte's molecular framework. Desirable ion-pairs may also become available when the analyte/IS pair, possessing active functional groups, are derivatized with an appropriate derivatization group [3].

Sections included in this chapter illustrate the generation and selection of favorable ion-pairs using various derivatization groups and ILAs. Also illustrated are methods used to evaluate the quality of the ion-pairs of interest. *Full-scan mass spectra* of the analytes (and their ILAs) studied, in various derivatization forms, are compiled in **Part Two** (**Appendix One**, pp 31–371), while the *CC data* evaluated for potentially favorable ion-pairs are shown in **Part Three** (**Appendix Two**, pp 373–492) of this book. Information included in Parts Two and Three should be of valuable reference to those who are interested in the analysis of drugs in various forms.

I. CHEMICAL DERIVATIZATION

Chemical derivatization was traditionally incorporated into the sample preparation process to convert the analyte to a form that is more *compatible to the chromatographic environment. Creating or optimizing separation* and *enhancing detection and structural elucidation efficiency* have later become the reasons for practicing chemical derivatization prior to the instrumental measurement step [4]. The primary objective of chemical derivatization discussed in this chapter is the generation of favorable ion-pairs for designating the analyte and the ILA serving as the IS. Specifically, it is hoped that the chemically derivatized analyte and IS would produce ion-pairs that are free of (or with minimal) CCs between the intensities of ions designating the analyte and the IS.

A. Production of Most Favorable Ion-Pairs for Drug Quantitation

Approaches that may potentially help generate favorable ion-pairs for a specific analyte/ILA IS system include:

    a. Positioning the isotopic atoms at the most desirable positions in the molecular framework;

    b. Experimenting various chemical derivatization alternatives; and

    c. Selecting an alternate ionization method, such as chemical ionization.

All three approaches have their limitations. The *number and positions of the isotopic atoms* are determined in the manufacturing process and cannot be altered by the laboratory analyst, while positioning isotopic atoms in the analyte is constrained by the availability of practical synthesis routes. *Chemical derivatization* approach is limited by existing functional groups in the analyte and reagents that are available. For practical purposes, *electron impact* and *chemical ionization* are the only ionization

options. Chemical ionization procedure, typically yielding one ion for the designation of the drug and one for the IS, has been criticized for providing inferior discriminating power for definite identification under routine high throughput test environment [5].

In practice, the analyst can explore the best combination of ILA (serving as the IS) and derivatization options through a comprehensive evaluation process. This process includes the examination of empirical CC data of various derivatization products resulting from all combinations of available ILA and amenable derivatization alternatives.

B. Exemplar Studies

Fragmentation in the ion source of a mass spectrometer often includes $[M - H_n]$ processes resulting in series of "cluster ions", where n is the number of H-atom involved in the process. When this occurs, $[M - {}^2H_n]$ processes originated from a ${}^2H$-analog of the analyte are likely to generate ions that may contribute to the intensities of ions designating the analyte. Thus, ${}^{13}C$-labeled analogs may cause less interference of this nature. A very limited number of ${}^{13}C$-labeled drug analogs are now commercially available and two of them have been thoroughly studied along with their ${}^2H$-counterparts [6–8]. The resulting CC data, as determined by various procedures for the secobarbital/secobarbital-d_5 (SB/SB-d_5), secobarbital/secobarbital-${}^{13}C_4$ (SB/SB-${}^{13}C_4$) systems, indeed confirmed this rationale.

With limited availability of ${}^{13}C$-analogs, ${}^2H$-analogs are more commonly used as the ISs in routine analysis and related studies. For example, various ${}^2H$-labeled analogs of amphetamine and methamphetamine, with different number of ${}^2H$-atoms and positions, were evaluated for their suitability in serving as the ISs for the analysis of these two drugs [9]. Suitability is judged primarily by the availability of three ions that are *interference-free, retaining some of the drug's structural feature*, and with *relatively high mass and sufficient intensity*. Another study on this subject matter focused on evaluating the contribution (interference) of ions m/z 91 and 118, derived from the heptafluorobutyryl derivative of methamphetamine-d_5, -d_8, and -d_{11}, to the intensities of these ions designating the analyte [10].

Since the analyte's fragmentation pathways and the resulting ions may be altered by the attached derivatization group, derivatization agents can play an important role in generating more favorable ion-pairs. Methamphetamine and several ^{2}H-analogs were used as exemplar system to evaluate the effect of the derivatization factor [3]. This study concluded that ion-pairs with the most favorable CC varied with the ILA serving as the IS and the derivatization group selected in the assay protocol. Thus, if a specific derivatization procedure is preferred, a specific ILA that may generate the most favorable ion-pairs should be selected, and vice versa.

Studies cited above [3,6–10] focused on examining the extent of CC between the ions designating the analyte and the IS. Studies examining the *derivatization yield*, that may affect the achievable limits of detection and quantitation, have also been reported [11,12]. These are also important factors affecting the effectiveness of a quantitation protocol, but they are not within the scope of this chapter.

C. Isotopically Labeled Analogs and Chemical Derivatization Groups

Part Two (pp 31–371) of this book is a systematic collection of reference mass spectra representing seven categories of drugs and their ILAs in various forms of derivatives. Since it is the intention of this project to empirically examine the combinations of available ILA serving as the ISs and amenable derivatization routes for the generation of the most favorable ion-pairs, all drugs with commercially available ILAs and all reported derivatization alternatives for these drugs are within the scope of this project. However, for the reasons listed below, there are omissions within this scope: (a) certain ILAs that are available through sources unknown to the authors or become available after the conclusion of the laboratory work devoted to this project; (b) certain

derivatization routes that have been reported but were found to result in exceptionally low yield in our laboratories; and (c) inclusion of all derivatization groups in a series of analogs, such as methyl-, ethyl-, propyl-, and butyl-, may only add information of limited value.

All drugs and these drugs' ILAs included in this project were purchased from Cerilliant Corp. (Austin, TX) except those listed in **Table 2-1**. Chemical derivatization reagents used in this project and their sources are shown in **Table 2-2**. Shown in **Table 2-3** are the protocols adapted to generate various forms of derivatives included in the mass spectra (Part Two) and ion CC tables (Part Three) sections of this book. The names of the drugs appeared in the titles of the references cited in Table 2-3 may appear irrelevant, but the derivatization procedures described in those publications were adapted in this study.

II. ION INTENSITY CROSS-CONTRIBUTION DATA

Through its intrinsic ion fragmentation mechanisms, an ILA with adequate isotopic purity and mass difference from the analyte, meeting the requirements studied in Chapter 1, may still generate ions contributing to the intensities of ions designating the analyte, and vice versa. Understanding the fragmentation pathways of a compound serves well for selecting appropriate positions in where the labeling isotopes (^{2}H or ^{13}C) are placed in the ILA synthesis process. It does not, however, always help identify the sources of the cross-contributing ions. Cross-contributing ions may have low intensities and pathways leading to the generation of these ions are, in general, not well understood. Cross-contribution data for a specific analyte/ILA system requires empirical evaluation.

Table 2-1. Analytes and their isotopically labeled analogs obtained from sources other than Cerilliant Corp.

Source	Compound
Research Triangle Institute (Research Triangle Park, NC)	Cannabinol/cannabinol-d$_3$; tetrahydrocannabinol/tetrahydrocannabinol-d$_3$; THC-OH/ THC-OH-d$_3$; THC-COOH/THC-COOH-d$_3$/THC-COOH-d$_9$
Sigma-Aldrich Fine Chemicals (Saint Louis, MO)	Methylphenidate/methylphenidate-d$_3$; ritalinic acid/ritalinic acid-d$_5$;butalbital/butalbital-d$_5$/ butalbital-^{13}C$_4$; codeine/codeine-d$_3$/codeine-d$_6$/codeine-^{13}C$_4$; secobarbital/secobarbital-d$_5$/ secobarbital-^{13}C$_4$; norbuprenorphine
Alltech Associates (Deerfield, IL)	Amitriptyline/amitriptyline-d$_3$
Cambridge Isotope Laboratories (Andover, MA)	Chloramphenicol/chloramphenicol-d$_5$; clonidine/clonidine-d$_4$
Lipomed, Inc. (Cambridge, MA)	MDA/MDA-d$_5$; MDMA/MDMA-d$_5$; diazepam/diazepam-d$_3$/diazepam-d$_5$; flunitrazepam/ flunitrazepam-d$_3$/flunitrazepam-d$_7$; anhydroecgonine methyl/anhydroecgonine methyl-d$_3$; 6-acetylcodeine/6-acetylcodeine-d$_3$; dihydrocodeine/dihydrocodeine-d$_3$/ dihydrocodeine-d$_6$

Table 2-2. Derivatization group and related information

Derivatization group	Derivatization reagent and source[a]	Group attached	Formula wt.[b]
1. None (H)	None	None (Parent drug)	1.00794
2. Methyl	Iodomethane[1]	$-CH_3$	15.03452
3. Ethyl	Iodoethane[1]	$-C_2H_5$	29.06110
4. Propyl	Iodopropane[1]	$-C_3H_7$	43.08768
5. Butyl	Iodobutane[1]	$-C_4H_9$	57.11426
6. Pentafluoro-1-propoxy (PFPoxy)	2,2,3,3,3-Pentafluoro-1-propanol (PFP-OH)[2]	$-OCH_2(C_2F_5)$	149.03940
7. Hexafluoro-2-propoxy (HFPoxy)	1,1,1,3,3,3-Hexafluoro-2-propanol (HFP-OH)[2]	$-OCH(CF_3)_2$	167.02986
8. Acetyl	Acetic anhydride (AA)[3]	$-CO(CH_3)$	43.04462
9. Trichloroacetyl (TCA)	Trichloroacetic anhydride (TCAA)[2]	$-CO(CCl_3)$	146.37890
10. Trifluoroacetyl (TFA)	N-Methyl-bis(trifluoroacetamide) (MBTFAA)[2]	$-CO(CF_3)$	97.01601
11. Trifluoroacetyl (TFA)	Trifluoroacetic anhydride (TFAA)[2]	$-CO(CF_3)$	97.01601
12. Propionyl	Propionic anhydride (PA)[1]	$-CO(C_2H_5)$	57.07120
13. Pentafluoropropionyl (PFP)	Pentafluoropropionic anhydride (PFPA)[2]	$-CO(C_2F_5)$	147.02357
14. Heptafluorobutyryl (HFB)	Heptafluorobutyric anhydride (HFBA)[2]	$-CO(C_3F_7)$	197.03102
15. 4-Carboethoxyhexafluorobutyryl (4-CB)	4-Carboethoxyhexafluorobutyryl chloride (4-CBC)[4]	$-CO[C_3F_6CO(OC_2H_5)]$	251.10322
16. Pentafluorobenzoyl (PFB)	2,3,4,5,6-Pentafluorobenzoyl chloride (PFBC)[2]	$-CO(C_6F_5)$	195.06632
17. Propylformyl	N-Propyl chloroformate[2]	$-CO(OC_3H_7)$	87.09718
18. l-N-(Trifluoroacetyl)prolyl	(S)-(-)-N-(Trifluoroacetyl)prolyl chloride (l-TPCC)[5]		187.07564
a. l-TPC-l			
b. l-TPC-d			
19. l-α-Methoxy-α-trifluoromethyl-phenylacetyl	(–)-α-Methoxy-α-trifluoromethylphenylacetic acid (l-MTPAA)[5]	$-CO[C(CF_3)(C_6H_5)OCH_3]$	217.16453
a. l-MTMP-l			
b. l-MTMP-d			
20. Trimethylsilyl (TMS)	N-Methyl-N-trimethylsilyltrifluoroacetamide (MSTFA) with 1% trimethylchlorosilane (TMCS)[6]	$-Si(CH_3)_3$	73.18906
21. t-Butyldimethylsilyl (t-BDMS)	N-Methyl-N-(t-butyldimethylsilyl) trifluoroacetamide (MTBSTFA) with 1% t-butyldimethylchlorosilane (TBDMCS)[6]	$-Si(CH_3)_2C(CH_3)_3$	115.26880
22. Methoxyimino (MA)	Methoxyamine[7]	$=NOCH_3$	45.04066
23. Hydroxylimino (HA)	Hydroxylamine[7]	$=NOH$	31.01408
24. Methyl/trifluoroacetyl			
25. Ethyl/acetyl			
26. Ethyl/trimethylsilyl			
27. Ethyl/t-butyldimethylsilyl			
28. Propyl/trimethylsilyl			

For entry 18, the group attached is depicted as a structural diagram of an N-(trifluoroacetyl)prolyl group.

29. Butyl/trimethylsilyl
30. Pentafluoro-1-propoxy/pentafluoropropionyl
31. Hexafluoro-2-propoxy/trifluoroacetyl
32. Hexafluoro-2-propoxy/heptafluorobutyryl
33. Acetyl/trimethylsilyl
34. Acetyl/t-butyldimethylsilyl
35. Trifluoacetyl/trimethylsilyl
36. Trifluoacetyl/t-butyldimethylsilyl
37. Pentafluoropropionyl/trimethylsilyl
38. Pentafluoropropionyl/t-butyldimethylsilyl
39. Heptafluorobutyryl/trimethylsilyl
40. Heptafluorobutyryl/t-butyldimethylsilyl
41. Methoxyimino/ethyl
42. Methoxyimino/acetyl
43. Methoxyimino/propionyl

44. Methoxyimino/heptafluorobutyryl
45. Methoxyimino/trimethylsilyl
46. Methoxyimino/t-butyldimethylsilyl
47. Methoxyimino/ethyl/propionyl
48. Methoxyimino/ethyl/trimethylsilyl
49. Methoxyimino/ethyl/t-butyldimethylsilyl
50. Methoxyimino/acetyl/trimethylsilyl
51. Methoxyimino/propionyl/trimethylsilyl
52. Hydroxylimino/di-propionyl
53. Hydroxylimino/trimethylsilyl
54. Hydroxylimino/ethyl/propionyl
55. Hydroxylimino/di-ethyl/propionyl
56. Hydroxylimino/di-ethyl/trimethylsilyl
57. Hydroxylimino/di-ethyl/t-butyldimethylsilyl

[a] Sources of reagents: [1]Acros Organic (Fairlawn, NJ); [2]Acros Organic (Geel, Belgium); [3]Ajax Finechem (Seven Hills, Australia); [4]Harris Specialty Chemicals PCR (Gainesville, FL); [5]Sigma-Aldrich (St. Louis, MO); [6]Pierce Chemical (Rockford, IL); [7]Yakuri Pure Chemical (Osaka, Japan).

[b] All formula weights were calculated based on the information listed below: H: 1.00794; D: 2.01400; C: 12.0107; [13]C: 13.00335; N: 14.00674; O: 15.99940; F: 18.99840; Si: 28.08550; Cl: 35.4527.

Table 2-3. Procedure adapted for chemical derivatization (CD) of drugs and their isotopically labeled analogs

CD group[a]	Procedure[a]
	General step: To a 16 × 100-mm glass tube, add 5 µL analyte (1 mg/mL) or 50 µL internal standard (0.1 mg/mL). Evaporate the solvent to dryness under a stream of nitrogen at 50 °C.

Derivatization with one derivatization group

Methyl	To the dried analyte (*see* the procedure described in the "General step" entry), add 100 µL freshly prepared TMAH/DMSO (1:20) solution and, 2 min later, 100 µL iodomethane. Vortex-mixed briefly, and incubate for 10 min at 40 °C in a heating block. Add 2 mL 0.1-N NaOH and 2 mL *n*-hexane. Mix thoroughly and centrifuge at 1500 rpm. Isolate the organic phase by decanting after freezing the lower aqueous layer in liquid nitrogen. Dry the organic phase at 50 °C under nitrogen. Reconstitute with ethyl acetate for GC-MS analysis [13,14].
Ethyl	Same as "Methyl" procedure, except iodoethane was used as the derivatization reagent [13,15].
Propyl	Same as "Methyl" procedure, except iodopropane was used as the derivatization reagent [13,15].
Butyl	Same as "Methyl" procedure, except iodobutane was used as the derivatization reagent [13,15].
PFPoxy	To the dried analyte (*see* the procedure described in the "General step" entry), add 50 µL PFP-OH. Cap the tube, mix and incubate for 30 min at 60 °C in a dry heating block. Cool the mixture, then evaporate to dryness at 50 °C under nitrogen. Reconstitute with ethyl acetate for GC/MS analysis [16–21].
HFPoxy	Same as "PFPoxy" procedure, except HFP-OH was used as the derivatization reagent [16–21].
Acetyl	To the dried analyte (*see* the procedure described in the "General step" entry), add 50 µL AA and 200 µL pyridine. Cap the tube, mix, and incubate for 20 min at 80 °C in a heating block. Evaporate the solvent to dryness at 50 °C under nitrogen. Reconstitute with ethyl acetate for GC-MS analysis [13,15,22].
TCA	To the dried analyte (*see* the procedure described in the "General step" entry), add 150 µL 0.1 mg/mL dimethylaminopyridinein acetoneand 75 µL TCAA, then vortex-mixed for approximately 30 s. Wash the derivatized specimens with a solution of water, 1.5-M carbonate buffer (pH = 9.5), and 1-N NaOH (1:0.5:0.4 by volume). Cap the tubes, vortex-mixed for 30 s, then incubate for 45 min at 60 °C in a heating block. Cool the mixture, then centrifuge for 5 min. Freeze the lower aqueous phase and decant the organic phase to a clean tube and evaporate to dryness under a stream of nitrogen at 50 °C. The residue was reconstituted with ethyl acetate for GC-MS analysis [3,23].
TFA	To the dried analyte (*see* the procedure described in the "General step" entry), add 100 µL TFAA. Cap the tube, mix, and incubate for 20 min at 80 °C in a heating block. Evaporate the solvent to dryness at 50 °C under nitrogen. Cool the mixture for GC-MS analysis [9,15,23–26].
Propionyl	To the dried analyte (*see* the procedure described in the "General step" entry), add 30 µL PA. Cap the tube, mix, and incubate for 15 min at 56 °C in a heating block. Add 1 mL hexane/chloroform (3:1) and 100 µL 50% ammonium hydroxide. Mix thoroughly and centrifuge at 1500 rpm. Isolate the organic phase by decanting after freezing the lower aqueous layer in liquid nitrogen. Evaporate the solvent to dryness at 50 °C under nitrogen. Reconstitute with ethyl acetate for GC-MS analysis [15].
PFP	To the dried analyte (*see* the procedure described in the "General step" entry), add 100 µL PFPA. Cap the tube, mix, and incubate for 20 min at 80 °C in a heating block. Evaporate the solvent to dryness at 50 °C under nitrogen. Cool the mixture for GC-MS analysis [3,9,15,24,26–28].
HFB	To the dried analyte (*see* the procedure described in the "General step" entry), add 100 µL HFBA. Cap the tube, mix, and incubate for 20 min at 80 °C in a heating block. Add 1 mL hexane/chloroform (3:1), and 100 µL 50% ammonium hydroxide. Mix thoroughly and centrifuge at 1500 rpm. Isolate the organic phase by decanting after freezing the lower aqueous layer in liquid nitrogen. Evaporate the solvent to dryness at 50 °C under nitrogen. Reconstitute with ethyl acetate for GC-MS analysis [3,9,15,23–30].
PFB	To the dried analyte (*see* the procedure described in the "General step" entry), add 100 µL acetone and 100 µL PFBC and 1 mL 1-chlorobutane. Cap the tube, mix, and incubate for 30 min at 80 °C in a heating block. Evaporate the solvent to dryness at 50 °C under nitrogen. Reconstitute with ethyl acetate for GC-MS analysis [27,28].
4-CB	To the dried analyte (*see* the procedure described in the "General step" entry), add 200 µL of 4-CBC solution (1:100 in 1-chlorobutane, v/v, prepared fresh daily). Cap the mixture, vortex-mixed, and incubate for 30 min at 50–60 °C. Convert excess acid chloride to diethyl ester by adding 0.3 mL of anhydrous ethanol, vortex-mixed, and incubate for 30 min at 50–60 °C. Carefully evaporate to dryness (to avoid evaporative loss of derivatives). Reconstitute the residue with ethyl acetate for GC-MS analysis [9,31].
Propyl-formyl	To the dried analyte (*see* the procedure described in the "General step" entry), add 100 µL propyl chloroformate. Cap the tube, mix, and incubate for 20 min at 80 °C in a heating block. Evaporate the solvent to dryness at 50 °C under nitrogen. Cool the mixture for GC-MS analysis [23–26].
l-TPC	To the dried analyte (*see* the procedure described in the "General step" entry), add 0.5 mL saturated potassium carbonate.4 mL *n*-hexane and 50 µL *l*-TPCC. Mix thoroughly and centrifuge at 3000 rpm. Isolate the organic phase by decanting after freezing the lower aqueous layer in liquid nitrogen. Evaporate the solvent to dryness at 50 °C under nitrogen. Reconstitute with ethyl acetate for GC-MS analysis [22,29].

Table 2-3. (Continued)

CD group[a]	Procedure[a]
l-MTPA	To the dried analyte (*see* the procedure described in the "General step" entry), add 50 μL *N,N*-dicyclohexycarbodiimide and 100 μL *l*-MTPA. Cap the tube, mix, and incubate for 20 min at 70 °C in a heating block. Evaporate the solvent to dryness at 50 °C under nitrogen. Reconstitute with ethyl acetate for GC-MS analysis [22].
TMS	To the dried analyte (*see* the procedure described in the "General step" entry), add 50 μL MSTFA (with 1% TMCS). Cap the tube, mix, and incubate for 20 min at 90 °C in a heating block. Cool the mixture, then add ethyl acetate for GCMS analysis [3,6,15,32–38].
t-BDMS	To the dried analyte (*see* the procedure described in the "General step" entry), add 50 μL acetonitrile and 50 μL MTBSTFA (with 1% TBDMCS). Cap the tube, mix, and incubate for 20 min at 90 °C in a heating block. Cool the mixture for GC-MS analysis [3,15,23,38,39].
Methoxyimino	To the dried analyte (*see* the procedure described in the "General step" entry), add 30 μL pyridine containing 2% methoxyamine HCl. Keep the tube at room temperature for 15 min. Evaporate the solvent to dryness under a stream of nitrogen at 50 °C. Reconstitute with ethyl acetate for GC-MS analysis [15,37,40–45].
Hydroxylimino	To the dried analyte (*see* the procedure described in the "General step" entry), add 2 mL 0.1-M acetate buffer (pH = 4.5) and 0.5 mL 10% hydroxylamine. Cap the tube, vortex, and incubate at 60 °C with in a heating block for 1 hr. Add 800 μL NaHCO$_3$ to extract for 10 min and add 3 mL methyl dichloride. Mix thoroughly and centrifuge at 1500 rpm. Isolate the organic phase by decanting after freezing the lower aqueous layer in liquid nitrogen [35,36].

Derivatization with more than one derivatization group

Ethyl/acetyl	Follow the "Ethyl" procedure to the point where the product is dried, then proceed with the "Acetyl" procedure [13,15,22].
Ethyl/TMS	Follow the "Ethyl" procedure to the point where the product is dried, then proceed with the "TMS" procedure [3,13,15,32–38].
Ethyl/*t*-BDMS	Follow the "Ethyl" procedure to the point where the product is dried, then proceed with the "*t*-BDMS" procedure [3,13,15,23,38,39].
Propyl/TMS	Follow the "Propyl" procedure to the point where the product is dried, then proceed with the "TMS" procedure [3,6,13,15,32–38].
Butyl/TMS	Follow the "Butyl" procedure to the point where the product is dried, then proceed with the "TMS" procedure [3,6,13,15,32–38].
Acetyl/TMS	Follow the "Acetyl" procedure to the point where the product is dried, then proceed with the "TMS" procedure [3,6,13,15,16,18,32–38].
Acetyl/*t*-BDMS	Follow the "Acetyl" procedure to the point where the product is dried, then proceed with the "*t*-BDMS" procedure [3,13,15–23,38,39].
TFA/methyl	Follow the "TFA" procedure to the point where the product is dried, then proceed with the "Methyl" procedure [9,13–15,23–26,46,47].
TFA/HFPoxy	To the dried analyte (see the procedure described in the "General step" entry), add 50 μL TFAA and 50 μL HFP-OH. Cap the tube, mix and incubate for 30 min at 60 °C in a dry heating block. Cool the mixture, then evaporate to dryness at 50 °C under nitrogen. Reconstitute with ethyl acetate for GC/MS analysis [9,15,17–20,23–26,47].
TFA/TMS	Follow the "TFA" procedure to the point where the product is dried, then proceed with the "TMS" procedure [3,6,7,9,15,23–26,32–38].
TFA/*t*-BDMS	Follow the "TFA" procedure to the point where the product is dried, then proceed with the "*t*-BDMS" procedure [3,9,15,23–26,39].
PFP/PFPoxy	Same as "TFA/HFPoxy" procedure, except PFPA and PFP-OH were used as the derivatization reagents [3,9,15–20,24,26–28,47].
PFP/TMS	Follow the "PFP" procedure to the point where the product is dried, then proceed with the "TMS" procedure [3,6,9,15,21,24,26–28,32–37].
PFP/*t*-BDMS	Follow the "PFP" procedure to the point where the product is dried, then proceed with the "*t*-BDMS" procedure [3,9,15,21,23,24,26–29,38].
HFB/HFPoxy	Same as "TFA/HFPoxy" procedure, except HFBA and HFP-OH were used as the derivatization reagents [3,9,15,20,23–26,45,47].
HFB/TMS	Follow the "HFB" procedure to the point where the product is dried, then proceed with the "TMS" procedure [3,6,9,14,15,21,23–30,32–39,48].
HFB/*t*-BDMS	Follow the "HFB" procedure to the point where the product is dried, then proceed with the "*t*-BDMS" procedure [3,5,9,15,21,23–30,38].

Table 2-3. (Continued)

CD group[a]	Procedure[a]
Methoxyimino/ethyl	Follow the "Methoxyimino" procedure to the point where the product is dried, then proceed with the "Ethyl" procedure[13,15,37,40–45].
Methoxyimino/ethyl/propionyl	Follow the "Methoxyimino/ethyl" procedure to the point where the product is dried, then proceed with the "Propionyl" procedure [13,15,37,40–45].
Methoxyimino/ethyl/TMS	Follow the "Methoxyimino/ethyl" procedure to the point where the product is dried, then proceed with the "TMS" procedure [3,6,13,15,32–38,40–45].
Methoxyimino/acetyl/TMS	Follow the "Methoxyimino" procedure to the point where the product is dried, then proceed with the "Acetyl/TMS" procedure [3,6,13,15,22,32–38,40–45].
Methoxyimino/propionyl	Follow the "Methoxyimino" procedure to the point where the product is dried, then proceed with the "Propionyl" procedure [13,15,37,40–45].
Methoxyimino/propinoyl/TMS	Follow the "Methoxyimino/propionyl" procedure to the point where the product is dried, then proceed with the "TMS" procedure [3,6,13,15,32–38,40–45].
Methoxyimino/HFB	Follow the "Methoxyimino" procedure to the point where the product is dried, then proceed with the "HFB" procedure [3,9,15,23–30,37,40–45].
Methoxyimino/TMS	Follow the "Methoxyimino" procedure to the point where the product is dried, then proceed with the "TMS" procedure [3,6,15,32–38,40–45].
Methoxyimino/t-BDMS	Follow the "Methoxyimino" procedure to the point where the product is dried, then proceed with the "t-BDMS" procedure [3,15,23,33,37–45].
Hydroxylimino/ethyl	Follow the "Hydroxylimino" procedure to completion, then proceed with the "Ethyl" procedure [13,15,35,36].
Hydroxylimino/ethyl/propionyl	Follow the "Hydroxylimino/ethyl" procedure to the point where the product is dried, then proceed with the "Propionyl" procedure [13,15,35,36].
Hydroxylimino/ethyl/TMS	Follow the "Hydroxylimino/ethyl" procedure to the point where the product is dried, then proceed with the "TMS" procedure [3,6,13,15,32–38,42].
Hydroxylimino/propionyl	Follow the "Hydroxylimino" procedure to completion, then proceed with the "Propionyl" procedure [15,35,36].
Hydroxylimino/propionyl/TMS	Follow the "Hydroxylimino/propionyl" procedure to the point where the product is dried, then proceed with the "TMS" procedure [6,15,32–38].
Hydroxylimino/HFB	Follow the "Hydroxylimino" procedure to completion, then proceed with the "HFB" procedure [3,9,23–30, 35,36].
Hydroxylimino/TMS	Follow the "Hydroxylimino" procedure to completion, then proceed with the "TMS" procedure [35,36].

[a] Abbreviations for derivatization groups and derivatization reagents are shown in Table 2-2.

The CC data between the corresponding ions designating SB and its ^{2}H- and ^{13}C-analogs (SB-d_5 and SB-$^{13}C_4$)—all as butyl derivatives—have been studied thoroughly [8]. Using the protocol established in that study [8], SB and SB-d_5—all as butyl derivatives—are adapted as the exemplar system, in the following sections, to illustrate the determination and evaluation of CC data for the ion-pairs designating the analytes and the ISs of interests.

A. Full-Scan Mass Spectra

Full-scan mass spectra of SB and SB-d_5, are shown in **Figure 2-1**. Also shown in Figure 2-1 is the full-scan mass spectrum of pentobarbital, serving as an IS in a later application to determine the intensities of ions resulting from SB and SB-d_5. Since ion intensity data of full-scan spectra are inherently less accurate, they are not used directly to derive CC information. Rather, these data are used to preliminarily select analogous ion-pairs that are apparently free of (or with minimal) CC between SB and SB-d_5. Ion-pairs thus selected for CC studies for the SB/SB-d_5 system (all as butyl derivatives) are m/z 207/212, 224/229, 279/284, 321/326, 263/268, and 350/355. The relative intensities of these ions in their respective full-scan mass spectra are shown in the second column of **Table 2-4**. Selected ion monitoring (SIM) protocols are then used to collect ion intensity data for these ions.

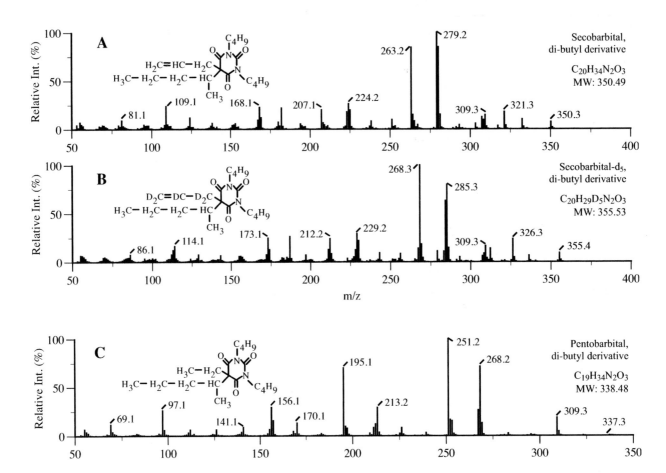

Figure 2-1. Mass spectra and structures of secobarbital (A); secobarbital-d₅ (B); and pentobarbital (C) — all as butyl-derivatives.

B. Selected Ion Monitoring and Calculation of Ion Cross-Contribution Data

1. Direct Measurement [8,13]

Raw SIM intensities of the selected ions for the SB/SB-d₅ system are shown in the third column of Table 2-4. Intuitively, contribution by the IS to the intensity of the ion designating the analyte (and vice versa) can be determined by a *direct measurement* method [13]. Specifically, the intensities of the ions designating the analyte and the IS are determined in two separate SIM runs; each run includes only a single and equal quantity of these two compounds. To determine the CC of the IS to the analyte, the intensity of the ion of interest (the ion designating the analyte) measured during the IS run is divided by the intensity of this ion measured during the analyte run. Similarly, to determine the CC of the analyte to the IS, the intensity of the ion of interest (the ion designating the IS) measured during the analyte run is divided by the intensity of this ion measured during the IS run.

Cross-contribution data derived from the *direct measurement* method [46] are listed inside parentheses following the SIM ion intensity data in the third column. For example, the SIM intensities (peak area) observed for *m/z* 207, an ion designating SB, were 925,096 and 38,264 when equal quantities of SB and SB-d₅ were analyzed. Thus, the CC of SB-d₅ to *m/z* 207 designating SB is 38,264/925,096 or 4.14%. Similarly, the contribution of SB to *m/z* 212 designating SB-d₅ is 38,464/744,493 or 5.17% (*see* the data shown in the lower section of the table).

Several factors may affect the accuracy of the CC data derived from the *direct measurement* method [46]. The *exact quantities* of the analyte and the ILA introduced into the GC/MS system, in two separate injections for ion intensity measurements, may not be exactly the same. This is due to variations in the following factors associated with the overall sample preparation process: (a) errors associated with pipetting process and the exact concentrations of the standards; (b) degrees of completeness in the chemical derivatization step; and (c)

Table 2-4. Selected ion monitoring (SIM) ion intensity and cross-contribution (inside parentheses in %) data for ions designating secobarbital (SB) and its $^2H_5$-analog (SB-d$_5$), all as butyl-derivatives — "Direct", "normalized direct", and "internal standard" methods

Ion (m/z)	Secobarbital run (raw data)		Secobarbital-d$_5$ run (raw data)		Normalizing the int. of m/z 355 to the int. of 350 (% CC by SB)[b]	Normalized based on the int. of m/z 251 (IS) in both runs (% CC by PB)[c,d]
	Full-scan (% int.)	SIM ion intensity (% CC by SB-d$_5$)[a,c]	Full-scan (% int.)	SIM ion intensity (% CC by SB)		
Ions designating secobarbital						
207	17.4	925,096 (4.14; 4.61; 4.44)	0.33	38,264	42,687	41,080
224	26.6	1,034,712 (3.02; 3.37; 3.24)	0.90	31,225	34,835	33,523
263	79.1	2,488,715 (0.24; 0.27; 0.26)	0.17	5,916	6,599	6,351
279	100	1,913,756 (15.4; 17.1; 16.5)	11.2	293,787	327,749	315,409
321	19.2	495,689 (7.02; 7.84; 7.54)	1.75	34,814	38,838	37,376
350	9.22	229,821 (0.00; 0.00; 0.00)	0.00	0	0	0
Ions designating secobarbital-d$_5$						
212	0.67	38,464	22.3	744,493 (5.17)	830,556 (4.63)	799,288 (4.81)
229	0.02	1,282	30.1	864,457 (0.15)	964,388 (0.13)	928,081 (0.14)
268	0.59	25,631	100	2,450,410 (1.05)	2,733,677 (0.94)	2,630,760 (0.97)
284	0.02	764	65.9	1,509,669 (0.051)	1,684,187 (0.045)	1,620,781 (0.047)
326	0.01	444	25.2	551,825 (0.080)	615,616 (0.072)	592,439 (0.075)
355	0.01	2,917	12.5	206,007 (1.42)	229,821 (1.27)	221,169 (1.32)

[a] The first, second, and third "% CC" data shown inside parentheses were derived, respectively, from raw data, normalizing the intensities of the molecular ion for SB-d$_5$ to that of SB, and normalizing the intensities of all ions for SB-d$_5$ based on the intensities of the selected ion (m/z 251) of the IS (pentobarbital) observed in both runs.

[b] The ion intensities of all ions were normalized based on the assumption that the intensity of the molecular ion (m/z 355) of SB-d$_5$ is the same as the molecular ion (m/z 350) of SB in two separate runs. New % CC data were calculated and included inside parentheses.

[c] SIM ion intensities of the IS (m/z 251) observed from the same amount of pentobarbital included in the SB and SB-d$_5$ runs were 3,965,624 and 3,693,741, respectively.

[d] These normalized SIM ion intensity data were obtained by multiplying the corresponding raw SIM ion intensity data by a factor of "3,965,624/ 3,693,741 = 1.0736". New % CC data were calculated and included inside parentheses.

losses due to adsorption and resolubitization. Secondly, the *GC/MS conditions* for measurement cannot be exactly reproduced in two separate experiments. Furthermore, the intensity of the "interfering" ion derived from the "interfering" isotopic analog is significantly lower than the intensity of the same ion derived from the isotopic analog accepting CC. Parameters, including threshold, peak width, peak position settings, that are required to generate the same level of accuracy for both low and high ion intensities, are not easily achievable [49]. Thus, other methods have also been studied for the derivation of CC data.

2. Normalized Direct Measurement [8,13]

The *normalized direct measurement* approach assumes that equimolar amounts of the isotopic analogs produce the same base-ion intensities. Thus, the intensities of ions observed from two separate experiments for the analyte/ ILA pair are first normalized with the assumption that the intensities of their respective molecular ions (m/z 350 and 355, in this case) are the same. These normalized data are

taken as true ion intensities generated by the same amount of the isotopic pair and used for the calculation of CC data.

Again, using the SB/SB-d$_5$ data shown in Table 2-4 as the example, the intensity of the molecular ion for SB (m/z 350) observed during the SIM run, was 229,821. On the other hand, the intensity for the corresponding ion (m/z 355) for SB-d$_5$, in the SB-d$_5$ run, was 206,007 (*see* data shown in the lower section of the table). For intensity normalization purpose, all ion intensity data derived from the SB-d$_5$ run were adjusted by a factor of 229,821/ 206,007 (or 1.1156) and shown in the 6th column in Table 2-4. For example, the normalized intensity for the ion m/z 207 collected during the run including only SB-d$_5$ is now 38,264 ✕ 1.1156 (or 42,687) and shown in the sixth column of the table. Thus, the CC of SB-d$_5$ to m/z 207 designating SB is 42,687/925,096 or 4.61% and shown in the third column inside the parentheses after the data derived from the direct measurement method. Similarly, the CC of SB to m/z 212 designating SB is 38,464/830,556 or 4.63% (*see* the data shown in the lower section of the table).

3. Internal Standard Method [8]

For the *internal standard* method, a set amount of a chromatographically resolved third compound is incorporated into the two separate experiments for the analyte/ILA pair to serve as the IS for ion intensity measurement. Variations in the GC-MS conditions and the sample preparation process are compensated for by normalizing the observed intensities of ions designating the analyte/ILA pair to the intensity of a selected ion derived from the IS.

The SB/SB-d_5 system was again adapted in this study using pentobarbital as the IS. Pentobarbital was selected as the IS for ion intensity measurement because of the similarity in the structural features and presumably, the resulting spectral and chromatographic characteristics.

For ion intensity normalization purpose, the intensity of the ion m/z 251 (designating pentobarbital) was adapted as the IS. Specifically, the intensities of ions observed in the SB-d_5 runs are multiplied by a correction factor. This correction factor is the ratio of the intensities of the ion (m/z 251) derived from the same amount of pentobarbital incorporated into the SB and the SB-d_5 runs. Since the intensities of m/z 251 observed in the SB and the SB-d_5 runs were 3,965,624 and 3,693,741, respectively (footnote c in Table 2-4), the correction factor is 3,965,624/3,693,741 (or 1.0736). Thus, the normalized intensity of the ion m/z 207 (designating SB) collected during the SB-d_5 run is: 38,264 × 1.0736 (or 41,080). Normalized data derived from the internal standard method are shown in the last column in Table 2-4. These data are then used for the calculation of CC data. For example, the CC of SB-d_5 to m/z 207 (designating SB) based on the normalized intensity data is: 41,080/925,096 or 4.44% and shown in the third column as the last entry inside the parentheses. Similarly, the CC of SB to m/z 212 (designating SB-d_5) is 38,464/799,288 or 4.81% (*see* the data shown in the lower section of the table).

4. Standard Addition Method

Again, data derived from the SB/SB-d_5 system (as butyl-derivatives) are used to illustrate the calculation of CC data by the *standard addition* method. The standard addition approach requires two sets of experiments for each analyte/ILA pair. Specifically, *two sets of experiments* required for evaluating the CC data between SB and SB-d_5 are: (a) the contribution of SB (interfering analog) to

the intensities of ions designating SB-d_5 (analog suffering interference); and (b) the contribution of SB-d_5 (interfering analog) to the intensities of ions designating SB (analog suffering interference).

Using set "(a)" experiments as an example, the purpose is to find out to what percentages the intensities of ions m/z 212, 229, 268, and 285 (ions designating SB-d_5—the analog suffering interference) are contributed by SB when the same amounts of SB and SB-d_5 are present. Since these ions are used to designate SB-d_5, their intensities derived from the presence of SB-d_5 are much higher, while their intensities due to the presence of the same amount of SB would be minimal if any. Furthermore, the intensity for each of these ions resulting from the presence of a set amount of SB-d_5 (or SB) will not be the same. Thus, the amount of each "addition" of the "standard" (SB-d_5) that is needed to make the optimal increase in the intensity (1/2 to 2 times of the original signal derived from 5 µg of SB) also varies with the ion (m/z 212, 229, 268, and 285) to be evaluated. For this specific example, the amounts of the analog suffering interference (SB-d_5) added in each of the four "addition" processes for each ion evaluated are shown in **Table 2-5**.

Data derived from set "(a)" experiments for the evaluation of CC data for m/z 212 are shown in **Table 2-6**. Intensity data resulting from these "addition" experiments were used in three different ways to estimate the intensities of these ions before the "addition"—the intensity of m/z 212 in 5 µg of SB. First, ion intensities observed from each series of "additions", which included the addition of none and four levels of SB-d_5, were used directly (Row "A" in Table 2-6). For the second method (Row "B" in Table 2-6), ion intensities were first normalized to the base-ion intensity (m/z 279) resulting from the presence of 5 µg SB in each run. For the third method (Row "C" in Table 2-6), ion intensities were first normalized to the intensity of an ion (m/z 251) resulting from a set amount of a third compound (pentobarbital)

Table 2-5. Quantities (µg) of SB-d_5 used (in the standard addition method) to determine SB's contribution to the intensities of ions (m/z) designating SB-d_5

Aliquot	Ion to be evaluated			
	212	229	268	285
1st	0	0	0	0
2nd	0.10	0.10	0.075	0.025
3rd	0.20	0.20	0.15	0.050
4th	0.30	0.30	0.25	0.075
5th	0.40	0.40	0.30	0.10

Table 2-6. Calculation of SB's contribution to *m/z* 212 (an ion designating SB-d$_5$) by the "standard addition" method using raw (A) and normalized (B and C) SIM ion intensity data

Method	Quantity (µg) of SB-d$_5$ added into 5 µg SB; Ion intensity observed with the addition of each aliquot					Linearity regression Equation	Coefficient	Calculated contribution
	0	0.10	0.20	0.30	0.40			
A	1.705 x 10^5	3.099 x 10^5	4.072 x 10^5	5.399 x 10^5	5.906 x 10^5	$y = 1077124x + 186834$	$r^2 = 0.9831$	3.47%
B	8.791 x 10^{-3}	1.431 x 10^{-2}	1.855 x 10^{-2}	2.778 x 10^{-2}	2.960 x 10^{-2}	$y = 0.05508x + 0.008787$	$r^2 = 0.9707$	3.19%
C	6.233 x 10^{-3}	1.012 x 10^{-2}	1.318 x 10^{-2}	1.896 x 10^{-2}	2.128 x 10^{-2}	$y = 0.03893x + 0.006169$	$r^2 = 0.9867$	3.17%

incorporated in each run. These normalized ion intensities data were then used for CC data evaluation for methods "B" and "C".

With the "standard addition" approach, the intensity (y) of the ion of interest observed in the test sample was plotted against the quantity (x) of the "standard" added. Least-square fit equation and correlation coefficient data for the exemplar set "(a)" experiments for ion *m/z* 212 are shown in the 3rd and the 2nd columns (from the right) in Table 2-6. The least-square fit equation was then used to calculate the equivalent quantity of the "standard" (SB-d$_5$) in the test sample (5 µg SB) prior to the addition of the "standard", i.e., the x value when y = 0 [50,51]. Resulting equivalent quantities of the "standard" are then divided by the quantity of the interfering compound (5 µg SB) and presented in percentage in the last column in Table 2-6. These are the percent CC data.

Data derived from these three versions of the standard addition method for ion *m/z* 212 in the exemplar set "(a)" experiments are plotted in **Figure 2-2**. The quality of the regression lines derived from methods "A' and "C" appear to be slightly better. This is also true for other data derived from the determination of CC data for others ions (data not shown).

C. Assessing the Accuracy of Empirically Determined Cross-Contribution Data [38]

Included in **Table 2-7** are summaries of CC data for ion-pairs that may potentially be used for designating SB and SB-d$_5$ in quantitative SIM GC/MS analytical protocols. These data were derived from *direct measurement, normalized direct measurement, internal standard*, and three variations of *standard addition* methods. All methods produce practically the same *order in magnitude* of CC data, among ion-pairs derived from each isotopic analog. Thus, all methods can be used to select the *best ion-pair within a selected analyte/ILA pair* for the intended quantitative analysis protocol [8]. However, which method would produce the most *accurate* CC data requires further investigation.

The six sets of CC data shown in Table 2-7 came from two sets of experiments. Specifically, the data derived from the direct measurement, normalized direct measurement and internal standard methods were obtained from one set of experiments, while those derived from the three versions of the standard addition method were obtained in another set of experiments. It is also noted that the three

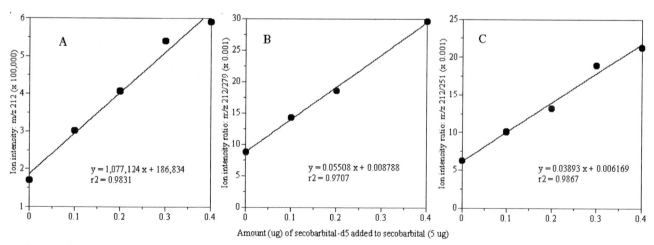

Figure 2-2. Evaluation of ion intensity derived from the interfering analog by standard addition method. Ion intensities adopted for calculation are based on: (A) *raw* intensity data; (B) data normalized to the *most intense ion of the interfering analog* and (C) data normalized to a *selected ion* (*m/z* 251) *derived from a reference compound* (pentobarbital).

Table 2-7. Cross-contribution (in %) to intensities of ions designating the analyte and its isotopic analog

Ion (m/z)	Direct	Normalized direct	Internal standard	Mean	Standard addition[a]			Mean
					A	B	C	
Ions designating secobarbital, but contributed by secobarbital-d₅								
207	4.14	4.63	4.43	4.40	0.85	1.07	0.89	0.94
224	3.02	3.38	3.23	3.21	2.70	2.64	2.72	2.69
263	0.24	0.27	0.25	0.25	0.072	0.046	0.049	0.056
280	2.14	2.41	2.20	2.25	1.55	1.57	1.49	1.54
321	7.02	7.87	7.51	7.47	—[b]	—	—	—
350	0.00	0.00	0.00	0.00	—	—	—	—
Ions designating secobarbital-d₅, but contributed by secobarbital								
212	4.90	4.34	4.76	4.67	3.47	3.19	3.17	3.28
229	0.19	0.16	0.18	0.18	0.068	0.049	0.070	0.062
268	1.22	1.08	1.19	1.16	0.13	0.15	0.16	0.15
285	0.062	0.055	0.060	0.059	0.030	0.094	0.086	0.070
326	0.035	0.031	0.034	0.033	—	—	—	—
355	0.59	0.52	0.57	0.56	—	—	—	—

[a] Ion intensity its used to derive the cross-contribution data were the observed value (for Method A), normalized to the base ion intensity of the analog receiving contribution (for Method B), and normalized to a selected ion (m/z 251) from a reference compound (pentobarbital) (for Method C).
[b] Not determined.

subsets of CC data derived from these two separate sets of experiments are similar. Thus, two sets of mean values are calculated and shown in columns 5 and 9 in Table 2-7.

A three-step process has been developed [32] to assess whether a set of empirically determined CC data for a specific ion-pair designating an analyte/IS system is accurate. Steps of this approach are first outlined, while details of each step will be further illustrated later: (a) a series of standard solutions are prepared and then analyzed to obtain a set of *empirically observed concentrations*; (b) the set of CC data (to be assessed) is used to derive a set of *theoretically calculated concentrations* for this set of standard solutions; and finally, (c) deviations of the empirically observed and the theoretically calculated concentrations from the expected concentrations of the set of standard solutions are compared. The closeness of these two sets of deviations is an indication of the accuracy of the set of CC evaluated.

The *empirically observed* concentrations of individual standards deviate from their respective true values as a result of the *true CC* imbedded in the adopted ion-pair designating the analyte and the IS. Whether the *theoretically calculated concentrations* would deviate from the respective true values, to the same extents as the empirically observed data, reflects the accuracy of the *empirically determined CC* values used in the calculation. Thus, if the set of CC data is accurate, deviations resulting

from the theoretically calculated data, using this set of CC, should coincide well with that derived from the empirically observed values (permitting random experimental errors). On the other hand, significant differences between these two sets of deviation data indicate existence of significant *random* and/or *systematic* errors in deriving this set of CC data under examination.

1. Empirically Observed Concentration

Data shown in Table 2-7 for the SB/SB-d₅ system are used as the example. First, a series of standard solutions containing 50–4000 ng/mL SB and 500 ng/mL SB-d₅ (as the IS) were prepared. Two sets of ion-pairs (m/z 207/212 and 263/268) with different levels of CC are adopted to illustrate (a) the effect of CC values on achievable linear range; and (b) how the accuracy of the empirically determined CC values is assessed. (It should be noted that, in a normal analytical protocol, the ion-pair with the most favorable CC, m/z 350/355 in this case, would be selected for quantitation purpose.)

As shown in Table 2-7, ion-pair m/z 263/268 exhibits minimal CC; thus, adopting this ion-pair for quantitation can generate high-quality data. Data are shown in the lower section of **Table 2-8** to serve as the "control", proving deviations resulting from the ion-pair m/z 207/212 are indeed caused by the significant CC imbedded in

Table 2-8. Comparison of empirically determined and theoretically calculated data derived from adapting ion-pairs with different levels of cross-contribution — Secobarbital/secobarbital-d$_5$ (as butyl-derivatives) example

| Theor. conc. | Empirically observed | | Theoretically calculated with CC derived from the mean of | | | |
| | | | Direct and normalized methods[a] | | Standard addition methods[b] | |
	Ion int. ratio	Observed conc. (% deviation)	Ion int. ratio	Calculated conc. (% deviation)	Ion int. ratio	Calculated conc. (% deviation)
m/z 207/212 (cross-contribution data: 4.40%/4.67%^{a}; 0.94%/3.28%^{b})						
50	0.1158	55.5 (+10.9)	0.1521	72.9 (+45.7)	0.1167	55.9 (+11.8)
80	0.1727	82.7 (+3.35)	0.2140	102.5 (+28.1)	0.1802	86.3 (+7.91)
100	0.2101	100.6 (+0.57)	0.2551	122.2 (+22.2)	0.2225	106.6 (+6.56)
200	0.4220	202.1 (+1.03)	0.4580	219.3 (+9.67)	0.4320	206.9 (+3.45)
300	0.6099	292.1 (−2.67)	0.6570	314.7 (+4.89)	0.6387	305.9 (+1.96)
500	1.044	500.0 (0.00)	1.044	500.0 (0.00)	1.044	500.0 (0.00)
800	1.688	808.5 (−1.03)	1.598	765.4 (−4.33)	1.632	781.8 (−2.28)
1,000	2.042	977.9 (−2.23)	1.951	934.4 (−6.56)	2.012	963.7 (−3.63)
1,300	2.648	1,268 (−2.49)	2.457	1,177 (−9.47)	2.564	1,228 (−5.54)
1,700	3.336	1,598 (−6.05)	3.093	1,481 (−12.9)	3.269	1,566 (−7.91)
2,000	3.822	1,831 (−8.50)	3.542	1,696 (−15.2)	3.775	1,808 (−9.60)
3,000	5.065	2,426 (−19.2)	4.892	2,343 (−21.9)	5.339	2,557 (−14.8)
4,000	6.507	3,116 (−22.1)	6.052	2,899 (−27.5)	6.737	3,226 (−19.3)
m/z 263/268 (cross-contribution data: 0.25%/1.16%^{a}; 0.056%/0.15%^{b})						
50	0.1050	50.2 (+0.46)	0.1080	51.7 (+3.33)	0.1052	50.3 (+0.63)
80	0.1684	80.6 (+0.74)	0.1711	81.8 (+2.31)	0.1679	80.3 (+0.43)
100	0.2096	100.3 (+0.31)	0.2131	101.9 (+1.95)	0.2097	100.3 (+0.34)
200	0.4173	199.7 (−0.17)	0.3423	202.2 (+1.08)	0.4187	200.3 (+0.17)
300	0.6201	296.7 (−1.10)	0.6310	301.9 (+0.63)	0.6276	300.3 (+0.097)
500	1.046	500.0 (0.00)	1.045	500.0 (0.00)	1.045	500.0 (0.00)
800	1.688	807.5 (+0.93)	1.659	793.7 (−0.78)	1.670	799.1 (−0.11)
1,000	2.043	977.4 (−2.26)	2.064	987.3 (−1.27)	2.086	998.2 (−0.18)
1,300	2.746	1,314 (+1.07)	2.663	1,274 (−1.97)	2.709	1,296 (−0.27)
1,700	3.578	1,712 (+0.71)	3.451	1,651 (−2.88)	3.589	1,693 (−0.39)
2,000	4.425	2,117 (+5.85)	4.032	1,929 (−3.54)	4.159	1,990 (−0.49)
3,000	6.367	3,047 (+1.55)	5.914	2,830 (−5.68)	6.220	2,976 (−0.79)
4,000	8.513	4,073 (+1.83)	7.715	3,692 (−7.71)	8.269	3,956 (−1.09)

^{a} Ion cross-contribution data used for theoretically calculation are means of the value derived from direct measurement, normalized direct measurement, and internal standard methods. The contribution of secobarbital-d$_5$ to the intensities of ion designating secobarbital were 4.40% for *m/z* 207 and 0.25% for *m/z* 263; the contribution of secobarbital to the intensities of ions designating secobarbital-d$_5$ were 4.67% for *m/z* 212 and 1.16% for *m/z* 268.

^{b} Ion cross-contribution data used for theoretically calculation are means of the value derived from standard addition methods. The contribution of secobarbital-d$_5$ to the intensities of ion designating secobarbital were 0.94% for *m/z* 207 and 0.056% for *m/z* 263; the contribution of secobarbital to the intensities of ions designating secobarbital-d$_5$ were 3.28% for *m/z* 212 and 0.15% for *m/z* 268.

ion intensity measurement. With significant CC, the ion-pair *m/z* 207/212 will generate quantitation data with significant deviations from their expected values; thus, making the following phenomena more apparent: (a) the effect of CC on achievable linear range and (b) whether the CCs of the ion-pair under evaluation are accurate.

The ion intensity ratios shown in the second column of Table 2-7 are the empirically observed values for the ion-pairs designating SB and SB-d$_5$. The concentrations shown in the third column are the empirically observed concentrations of these standard solutions based on the ratios shown in the second column, using the 500 ng/mL standard as the calibration standard.

The percentage figures shown inside parentheses in the third column are percentage deviations of the empirically observed from the true (or expected) concentrations. For the standard solutions containing 50 and 4000 ng/mL of SB, deviations of the empirically observed from the expected concentrations are much higher when the ion-pair *m/z* 207/212 are used to derive the quantitation data. This is an indication that the CC between SB and SB-d$_5$ is much more significant for the ion-pair *m/z* 207/212 than *m/z* 263/268. Thus, the achievable linear range would be able to reach a lower concentration level when the ion-pair *m/z* 263/268 is used to designate SB/SB-d$_5$.

2. Theoretically Calculated Concentration

As reported earlier [14], the CC phenomenon will cause the intensity ratio values, shown in the second column (Table 2-8), to deviate from a linear relationship when plotted against their respective concentrations. This non-linear relationship and the need for correction have also been emphasized by Duncan et al. [52].

A formula has been developed for calculating the theoretical ratio of a pair of ions designating the analyte and the IS [32]. This formula is further modified as shown in **Equation 2-1**.

The objective here is *to evaluate the accuracy of a set of CC data*. For this purpose, two sets of CC data for the ion-pair *m/z* 207/212 derived from the SB/SB-d$_5$ system (Table 2-7) are used as the example. The first set is the mean of the three values derived from the *direct measurement, normalized direct measurement*, and *internal standard* methods, while the another set is the mean of the three values derived from the three variations of the *standard addition* methods. For the first set, the CC of the SB-d$_5$ to SB and SB to SB-d$_5$ are 4.40% and 4.67%, respectively, while for the second set, the corresponding CC values are 0.94% and 3.28%.

Each set of the CC data is used to derive a set of theoretical ion intensity ratios for a series of standard solutions. The calculated intensity ratios were then used to derive the theoretically calculated concentrations for this series of standard solutions. With two sets of CC data, two sets of calculated ratio/concentration figures are derived. These two sets of data are shown in the fourth/fifth and the sixth/seventh columns in Table 2-8 as further described below.

Theoretically calculated concentrations are derived with the following stipulations and steps: (a) the intensities of the ions, designating the analyte and the IS, increase and decrease linearly with their concentrations; (b) the CC values (i.e., "analyte's contribution to the intensity of the ion designating the IS" and the "IS's contribution to the ion designating the analyte") as empirically determined, are applied to arrive a theoretical analyte/IS ion intensity ratio for a standard solution with a specific analyte concentration; and (c) the resulting theoretical analyte/IS intensity ratio is then used to derive the theoretical analyte concentration for that specific standard solution. Thus, *if the empirically determined CCs are inaccurate, the approach adapted in step "b" would embed a systematic error in the calculated concentrations*. This error would allow for assessing the trueness of the CC values as discussed in the next section.

With these stipulations and steps, a sample calculation, using the formula shown above and the ion-pair *m/z* 207/212 designating SB/SB-d$_5$ as the example, is shown below.

At 500 ng/mL, the average of the intensity ratio between *m/z* 207 (I_{207}) and *m/z* 212 (I_{212}) observed from repeated measurement is: $I_{207}/I_{212} = 1.0443/1$. Applying the first set of CC values (4.40% and 4.67%, mean values derived from the *direct measurement, normalized direct measurement*, and *internal standard* methods) into Equation 2-1, the theoretical ion intensity ratio (R_a) for the standard at 4,000 ng/mL is:

$$\frac{I_{207}}{I_{212}} = \frac{1.0443 \times [1 + \frac{(4{,}000 - 500)}{500} \times (1 - 0.044)]}{1 + [\frac{(4{,}000 - 500)}{500} \times 0.0467]} = 6.052$$

With this theoretically calculated ion intensity ratio, the resulting theoretically calculated concentration (X) of the analyte for the standard at 4,000 ng/mL can be calculated as follows:

$$1.0443 / 500 = 6.052 / X;$$
$$X = 6.052 \times 500 / 1.0443 = 2{,}899 \text{ ng/mL}$$

$$R_a = \frac{R_{cal} + \dfrac{(a - C_{cal})}{C_{cal}} \times (R_{cal} - R_{cal} \times x)}{1 + \dfrac{(a - C_{cal})}{C_{cal}} \times (1 \times y)} = \frac{R_{cal}\,[1 + \dfrac{(a - C_{cal})}{C_{cal}} \times (1 - x)]}{1 + \dfrac{(a - C_{cal})}{C_{cal}} \times y} \tag{2-1}$$

where

R_a = theoretically calculated anayte-to-IS intensity ratio when the analyte's concentration = a;

R_{cal} = anayte-to-IS intensity ratio when the analyte's concentration = the IS's concentration (as observed in the one-point calibration standard);

C_{cal} = concnetration of the anayte and the IS of the one-point calibration standard;

x = the contribution of the IS to the intensity of the ion designating the analyte (in %); and

y = the contribution of the analyte to the intensity of the ion designating the IS (in %).

Thus, the calculated concentration of the analyte is (2899 – 4000) / 4000, or –27.53%, lower than the expected value, 4,000 ng/mL. The theoretically calculated concentrations for the standards at other concentrations (and their deviations from their respectively expected values) are similarly calculated and placed in the fourth and fifth columns of Table 2-8.

The second set of CC values (0.94% and 3.28%, mean values derived from the three variations of the *standard addition* method) is also used to calculate the theoretical ion intensity ratios, then the resulting theoretical concentrations, for standards at various concentrations. The resulting data are shown in the sixth and the seventh columns of Table 2-8.

3. Comparing Empirically Observed and Theoretically Calculated Concentrations — Graphic Presentation

Shown in the third column of Table 2-8 are the empirically observed concentrations for the series of standard solutions under examination, while the theoretically calculated concentrations using two different sets of CC values are shown in the fifth and the seventh columns in the same table. Deviations (in %) of these concentrations from their respectively expected values are shown inside parentheses following the concentration data.

These deviations data shown in the third, fifth, and seventh columns (Table 2-8) are graphically presented in **Figure 2-3A** as lines "a", "b", and "c". Line "b" is signifi-

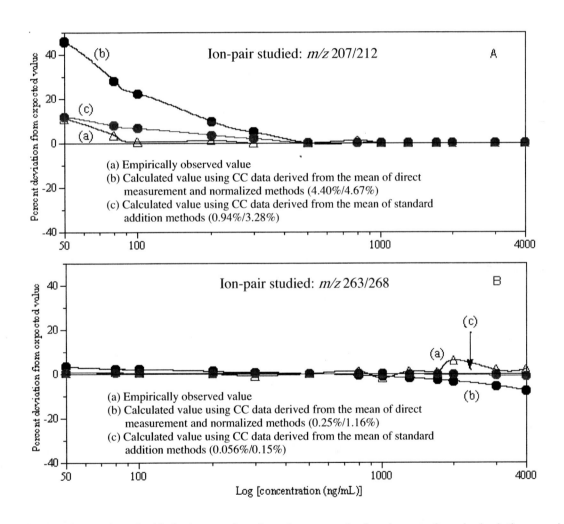

Figure 2-3. Deviations (in %) of secobarbital concentrations from the expected values in a set of standard solutions ranging from 50 to 4000 ng/mL: empirically observed (a) and theoretically calculated based on the mean of the cross-contribution data derived from *direct measurement, normalized direct measurement, and internal standard* methods (b); and means of the cross-contribution data derived from three version of the *standard addition* method (c). Part A (upper): Ions *m/z* 207 and 212 are adopted for designating secobarbital and secobarbital-d$_5$, respectively. For line "b", the cross-contributions of the IS to the analyte and the analyte to the IS are 4.40% and 4.67%, respectively. For line "c", the corresponding cross-contribution data are 0.94% and 3.28%, respectively. Part B (lower): Same as Part A, except the ions adopted for designating the analyte and the IS are *m/z* 263 and 268, respectively; the corresponding cross-contribution data are 0.25% and 1.16% for line "b" and 0.056% and 0.15% for line "c".

cantly different from line "a", especially at the lower concentration end. This is an indication that the 4.40% (contribution of SB-d_5 to the intensity of the ion, m/z 207, designating SB) used for theoretical calculation is too high. Similarly, the higher percentage of the theoretically calculated deviations at the higher concentration end can be attributed to the over-estimated contribution (4.67%) of SB to the intensity of the ion (m/z 212) designating SB-d_5. On the other hand, line "c" coincides well with line "a", indicating the second set of CC data (0.94% and 3.28%) is a good estimate of the true values.

Corresponding data for the ion-pair (m/z 263/268) with much smaller CC values are shown in Figure 2-3B. As expected, the empirically observed concentrations do not deviated significantly from the expected concentrations. Again, the deviations of line "b" from line "a" is more significant at the higher concentration end, indicating that the 1.16% (contribution of SB to the intensity of the ion, m/z 268, designating SB-d_5) used for theoretical calculation is too high.

4. Summary

In essence, the evaluation of a set of CC data constitutes the following steps: (a) preparing sets of standard solutions ranging from low to high analyte concentrations; (b) comparing the empirically observed concentrations against the expected concentrations for each of these standard solutions, then plotting "the difference (in %) between these two values at each concentration" against "the concentration of the standard"; (c) calculating the concentrations that should theoretically be "observed" when the CC factor is taking into account, then plotting "the difference (in %) between this theoretically observed concentration and the prepared concentration" against "the concentration of the standard"; and (d) Comparing the plots obtained in steps "(b)" and "(c)" to assess the accuracy of the CC data used in calculating the "theoretically observed concentration" used in step "(c)".

III. COMPILATION OF FULL-SCAN MASS SPECTRA AND ION INTENSITY CROSS-CONTRIBUTION TABLES

In addition to providing an overview on quantitation by the internal standard method using the analytes' ILAs as the ISs, the main objective of this book is the compilation of full-scan mass spectra and ion intensity CC data for all drugs with commercially available ILAs. Since the analysis of these drugs are accomplished by various derivatization approaches, the mass spectra and CC data are derived from various derivatives of these drugs. Derivatization reagents and the adapted derivatization procedures are first summarized in Tables 2-2 and 2–3. Information related to the generation and presentation of the mass spectra (Appendix One, pp 31–371) and CC data (Appendix Two, pp 373–492) are further described in the following sections.

A. Derivatization Procedures, Instrumentation, and Analytical Parameters

Derivatization procedures and reagents, as described in Table 2-3, are not necessarily the most effective ones for generating the intended products. They were merely used to produce the derivatives to generate the mass spectrometric information (full-scan spectra and SIM data) for evaluation.

GC-MS analyses were performed on the following two systems: (a) an Agilent 6890 gas chromatograph equipped with an Agilent (Wilmington, DE) HP-ULTRA-1 capillary column (crosslinked 100% methyl siloxane phase; 12-m, 0.20-mm ID, 0.33-μm film thickness) interfaced to an Agilent 5973 mass selective detector (Palo Alto, CA); (b) an Agilent 6890N gas chromatograph equipped with an Agilent (Wilmington, DE) HP-5 capillary column (crosslinked 5% phenyl methyl siloxane phase; 12-m, 0.20-mm ID, 0.33-m film thicknesses) interfaced to an Agilent 5975 mass selective detector (Palo Alto, CA). These systems were operated at 70 eV with helium as the carrier gas (flow rate: 1.0 mL/min). Injector, GC-MS interface, and the ion source temperatures were set at 250, 290, and 230 °C, respectively. Various GC oven temperature and programming parameters were adapted for the analyses of the analytes and their ISs in various derivatization status or forms. Since chromatographic resolution is not the emphasis of this study, temperature and programming rates are not critical parameters and are not provided for individual analyses.

B. Collection of Mass Spectrometric Data (Appendix One, pp 31–371)

As described in Section II of this chapter, typically, a full-scan mass spectrum of a drug of interest was obtained by injecting the drug (in various derivatization forms) into the GC-MS system. With a few exceptions, full-scan mass spectra were collected starting at m/z 40 or 50 and ended at a mass higher than the molecular weights of the derivatized products, rounded to the next 50 or 100. The drug's ILAs were analyzed separately. Information derived from these ion chromatograms (retention time and mass spectrometric data) was used to characterize the analytes and their ILAs. Full-scan mass spectrometric data from these two runs were reviewed to select ions that may be suited for designating the analytes and the ISs in routine GC-MS protocols.

The same drugs and their derivatized products were again injected (separately) into the GC-MS operated under SIM mode. Ions selected from the full-scan MS data for both the analyte and the IS were monitored. Mass spectrometric data derived from these SIM runs were then used to evaluate the CC data using the *normalized direct measurement* method as described in Section II of this chapter. Details of the methodology have been described in our earlier publications [8,13].

For presentation purpose, full-scan mass spectra were stored as digital data and then converted by the DeltaGraph software (Seattle, WA) into mass spectra of a more desirable format as shown in Part Two of this book.

The mass spectra in these figures (Appendix One) are organized as follows. All mass spectra of a drug and its ILAs are presented in one figure. The mass spectra resulting from the use of one derivatization group for a specific drug and its ILAs are grouped together, followed by the same set of mass spectra resulting from the use of a different derivatization group. The appearance order of these groups of mass spectra follows the order of the derivatization groups listed in Table 2-2. For example, the mass spectra for all forms of derivatization for *amphetamine* are included in **Figure I-1** (pp 39–56), where "I" is the designation of compound category, stimulant; and "1" is the designation of the first compound in this drug category, amphetamine. The mass spectra for the acetyl-derivatives of amphetamine and its ILAs (amphetamine-d_5, amphetamine-d_6, amphetamine-d_8, amphetamine-d_{10} and amphetamine-d_{11}) are grouped together and presented first as Figure I-1-A, where "A" is the designation of the "acetyl" derivatization group in this case. Corresponding mass spectra of these compounds, with other derivatization groups attached (such as TCA, TFA, TFA/t-BDMS, PFP, PFP/t-BDMS, HFB, HFB/t-BDMS, 4-CB, PFB, propylformyl, l-TPC, d-TPC, l-MTPA, d-MTPA, TMS and t-BDMS) are grouped

together, respectively, and presented in sequence as Figure I-1-B to Figure I-1-Q. The mass spectra for various forms of derivatives for *methamphetamine* are similarly grouped and presented in **Figure I-2** (pp 57–71).

Many of the mass spectra included in Appendix One have not been published in literature generally available to the scientific community. Certainly, they have not been systematically compiled as presented here and, therefore, should be of routine reference value to laboratories engaged in drug analysis.

C. Ion Intensity Cross-Contribution Data (Appendix Two, pp 373–492)

The second set of data (ion intensity CC data), as presented in Part Three of this book in table format, are pairs of ions with potential for designating the drugs and their ILAs. Cross-contribution data are calculated based on data derived from SIM runs using the *normalized direct measurement* method described in Table 2-4 of Section II in this chapter. Ion-pairs adopted for the normalization process are underlined.

Ion-pairs included in each table are limited at two levels. First, full-scan data showing an ion-pair having >10% CC or with <10% relative intensity are not included in SIM data collection. At the second level, ion-pairs with CC >5% (based on SIM data) are also excluded from the table. Common practices in choosing the linear model for calibration mandate the use of ion-pairs with low CC for quantitation (or as a criterion for qualitative confirmation purpose). This is especially true when the calibration is to be established for a reasonable concentration range (e.g., in three orders of magnitude), as demonstrated in an earlier study [14]. An ion-pair with significant CC can be noted by the presence of the cross-contributing ion in the full-scan mass spectrum of the corresponding isotopic analog.

The CC data may appear to be irrelevant in cases where the number of the isotopic atoms in the ILAs is so large that the analytes and the ISs are practically resolved chromatographically. However, these analogs (such as methamphetamine-d_{14}) are still included in this study for the following two reasons: (a) to optimize the analytical time and to keep the system clean, chromatography is normally conducted at high temperatures resulting in inadequate resolution between the analyte and the IS; and (b) the retention time window set for automatic integration of ion intensities may not always be properly adjusted.

The sequence and manner adapted to present CC data are different from that adapted for presenting the full-scan mass spectra in Appendix One. Specifically, CC data are presented in table format and the CC data for all chemical derivatization products for one specific isotopic analog are included in one table. For example, with six ILAs included in this project, *amphetamine's CC data* are presented in six tables, ranging from **Table I-1a** to **Table I-1f** (pp 381–388), where "I" is the designation of compound category, stimulant; "1" is the designation of the first compound in this category, amphetamine; and "a" to "f" are the designations of the six isotopically labeled analogs). These six tables summarize the CC data for the following analyte/ILA pairings: amphetamine/amphetamine-d_5 (Table I-1a), amphetamine/amphetamine-d_5(ring) (Table I-1b), amphetamine/amphetamine-d_6 (Table I-1c), amphetamine/amphetamine-d_8 (Table I-1e), amphetamine/amphetamine-d_{10} (Table I-1e), and amphetamine/amphetamine-d_{11} (Table I-1f). Similarly, *methamphetamine's CC data* are presented in **Table I-2a** to **Table I-12e** (pp 388–393).

Each table includes all derivatization groups that have been attempted. The order of appearance of these derivatization groups in each table is the same as those listed in Table 2-2 in this chapter.

CONCLUDING REMARKS

Following a review on: (a) the structural features of commonly encountered drugs; (b) commercially available isotopically labeled analogs of these drugs; and (c) commonly utilized chemical derivatization approaches, the authors have carried out a series of chemical derivatization experiments and collected a set of full-scan mass spectra and CC data for ion-pairs with potential for designating the analytes and their ILAs serving as the ISs. An approach for evaluating the accuracy of a set of specific CC data has been presented; however, the CC data summarized in Appendix Two have not been systematic validated.

Full-scan mass spectra compiled in Appendix One represent the most comprehensive collection of mass spectra for these drugs and their isotopic analogs in various chemical derivatization forms. Comprehensive listings of CC data shown in Appendix Two should save an enormous amount of time and efforts for practicing laboratories in their search for this analytical parameter to establish optimal quantitation protocols.

REFERENCES

1. SOFT/AAFS Forensic Toxicology Laboratory Guidelines (2006 Version); Section 8 (www.soft-tox.org).
2. Liu RH, Lin T-L, Chang W-T, Liu C, Tsay W-I, Li J-H, Kuo T-L: Isotopically labeled analogues for drug quantitation; *Anal Chem* 74:618A; 2002.
3. Lin DL, Chang WT, Kuo TL, Liu RH: Chemical derivatization and the selection of deuterated internal standard for quantitative determination — Methamphetamine example; *J Anal Toxicol* 24:275; 2000.
4. Lin DL, Wang SM, Liu RH: Chemical derivatization in drug analysis — A conceptual review; *J Food Drug Ana* 16:1; 2008.
5. Smith FP, Kidwell D: Commentary on minimal standards for the performance and interpretation of toxicology tests in legal proceedings; *J Forensic Sci* 45:237; 2000.
6. Chang WT, Lin DL, Low IA, Liu RH: $^{13}C_4$-Secobarbital as the internal standard for the quantitative determination of secobarbital — A critical evaluation; *J Forensic Sci* 45:659; 2000.
7. Chang WT, Liu RH: Mechanistic studies on the use of ^{2}H- and ^{13}C-analogs as internal standards in selected ion monitoring GC-MS quantitative determination — Butalbital example; *J Anal Toxicol* 25:659; 2001.
8. Chang WT, Lin DL, Liu RH: Isotopic analogs as internal standards for quantitative analyses by GC/MS — Evaluation of cross-contribution to ions designated for the analyte and the isotopic internal standard; *Forensic Sci Int* 121:174; 2001.
9. Valtier S, Cody JT: Evaluation of internal standards for the analysis of amphetamine and methamphetamine; *J Anal Toxicol* 19:375; 1995.
10. Urry FM, Kushnir M, Nelson G, McDowell M, Jennison T: Improving ion mass ratio performance at low concentration in methamphetamine GC-MS assay through internal standard selection; *J Anal Toxicol* 20:592; 1996.
11. Wu C-H, Huang M-H, Wang S-M, Lin C-C, Liu RH: Gas chromatography-mass spectrometry analysis of ketamine and its metabolites — A comparative study on the utilization of different derivatization groups; *J Chromatogr A* 1157:336; 2007.
12. Wu C-H, Yang S-C, Wang Y-S, Chen B-G, Lin C-C, Liu RH: Evaluation of various derivatization approaches for gas chromatography-mass spectrometry analysis of buprenorphine and norbuprenorphine; *J Chromatogr A* 1182:93; 2008.
13. Liu RH, Foster G, Cone EJ, Kumar SD: Selecting an appropriate isotopic internal standard for gas chromatography/mass spectrometry analysis of drugs of abuse — Pentobarbital example; *J Forensic Sci* 40:983; 1995.
14. Whiting TC, Liu RH, Chang W-T, Bodapati MR: Isotopic analogs as internal standards for quantitative analyses of drugs/metabolites by GC/MS — Non-linear calibration approaches; *J Anal Toxicol* 25:179; 2001.

15. Chen B-G, Wang S-M, Liu RH: GC-MS analysis of multiply-derivatized opioids in urine; *J Mass Spectrom* 42:1012; 2007.

16. Szirmai M, Beck O, Stephansson N, Halldin MM: A GC-MS study of three major acidic metabolites of delta-1-tetrahydrocannabinol; *J Anal Toxicol* 20:573; 1996.

17. Huang W, Moody DE, Andrenyak DM, Smith EK, Foltz RL, Huestis MA, Newton JF: Simultaneous determination of delta-9-tetrahydrocannabinol and 11-nor-9-carboxy-delta-9-tetrahydrocannabinol in human plasma by solid-phase extraction and gas chromatography-negative ion chemical ionization-mass spectrometry; *J Anal Toxicol* 25:531; 2001.

18. Bourland JA, Hayes EF, Kelly RC, Sweeney SA, Hatab MM: Quantitation of cocaine, benzoylecgonine, cocaethylene, methylecgonine, and norcocaine in human hair by positive ion chemical ionization (PICI) gas chromatography-tandem mass spectrometry; *J Anal Toxicol* 24:489; 2000.

19. Moore C, Guzaldo F, Donahue T: The determination of 11-nor-delta-9-tetrahydrocannabinol-9-carboxylic acid (THC-COOH) in hair using negative ion gas chromatography-mass spectrometry and high-volume injection; *J Anal Toxicol* 25:555; 2001.

20. Baptista MJ, Monsanto PV, Marques EGP, Bermejo A, Avila S, Castanheira AM, Margalho C, Barroso M, Vieira DN: Hair analysis for delta-9-THC, delta-9-THC-COOH, CBN and CBD, by GC/MS-EI comparison with GC/MS-NCI for Delta-9-THC-COOH; *Forensic Sci Int* 128:66; 2002.

21. Jurado C, Gimenez MP, Menendez M, Repetto M: Simultaneous quantitation of opiates, cocaine and cannabinoids in hair; *Forensic Sci Int* 70:165; 1995.

22. Toseland PA: Determination of amphetamine as its *N*-acetyl derivative by gas-liquid chromatography; *Clin Chem Acta* 25:75; 1969.

23. Hornbeck CL, Czarny RJ: Quantitation of methamphetamine and amphetamine in urine by capillary GC/MS Part. I. Advantages of trichloroacetyl derivatization; *J Anal Toxicol* 13:144; 1989.

24. Elian AA: Detection of low levels of flunitrazepam and its metabolites in blood and bloodstains; *Forensic Sci Int* 101:107; 1999.

25. Hornbeck CL, Carrig JE, Czarny RJ: Detection of a GC/MS artifact peak as methamphetamine; *J Anal Toxicol* 17:257; 1993.

26. Reagent insert; Pierce Biotechnology Inc: Rockford, IL; 2003.

27. Gilbert RB, Peng PI, Wong D: A labetalol metabolite with analytical characteristics resembling amphetamines; *J Anal Toxicol* 19:84; 1995.

28. Gan BK, Baugh D, Liu RH, Walia AS: Simultaneous analysis of amphetamine, methamphetamine, and 3,4-methylene-dioxymethamphetamine (MDMA) in urine samples by solid-phase extraction, derivatization, and gas chromatography/mass spectrometry; *J Forensic Sci* 36:1331; 1991.

29. Cody JT, Schwarzhoff R: Interpretation of methamphetamine and amphetamine enantiomer data; *J Anal Toxicol* 17:321; 1993.

30. Jones JB, Mell LD: A simple wash procedure for improving chromatography of HFAA derivatized amphetamine extracts for GC/MS analysis; *J Anal Toxicol* 17:447; 1993.

31. Czarny RJ, Hornbeck CL: Quantitation of methamphetamine and amphetamine in urine by GC/MS Part. II. Derivatization with 4-carbethoxyhexafluorobutyl chloride; *J Anal Toxicol* 13:257; 1989.

32. Ropero-Miller JD, Lambing MK, Winecker RE: Simultaneous quantitation of opioids in blood by GC-EI-MS analysis following deproteination, detautomerization of keto analytes, solid-phase extraction, and trimethylsilyl derivatization; *J Anal Toxicol* 26:524; 2002.

33. Wang WL, Darwin WD, Cone EJ: Simultaneous assay of cocaine, heroin and metabolites in hair, plasma, saliva and urine by gas chromatography-mass spectrometry; *J Chromatogr B Biomed Appl* 660:279; 1994.

34. Chen BH, Taylor EH, Pappas AA: Comparison of derivatives for determination of codeine and morphine by gas chromatography/mass spectrometry; J Anal Toxicol 14:12; 1990.

35. Broussard LA, Presley LC, Pittman T, Clouette R, Wimbish GH: Simultaneous identification and quantitation of codeine, morphine, hydrocodone, and hydromorphone in urine as trimethylsilyl and oxime derivatives by gas chromatography-mass spectrometry; *Clin Chem* 43:1029; 1997.

36. Cremese M, Wu AHB, Cassella G, O'Connor E, Rymut K, Hill DW: Improved GC/MS analysis of opiates with use of oxime-TMS derivatives; *J Forensic Sci* 43:1220; 1998.

37. Nowatzke W, Zeng J, Sauders A, Bohrer A, Koenig J, Turk J: Distinctttttttion among eight opiate drugs in urine by gas chromatography-mass spectrometry; *J Pharm Biomed Anal* 20:829; 1999.

38. Chen BG, Chang CD, Wang CT, Chang WT, Wang SM, Liu RH: A novel approach to evaluate the extent of cross-contribution to the intensity of ions designating the analyte and the internal standard in quantitative GC-MS analysis; *J Am Soc Mass Spectrom* 19:598; 2008.

39. Jones J, Tomlinson K, Moore C: The simultaneous determination of codeine, morphine, hydrocodone, hydromorphone, 6-acetylmorphine, and oxycodone in hair and oral fluid; *J Anal Toxicol* 26:171; 2002.

40. Meatherall R: GC-MS confirmation of codeine, morphine, 6-acetylmorphine, hydrocodone, hydromorphone, oxycodone, and oxymorphone in blood; *J Anal Toxicol* 29:301; 2005.

41. Fenton J, Mummert J, Childers M: Hydromorphone and hydrocodone interference in GC/MS assays for morphine and codeine; *J Anal Toxicol* 18:159; 1994.

42. Meatherall R: GC-MS confirmation of codeine, morphine, 6-acetylmorphine, hydrocodone, hydromorphone, oxycodone, and oxymorphone in urine; *J Anal Toxicol* 23:177; 1999.

Chapter 2 — Chemical Derivatization and Mass Spectrometric Data Collection

43. Smith ML, Hughes RO, Levine B, Dickerson S, Darwin WD, Cone EJ: Forensic drug testing for opiates. VI. Urine testing for hydromorphone, hydrocodone, oxymorphone, and oxucodone with commercial opiate immunoassays and gas chromatography-mass spectrometry; *J Anal Toxicol* 19:18; 1995.

44. Broussard LA, Presley LC, Tanous M, Queen C: Improved gas chromatography-mass spectrometry method for simultaneous identification and quantitation of opiates in urine as propionyl and oxime derivatives; *Clin Chem* 47:127; 2001.

45. Melgar R, Kelly RC: A novel GC/MS derivatization method for amphetamines; *J Anal Toxicol* 17:399; 1993.

46. Yoo YC, Chung HS, Kim IS, Jin WT, Kim MK: Determination of nalbuphine in drug abusers' urine; *J Anal Toxicol* 19:120; 1995.

47. Bioaeronautical Sciences Research Laboratory: *Laboratory Operation Manual*; U.S. FAA Civil Aerospace Medical Institute: Oklahoma City, OK; 2004.

48. Valentine JL, Middleton R: GC-MS identification of sympathomimetic amine drugs in urine: rapid methodology applicable for emergency clinical toxicology; *J Anal Toxicol* 24:211; 2000.

49. Low IA, Liu RH, Barker SA, Fish F, Settine RL, Piotrowski EG, Damert WC, Liu JY: Selected ion monitoring mass spectrometry: parameters affecting quantitative determination; *Biomed Mass Spectrom* 12:633; 1985.

50. Willard HH, Merritt LL, Dean JA, Settle FA: *Instrumental Methods of Analysis*, 7th ed; Wadsworth Publishing: Belmont, CA; p. 32; 1988.

51. Krull I, Swartz M: Quantitation in method validation. *LC•GC* 16:1984; 1998.

52. Duncan MW, Gale PJ, Yergey AL: *The Principles of Quantitative Mass Spectrometry*; Rockpool Productions: Denver, CO; p. 97; 2006.

PART TWO

MASS SPECTRA OF COMMONLY ABUSED DRUGS AND THEIR ISOTOPICALLY LABELED ANALOGS IN VARIOUS DERIVATIZATION FORMS

Appendix One

Mass Spectra of Commonly Abused Drugs and Their Isotopically Labeled Analogs in Various Derivatization Forms

Table of Contents for Appendix One

Summary of Drugs, Isotopic Analogs, and Chemical Derivatization Groups Included in Figure I (Stimulants)

Compound	Isotopic analog	Chemical derivatization group (no. of spectra)	Figure #
Amphetamine	d_5, d_5 (ring), d_6, d_8, d_{10}, d_{11}	None, Acetyl, TCA, TFA, PFP, HFB, 4-CB, PFB, propylformyl, *l*-TPC, *d*-TPC, *l*-MTPA, *d*-MTPA, TMS, *t*-BDMS, TFA/*t*-BDMS, PFP/*t*-BDMS, HFB/*t*-BDMS (126)	I-1
Methamphetamine	d_5, d_8, d_9, d_{11}, d_{14}	None, acetyl, TCA, TFA, PFP, HFB, 4-CB, PFB, propylformyl, *l*-TPC, *d*-TPC, *l*-MTPA, *d*-MTPA, TMS, *t*-BDMS (90)	I-2
Ephedrine	d_3	None, acetyl, TCA, [TFA]₂, [PFP]₂, [HFB]₂, 4-CB, PFB, propylformyl, *d*-TPC, *d*-MTPA, [TMS]₂ (24)	I-3
Phenylpropanolamine	d_3	None, acetyl, TCA, [TFA]₂, [PFP]₂, [HFB]₂, 4-CB, PFB, *l*-TPC, *d*-TPC, *l*-MTPA, *d*-MTPA, [TMS]₂, *t*-BDMS, [*t*-BDMS]₂, TFA/[*t*-BDMS]₂, PFP/[*t*-BDMS]₂, HFB/[*t*-BDMS]₂ (36)	I-4
MDA	d_5	None, acetyl, TCA, TFA, PFP, HFB, 4-CB, PFB, propylformyl, *l*-TPC, *d*-TPC, *l*-MTPA, *d*-MTPA, TMS, TFA/*t*-BDMS, PFP/*t*-BDMS, HFB/*t*-BDMS (34)	I-5
MDMA	d_5	None, acetyl, TCA, TFA, PFP, HFB, 4-CB, PFB, propylformyl, *l*-TPC, *d*-TPC, *l*-MTPA, *d*-MTPA, TMS (28)	I-6
MDEA	d_5, d_6	None, acetyl, TCA, TFA, PFP, HFB, 4-CB, PFB, propylformyl, *l*-TPC, *d*-TPC, *l*-MTPA, *d*-MTPA, TMS (42)	I-7
MBDB	d_5	None, acetyl, TCA, TFA, PFP, HFB, 4-CB, PFB, propylformyl, *l*-TPC, *d*-TPC, *l*-MTPA, *d*-MTPA, TMS (28)	I-8
Selegiline	d_8	None (2)	I-9
N-Desmethylselegiline	d_{11}	None, acetyl, TCA, TFA, PFP, HFB, 4-CB, TMS (16)	I-10
Fenfluramine	d_{10}	None, acetyl, TCA, TFA, PFP, HFB, 4-CB (14)	I-11
Norcocaine	d_4	None, TFA, PFP, HFB, TMS (10)	I-12
Cocaine	d_3	None (2)	I-13
Cocaethylene	d_3, d_8	None (3)	I-14
Ecgonine methyl ester	d_3	None, TFA, PFP, HFB, TMS, *t*-BDMS (12)	I-15
Benzoylecgonine	d_3, d_8	Methyl, ethyl, propyl, butyl, PFPoxy, HFPoxy, TMS, *t*-BDMS (24)	I-16
Ecgonine	d_3	[TMS]₂, [*t*-BDMS]₂, HFPoxy/TFA, PFPoxy/PFP, HFPoxy/HFB (10)	I-17
Anhydroecgonine methyl ester	d_3	None (2)	I-18
Caffeine	$^{13}C_3$	None (2)	I-19
Methylphenidate	d_3	None, TFA, PFP, HFB, 4-CB, TMS (12)	I-20
Ritalinic acid	d_5	4-CB, [TMS]₂, *t*-BDMS (6)	I-21

Total no. of mass spectra: 523

Figure I — Stimulants

Appendix One — Figure I
Mass Spectra of Commonly Abused Drugs and Their Isotopically Labeled Analogs
in Various Derivatization Forms — Stimulants

Figure I — Stimulants

Figure I-1. Mass spectra of amphetamine (AM) and its deuterated analogs (AM-d$_5$ [ring], -d$_5$ [side chain], -d$_6$, -d$_8$, -d$_{10}$, -d$_{11}$): (A) underivatized; (B) acetyl-derivatized; (C) TCA-derivatized; (D) TFA-derivatized; (E) PFP-derivatized; (F) HFB-derivatized; (G) 4-CB-derivatized; (H) PFB-derivatized; (I) propylformyl-derivatized; (J,K) *l*-TPC-derivatized; (L,M) *l*-MTPA-derivatized; (N) TMS-derivatized; (O) *t*-BDMS-derivatized; (P) TFA/*t*-BDMS-derivatized; (Q) PFP/*t*-BDMS-derivatized; (R) HFB/*t*-BDMS-derivatized.

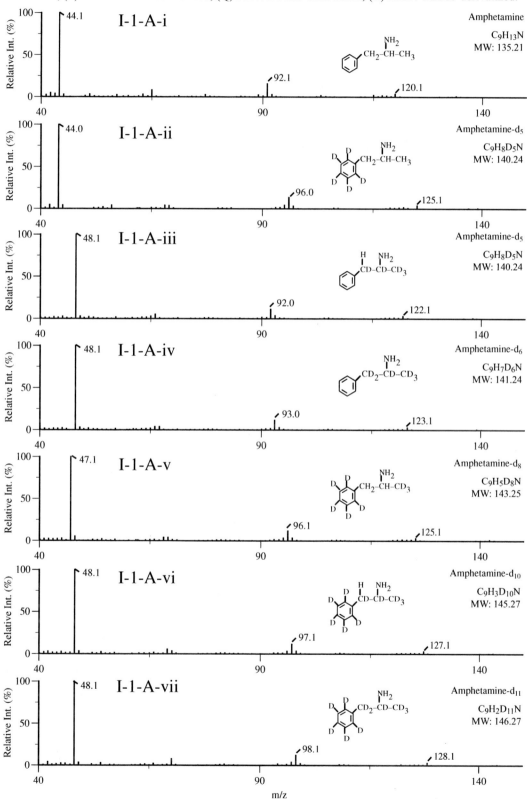

Figure I — Stimulants

40

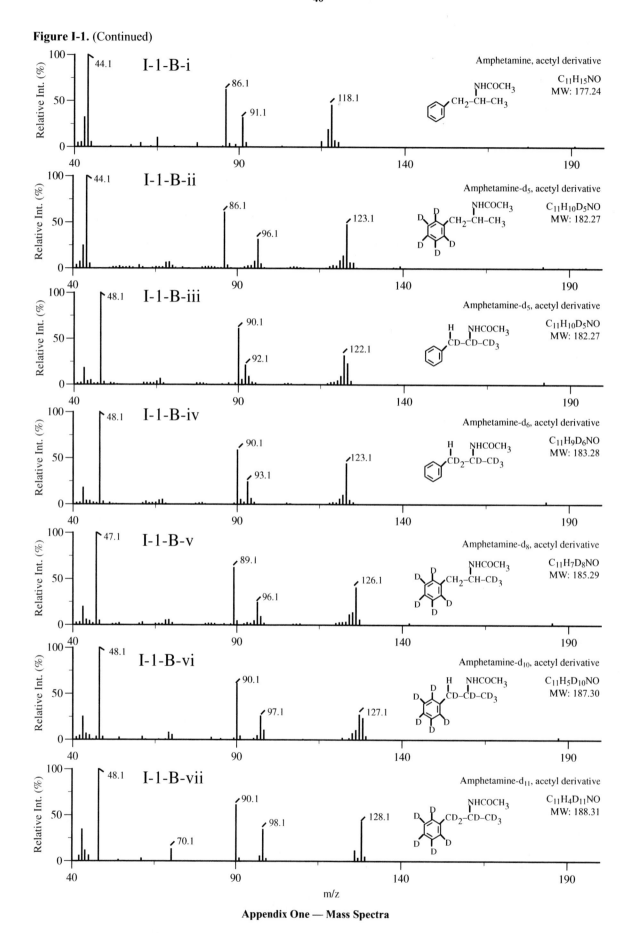

Figure I-1. (Continued)

Appendix One — Mass Spectra

Figure I-1. (Continued)

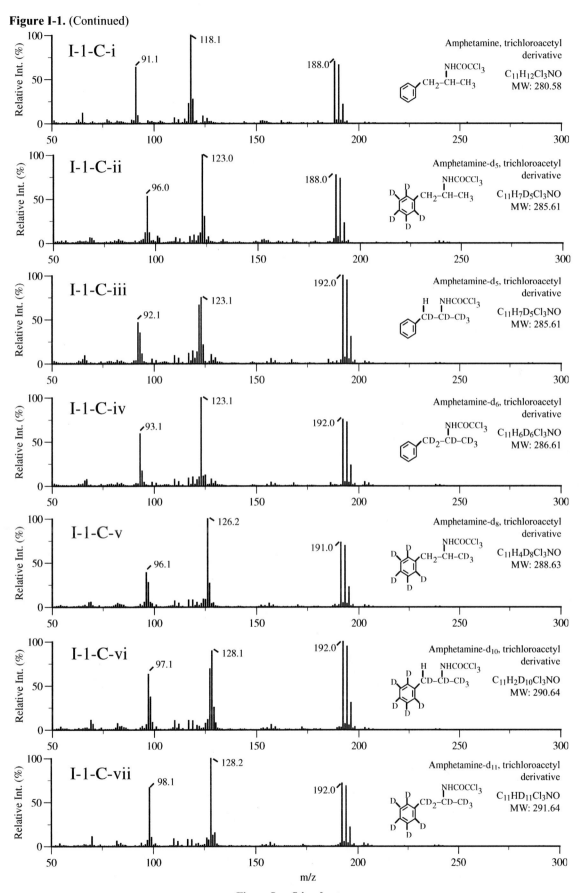

Figure I — Stimulants

Figure I-1. (Continued)

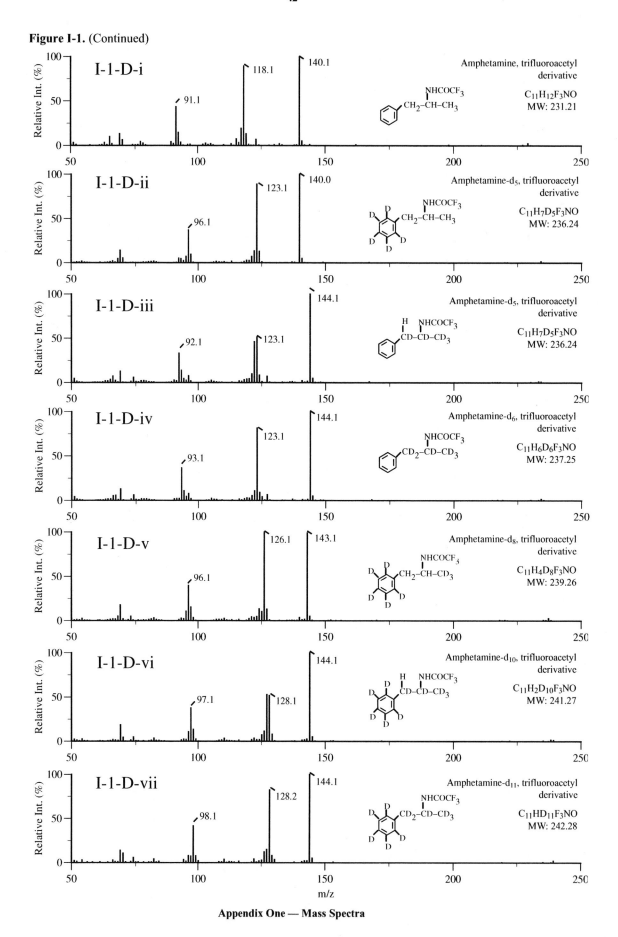

Figure I-1. (Continued)

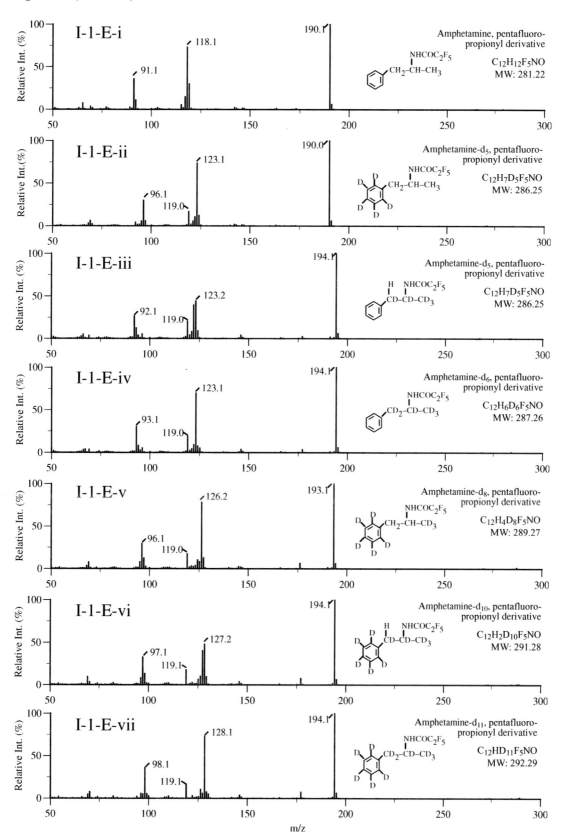

Figure I — Stimulants

Figure I-1. (Continued)

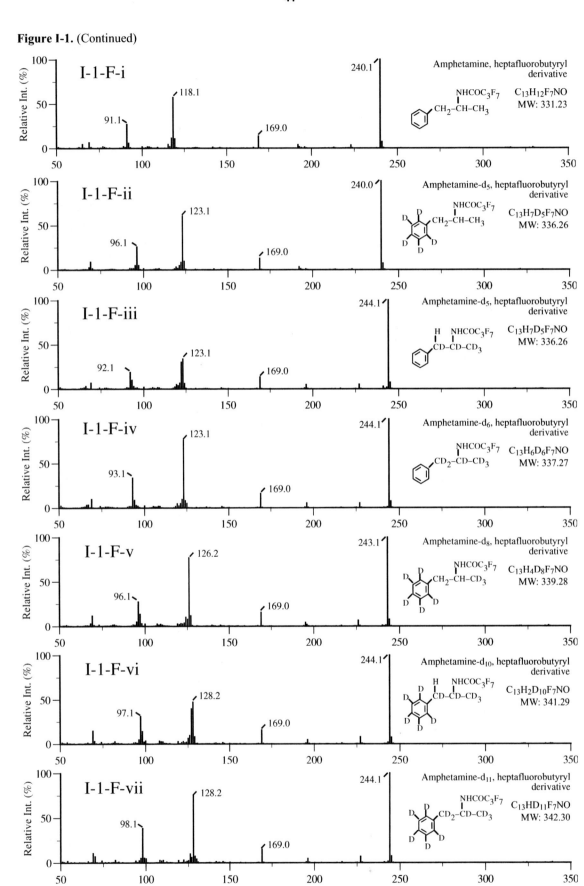

Figure I-1. (Continued)

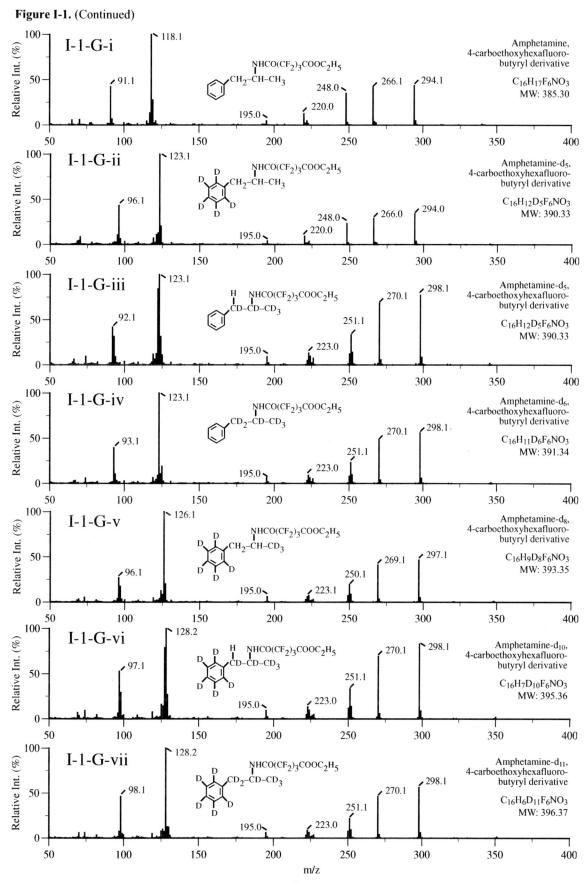

Figure I — Stimulants

Figure I-1. (Continued)

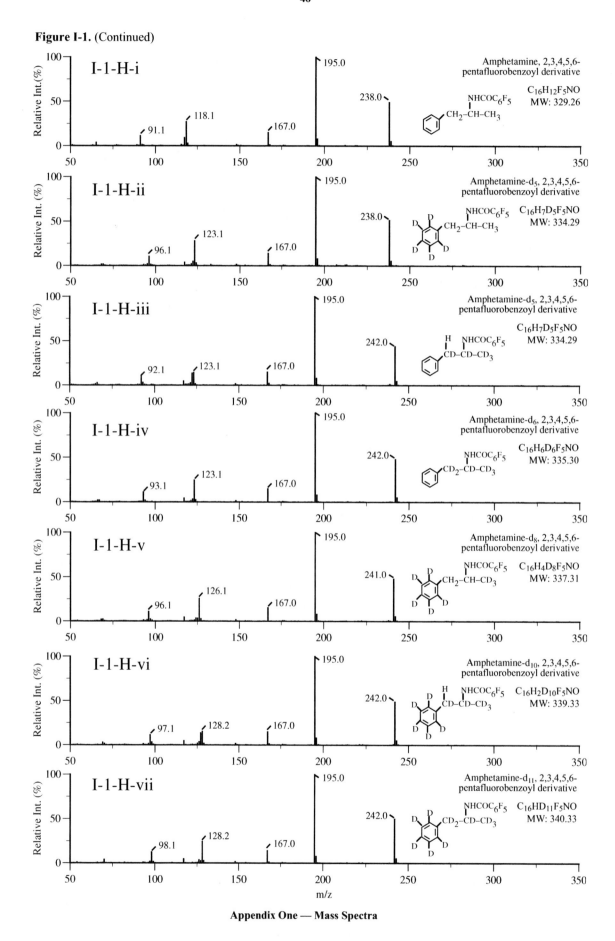

Figure I-1. (Continued)

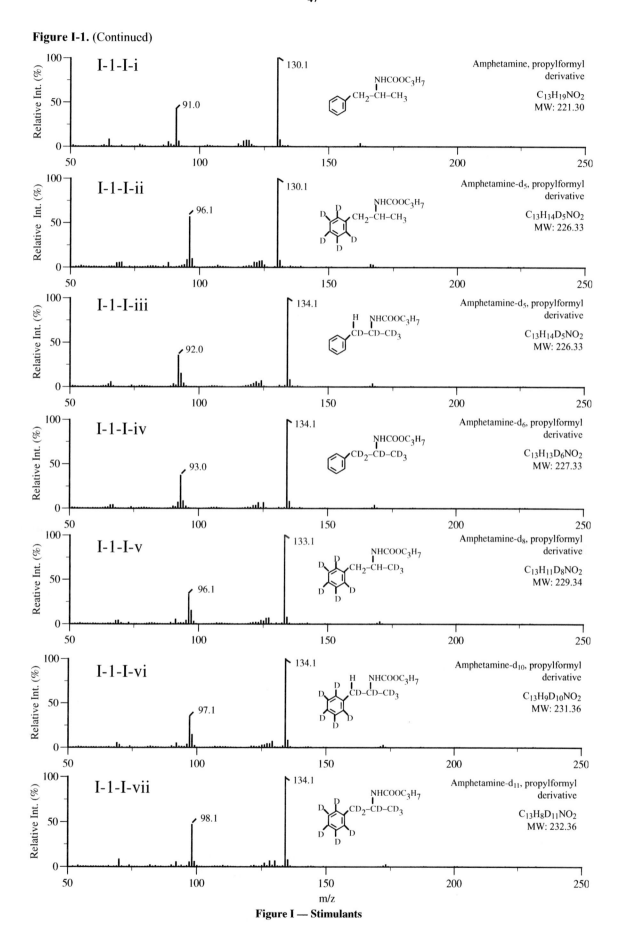

Figure I — Stimulants

Figure I-1. (Continued)

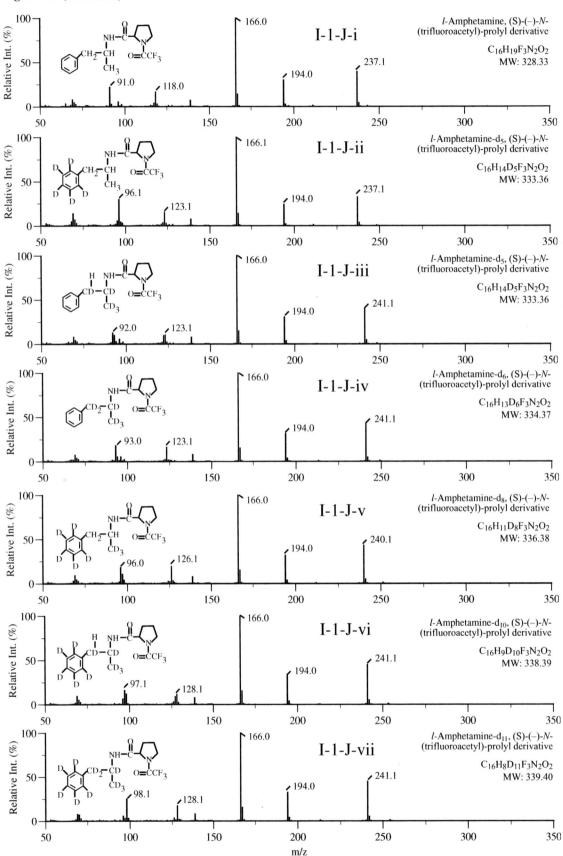

49

Figure I-1. (Continued)

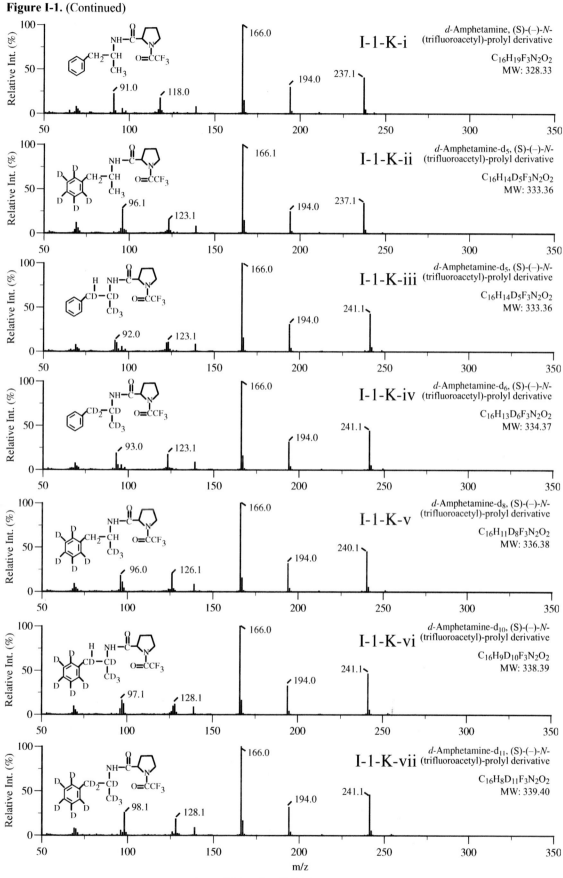

I-1-K-i — *d*-Amphetamine, (S)-(–)-*N*-(trifluoroacetyl)-prolyl derivative — C$_{16}$H$_{19}$F$_3$N$_2$O$_2$ — MW: 328.33

I-1-K-ii — *d*-Amphetamine-d$_5$, (S)-(–)-*N*-(trifluoroacetyl)-prolyl derivative — C$_{16}$H$_{14}$D$_5$F$_3$N$_2$O$_2$ — MW: 333.36

I-1-K-iii — *d*-Amphetamine-d$_5$, (S)-(–)-*N*-(trifluoroacetyl)-prolyl derivative — C$_{16}$H$_{14}$D$_5$F$_3$N$_2$O$_2$ — MW: 333.36

I-1-K-iv — *d*-Amphetamine-d$_6$, (S)-(–)-*N*-(trifluoroacetyl)-prolyl derivative — C$_{16}$H$_{13}$D$_6$F$_3$N$_2$O$_2$ — MW: 334.37

I-1-K-v — *d*-Amphetamine-d$_8$, (S)-(–)-*N*-(trifluoroacetyl)-prolyl derivative — C$_{16}$H$_{11}$D$_8$F$_3$N$_2$O$_2$ — MW: 336.38

I-1-K-vi — *d*-Amphetamine-d$_{10}$, (S)-(–)-*N*-(trifluoroacetyl)-prolyl derivative — C$_{16}$H$_9$D$_{10}$F$_3$N$_2$O$_2$ — MW: 338.39

I-1-K-vii — *d*-Amphetamine-d$_{11}$, (S)-(–)-*N*-(trifluoroacetyl)-prolyl derivative — C$_{16}$H$_8$D$_{11}$F$_3$N$_2$O$_2$ — MW: 339.40

m/z

Figure I — Stimulants

Figure I-1. (Continued)

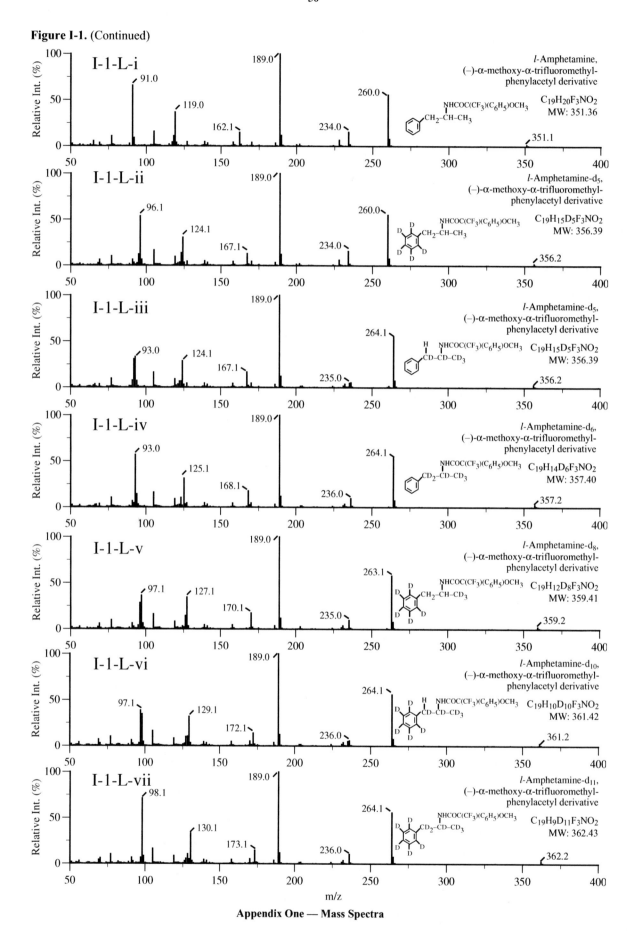

Figure I-1. (Continued)

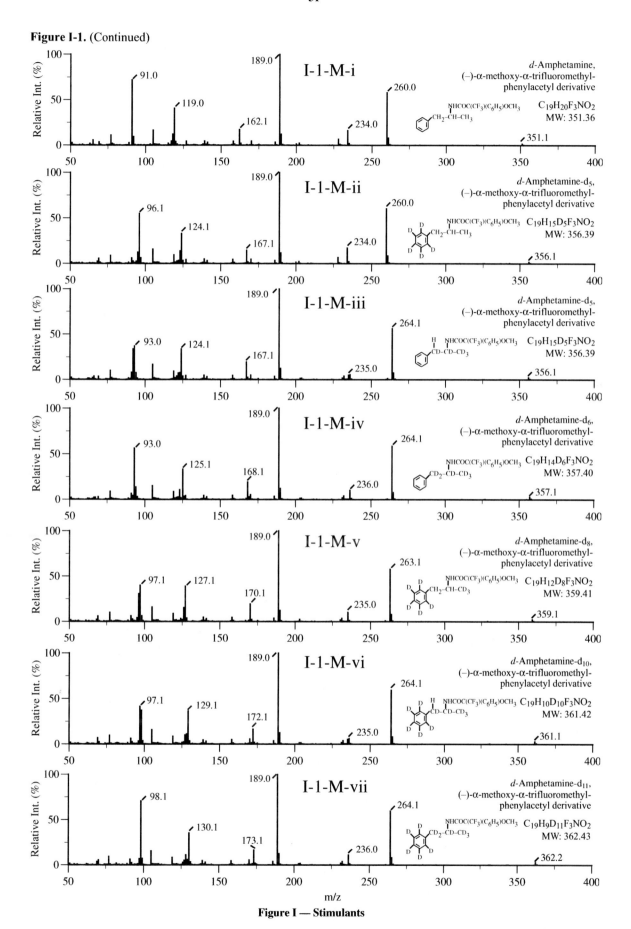

Figure I — Stimulants

Figure I-1. (Continued)

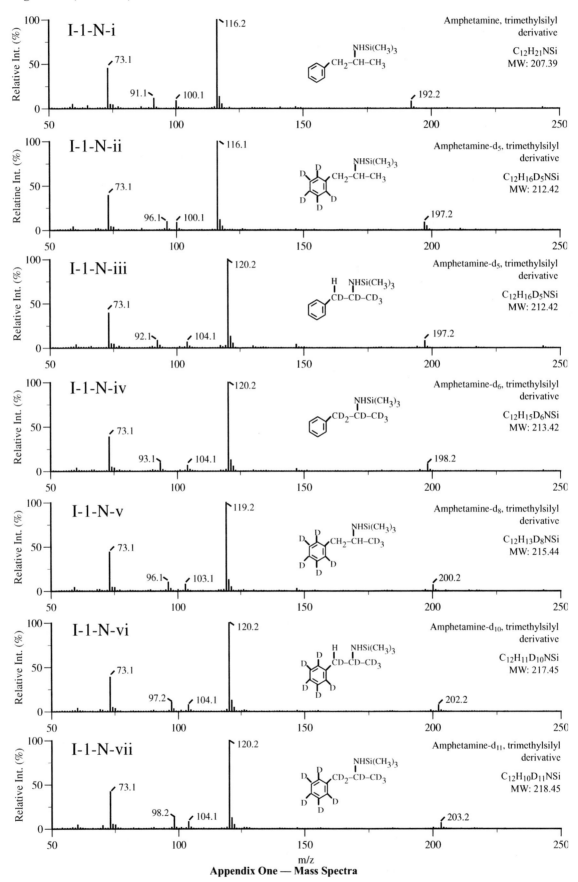

Figure I-1. (Continued)

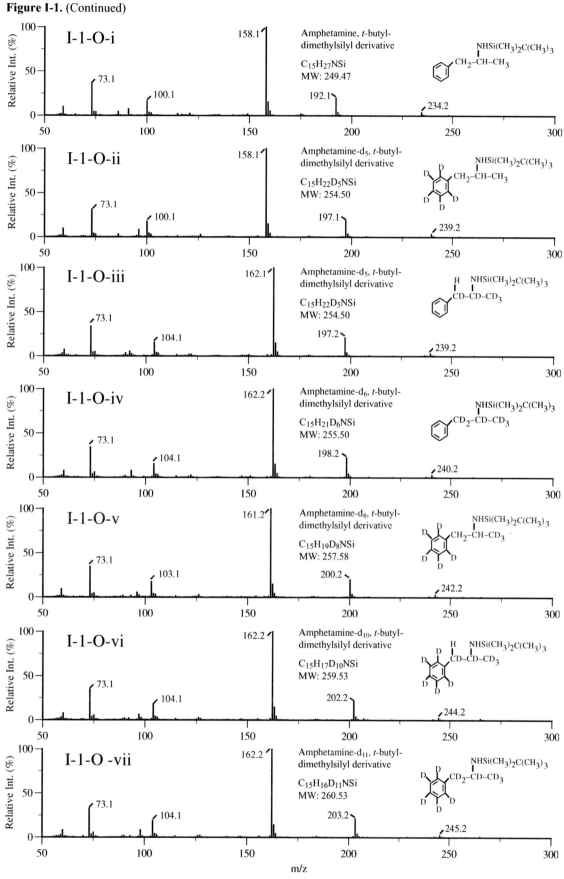

Figure I — Stimulants

Figure I-1. (Continued)

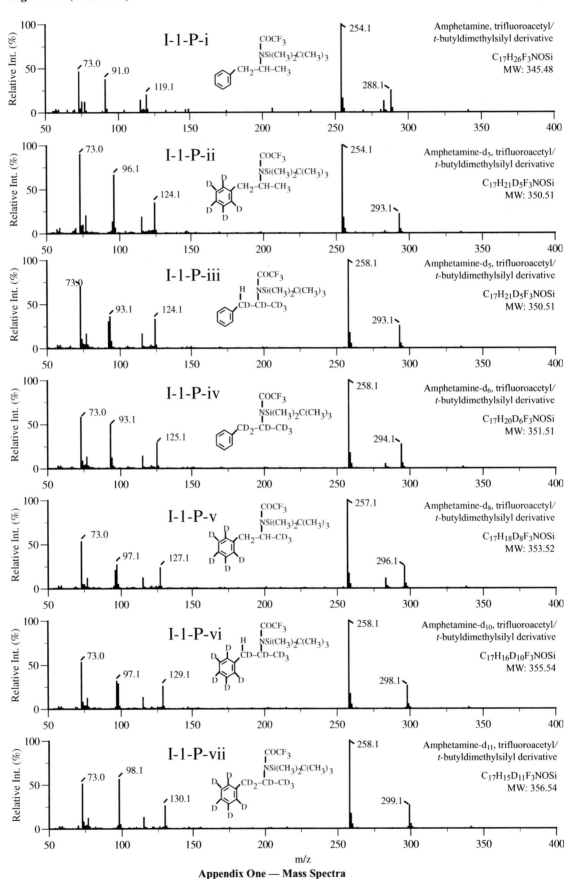

Figure I-1. (Continued)

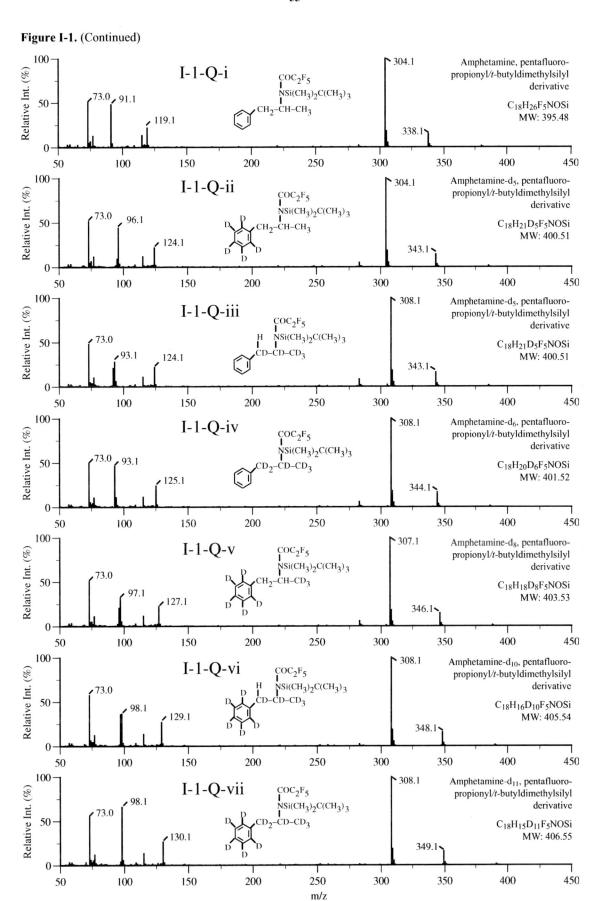

I-1-Q-i

Amphetamine, pentafluoro-propionyl/*t*-butyldimethylsilyl derivative

$C_{18}H_{26}F_5NOSi$
MW: 395.48

I-1-Q-ii

Amphetamine-d$_5$, pentafluoro-propionyl/*t*-butyldimethylsilyl derivative

$C_{18}H_{21}D_5F_5NOSi$
MW: 400.51

I-1-Q-iii

Amphetamine-d$_5$, pentafluoro-propionyl/*t*-butyldimethylsilyl derivative

$C_{18}H_{21}D_5F_5NOSi$
MW: 400.51

I-1-Q-iv

Amphetamine-d$_6$, pentafluoro-propionyl/*t*-butyldimethylsilyl derivative

$C_{18}H_{20}D_6F_5NOSi$
MW: 401.52

I-1-Q-v

Amphetamine-d$_8$, pentafluoro-propionyl/*t*-butyldimethylsilyl derivative

$C_{18}H_{18}D_8F_5NOSi$
MW: 403.53

I-1-Q-vi

Amphetamine-d$_{10}$, pentafluoro-propionyl/*t*-butyldimethylsilyl derivative

$C_{18}H_{16}D_{10}F_5NOSi$
MW: 405.54

I-1-Q-vii

Amphetamine-d$_{11}$, pentafluoro-propionyl/*t*-butyldimethylsilyl derivative

$C_{18}H_{15}D_{11}F_5NOSi$
MW: 406.55

m/z

Figure I — Stimulants

Figure I-1. (Continued)

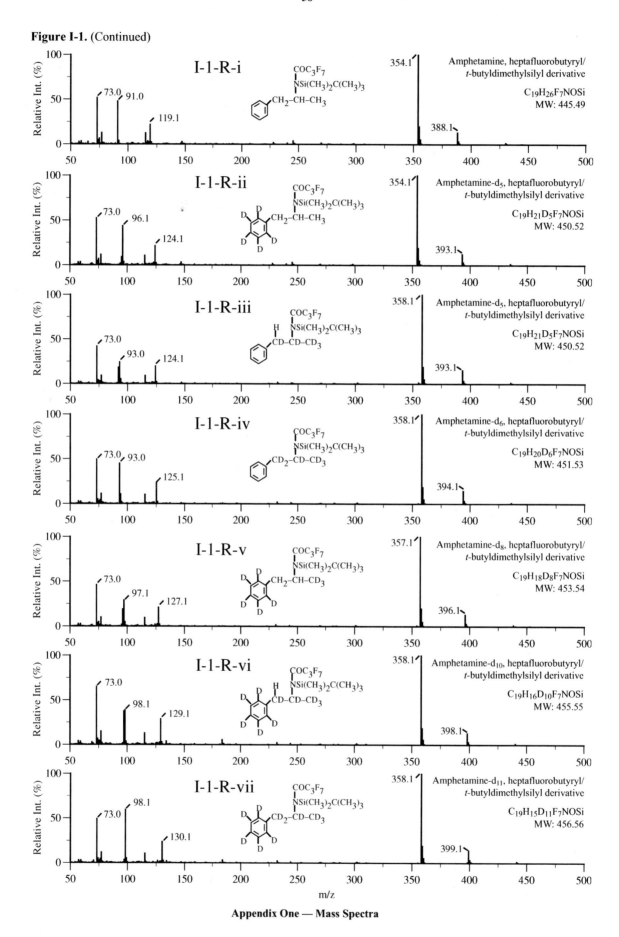

Appendix One — Mass Spectra

Figure I-2. Mass spectra of methamphetamine (MA) and its deuterated analogs (MA-d₅, -d₈, -d₉, -d₁₁, -d₁₄): (A) underivatized; (B) acetyl-derivatized; (C) TCA-derivatized; (D) TFA-derivatized; (E) PFP-derivatized; (F) HFB-derivatized; (G) 4-CB-derivatized; (H) PFB-derivatized; (I) propylformyl-derivatized; (J,K) *l*-TPC-derivatized; (L,M) *l*-MTPA-derivatized; (N) TMS-derivatized; (O) *t*-BDMS-derivatized.

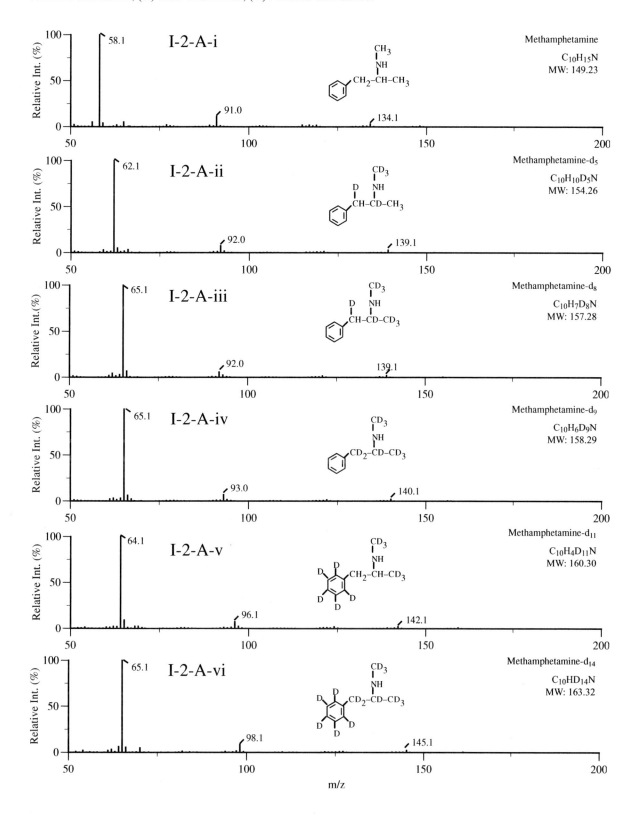

Figure I — Stimulants

Figure I-2. (Continued)

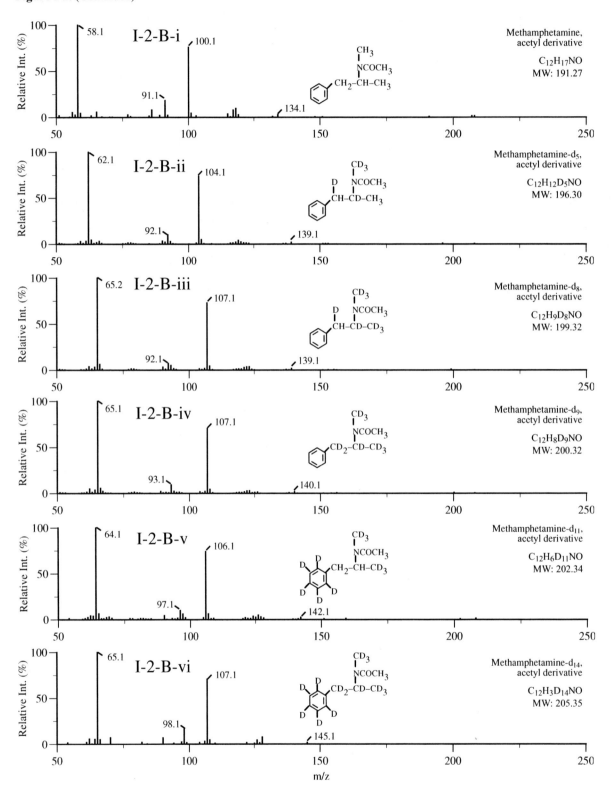

Figure I-2. (Continued)

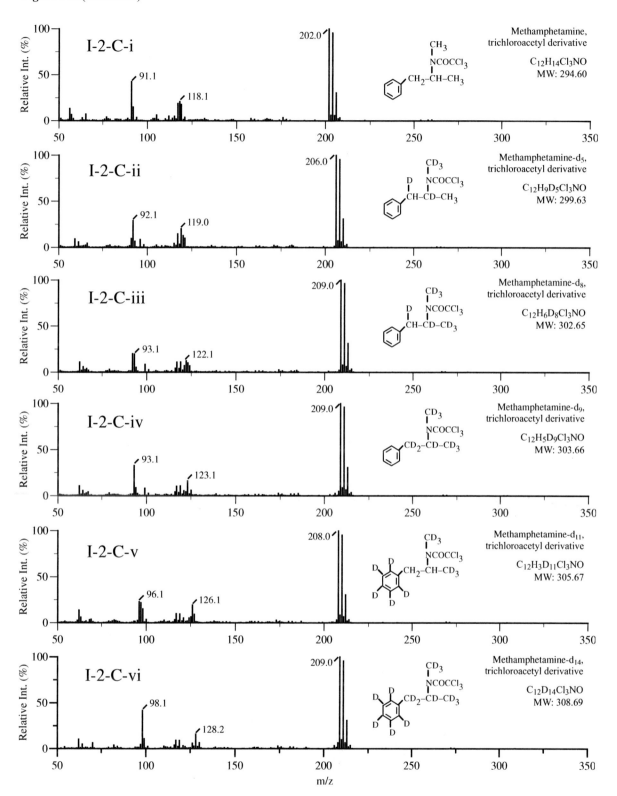

Figure I — Stimulants

Figure I-2. (Continued)

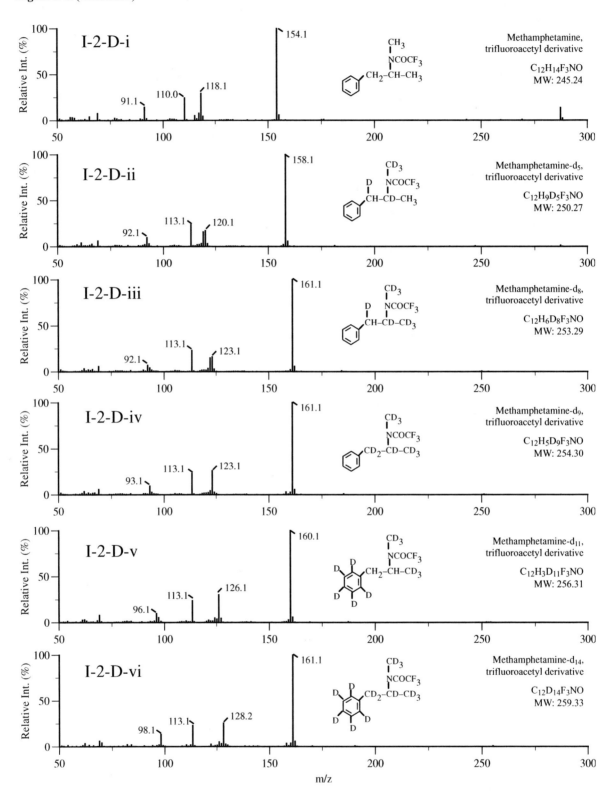

Figure I-2. (Continued)

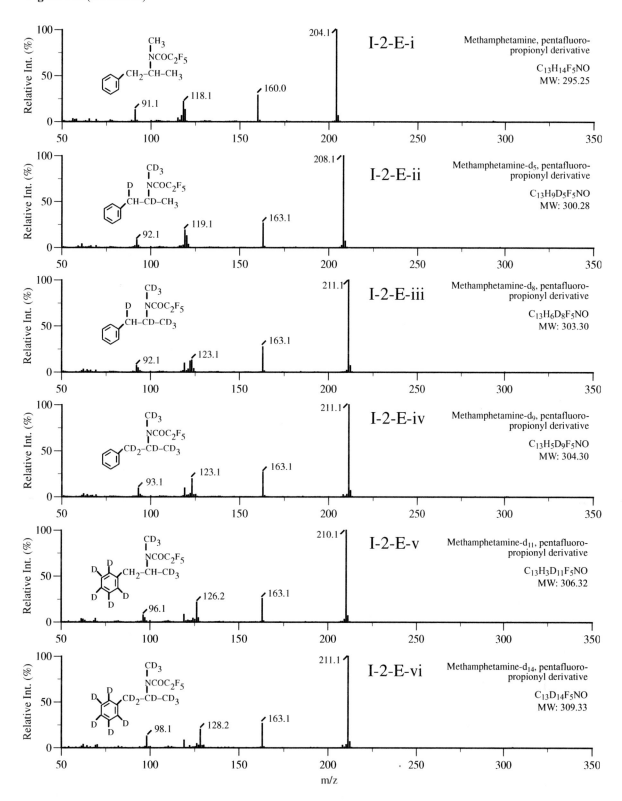

I-2-E-i — Methamphetamine, pentafluoropropionyl derivative; $C_{13}H_{14}F_5NO$; MW: 295.25

I-2-E-ii — Methamphetamine-d5, pentafluoropropionyl derivative; $C_{13}H_9D_5F_5NO$; MW: 300.28

I-2-E-iii — Methamphetamine-d8, pentafluoropropionyl derivative; $C_{13}H_6D_8F_5NO$; MW: 303.30

I-2-E-iv — Methamphetamine-d9, pentafluoropropionyl derivative; $C_{13}H_5D_9F_5NO$; MW: 304.30

I-2-E-v — Methamphetamine-d11, pentafluoropropionyl derivative; $C_{13}H_3D_{11}F_5NO$; MW: 306.32

I-2-E-vi — Methamphetamine-d14, pentafluoropropionyl derivative; $C_{13}D_{14}F_5NO$; MW: 309.33

Figure I — Stimulants

Figure I-2. (Continued)

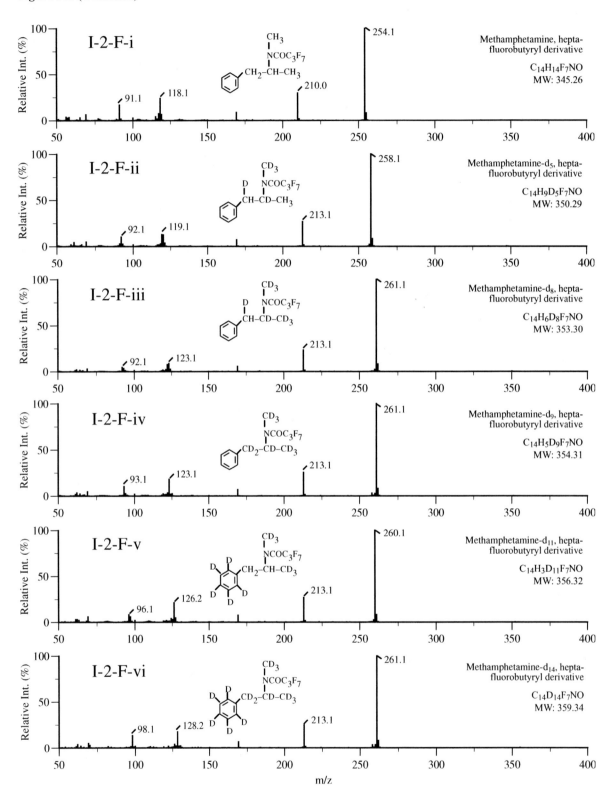

Figure I-2. (Continued)

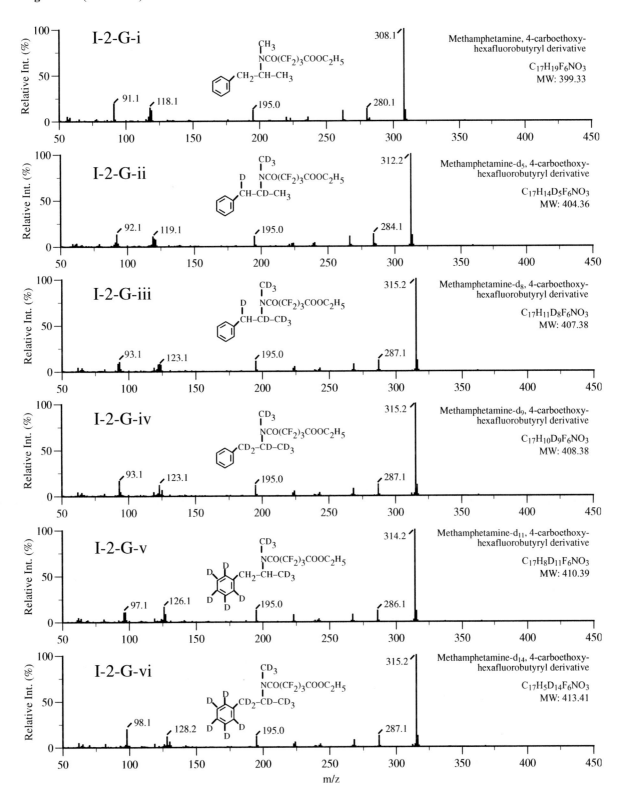

Figure I — Stimulants

Figure I-2. (Continued)

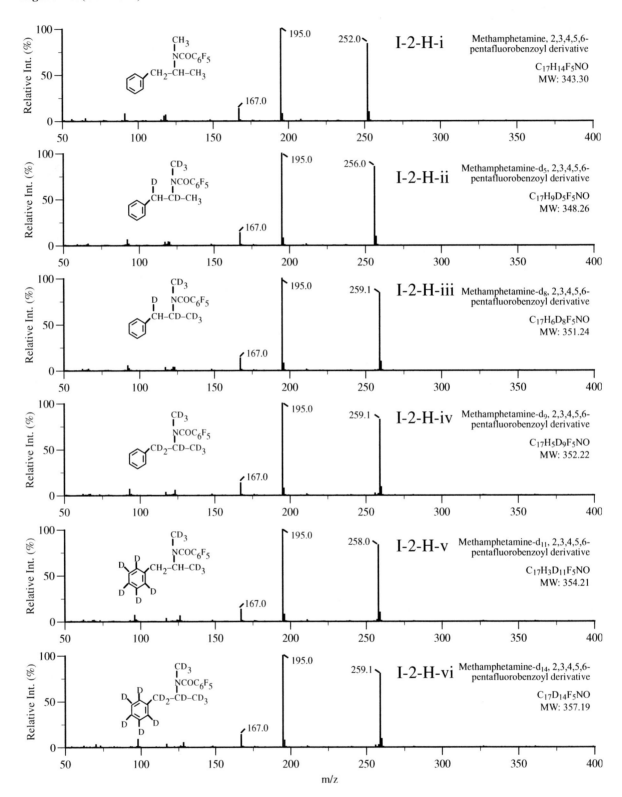

I-2-H-i — Methamphetamine, 2,3,4,5,6-pentafluorobenzoyl derivative, $C_{17}H_{14}F_5NO$, MW: 343.30

I-2-H-ii — Methamphetamine-d_5, 2,3,4,5,6-pentafluorobenzoyl derivative, $C_{17}H_9D_5F_5NO$, MW: 348.26

I-2-H-iii — Methamphetamine-d_8, 2,3,4,5,6-pentafluorobenzoyl derivative, $C_{17}H_6D_8F_5NO$, MW: 351.24

I-2-H-iv — Methamphetamine-d_9, 2,3,4,5,6-pentafluorobenzoyl derivative, $C_{17}H_5D_9F_5NO$, MW: 352.22

I-2-H-v — Methamphetamine-d_{11}, 2,3,4,5,6-pentafluorobenzoyl derivative, $C_{17}H_3D_{11}F_5NO$, MW: 354.21

I-2-H-vi — Methamphetamine-d_{14}, 2,3,4,5,6-pentafluorobenzoyl derivative, $C_{17}D_{14}F_5NO$, MW: 357.19

Appendix One — Mass Spectra

Figure I-2. (Continued)

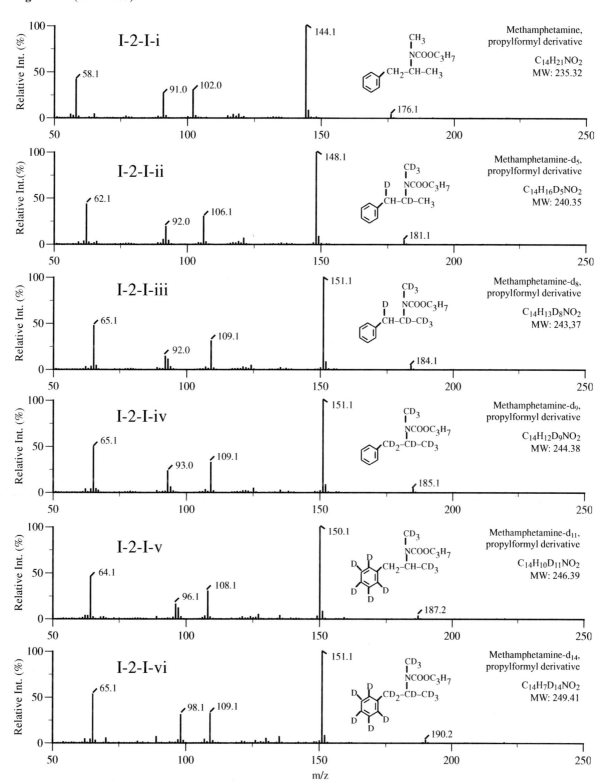

I-2-I-i — Methamphetamine, propylformyl derivative, $C_{14}H_{21}NO_2$, MW: 235.32. Fragments: 58.1, 91.0, 102.0, 144.1, 176.1.

I-2-I-ii — Methamphetamine-d_5, propylformyl derivative, $C_{14}H_{16}D_5NO_2$, MW: 240.35. Fragments: 62.1, 92.0, 106.1, 148.1, 181.1.

I-2-I-iii — Methamphetamine-d_8, propylformyl derivative, $C_{14}H_{13}D_8NO_2$, MW: 243.37. Fragments: 65.1, 92.0, 109.1, 151.1, 184.1.

I-2-I-iv — Methamphetamine-d_9, propylformyl derivative, $C_{14}H_{12}D_9NO_2$, MW: 244.38. Fragments: 65.1, 93.0, 109.1, 151.1, 185.1.

I-2-I-v — Methamphetamine-d_{11}, propylformyl derivative, $C_{14}H_{10}D_{11}NO_2$, MW: 246.39. Fragments: 64.1, 96.1, 108.1, 150.1, 187.2.

I-2-I-vi — Methamphetamine-d_{14}, propylformyl derivative, $C_{14}H_7D_{14}NO_2$, MW: 249.41. Fragments: 65.1, 98.1, 109.1, 151.1, 190.2.

Figure I — Stimulants

Figure I-2. (Continued)

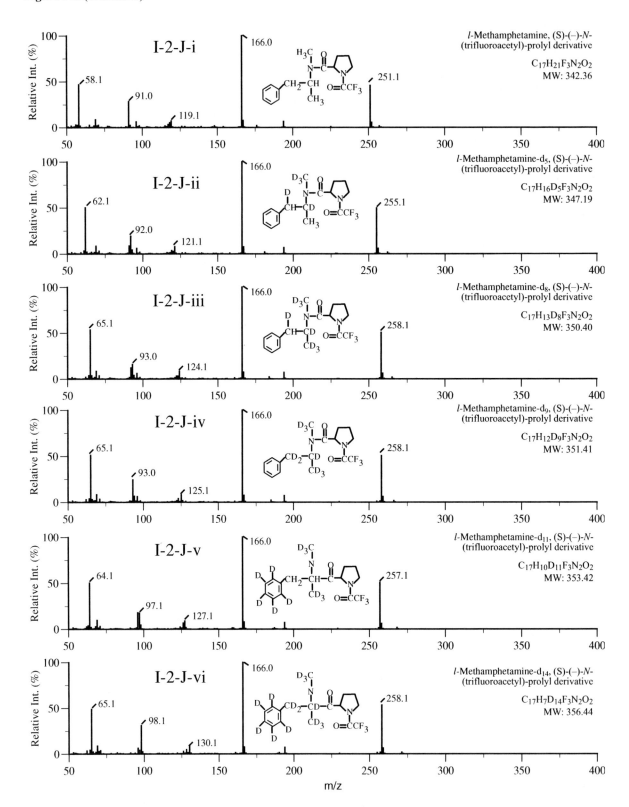

Figure I-2. (Continued)

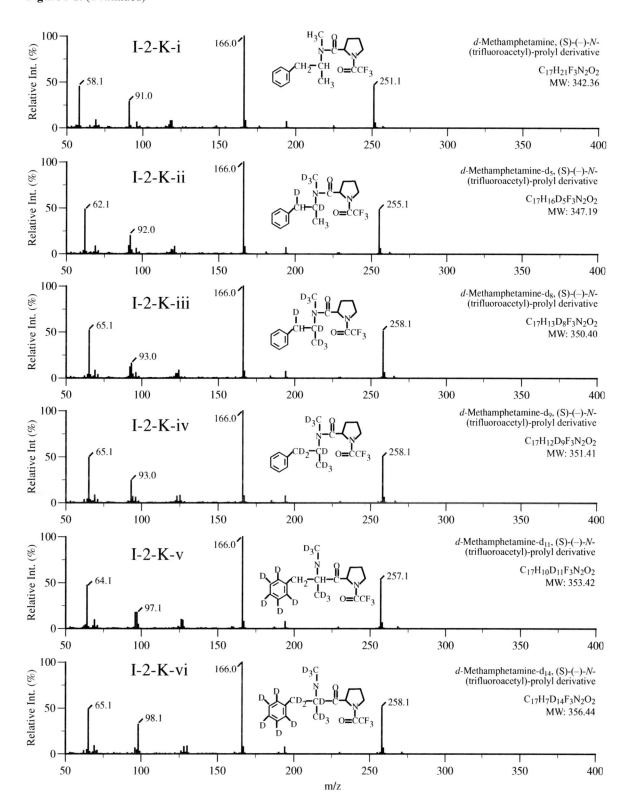

Figure I — Stimulants

Figure I-2. (Continued)

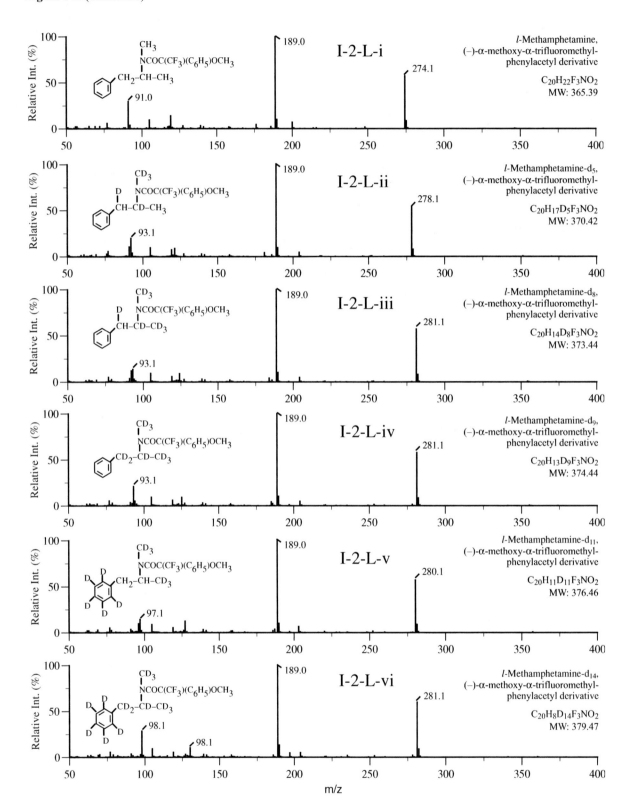

Figure I-2. (Continued)

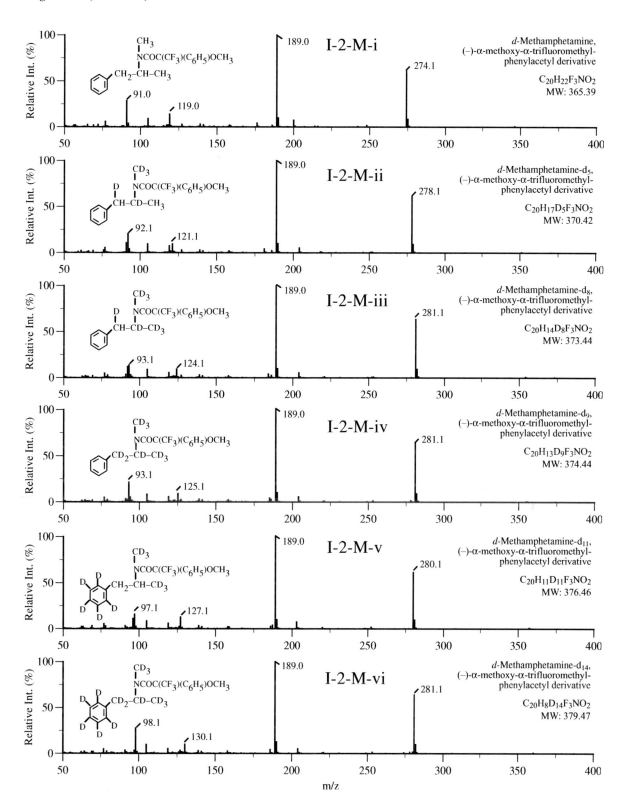

Figure I — Stimulants

Figure I-2. (Continued)

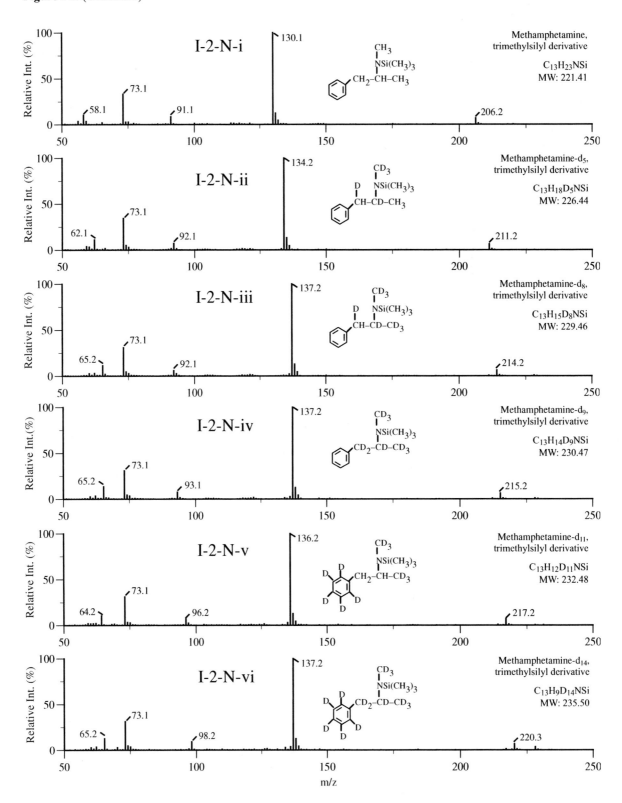

Figure I-2. (Continued)

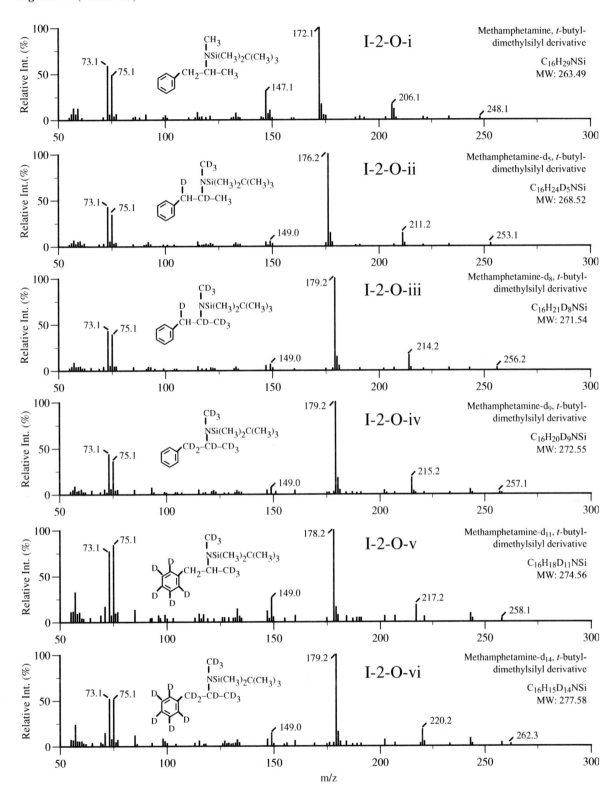

I-2-O-i — Methamphetamine, *t*-butyl-dimethylsilyl derivative — $C_{16}H_{29}NSi$ — MW: 263.49

I-2-O-ii — Methamphetamine-d_5, *t*-butyl-dimethylsilyl derivative — $C_{16}H_{24}D_5NSi$ — MW: 268.52

I-2-O-iii — Methamphetamine-d_8, *t*-butyl-dimethylsilyl derivative — $C_{16}H_{21}D_8NSi$ — MW: 271.54

I-2-O-iv — Methamphetamine-d_9, *t*-butyl-dimethylsilyl derivative — $C_{16}H_{20}D_9NSi$ — MW: 272.55

I-2-O-v — Methamphetamine-d_{11}, *t*-butyl-dimethylsilyl derivative — $C_{16}H_{18}D_{11}NSi$ — MW: 274.56

I-2-O-vi — Methamphetamine-d_{14}, *t*-butyl-dimethylsilyl derivative — $C_{16}H_{15}D_{14}NSi$ — MW: 277.58

Figure I — Stimulants

Figure I-3. Mass spectra of ephedrine and its deuterated analog (ephedrine-d₃): (A) underivatized; (B) acetyl-derivatized; (C) TCA-derivatized; (D) [TFA]₂-derivatized; (E) [PFP]₂-derivatized; (F) [HFB]₂-derivatized; (G) 4-CB-derivatized; (H) PFB-derivatized; (I) propylformyl-derivatized; (J) *d*-TPC-derivatized; (K) *d*-MTPA-derivatized; (L) [TMS]₂-derivatized.

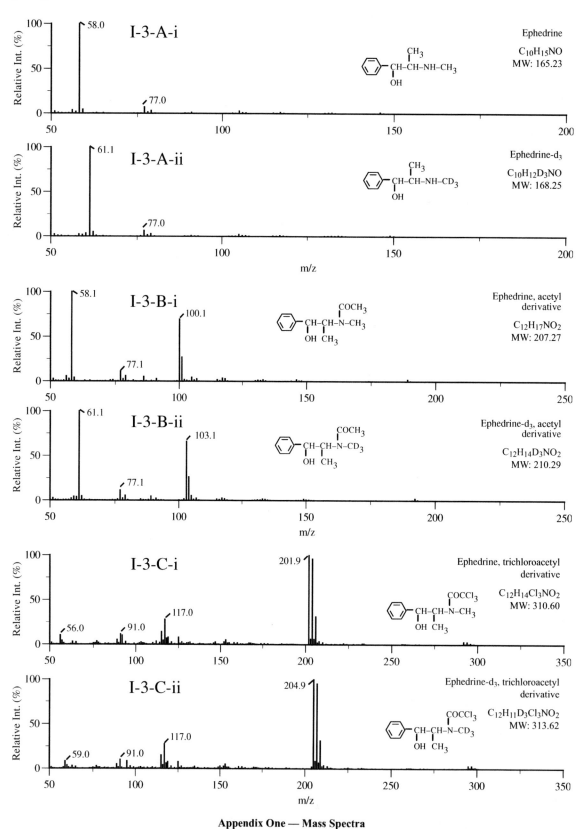

Figure I-3. (Continued)

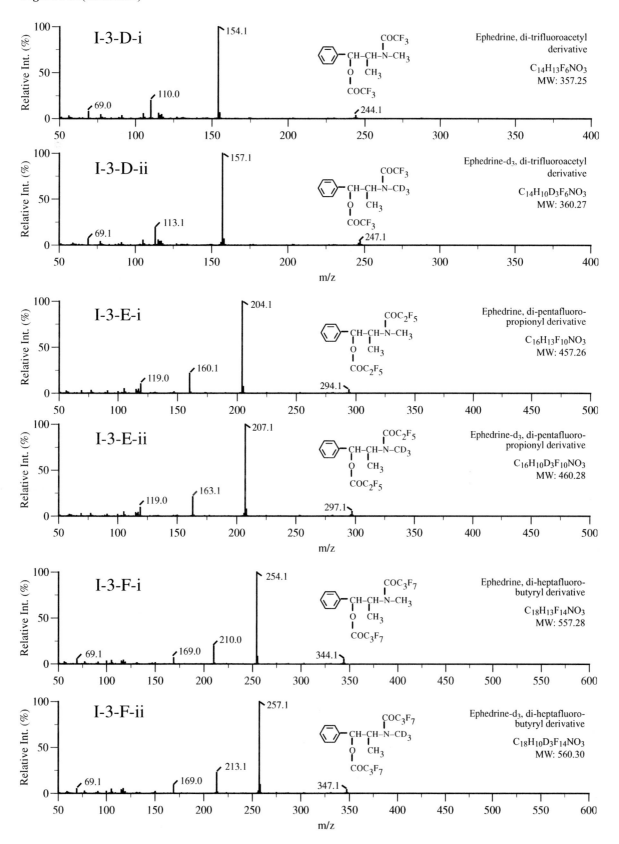

Figure I — Stimulants

Figure I-3. (Continued)

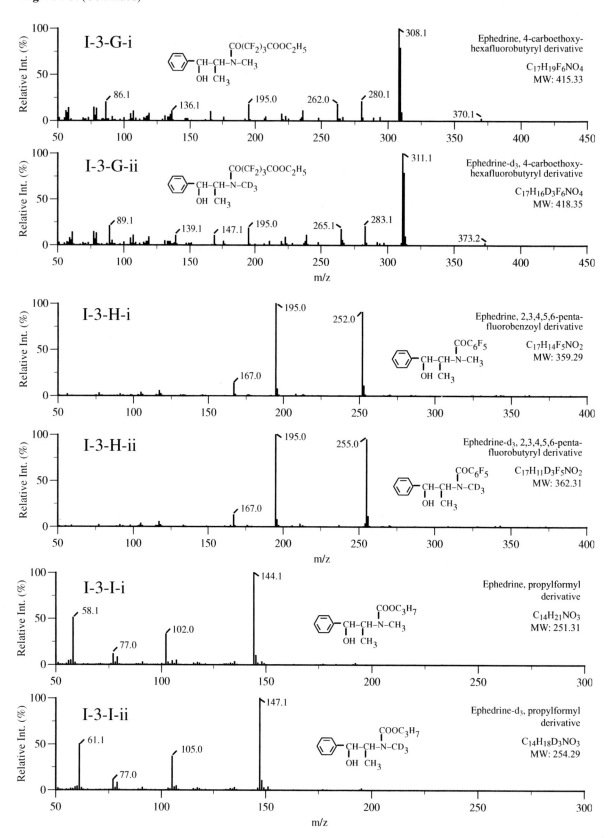

Appendix One — Mass Spectra

Figure I-3. (Continued)

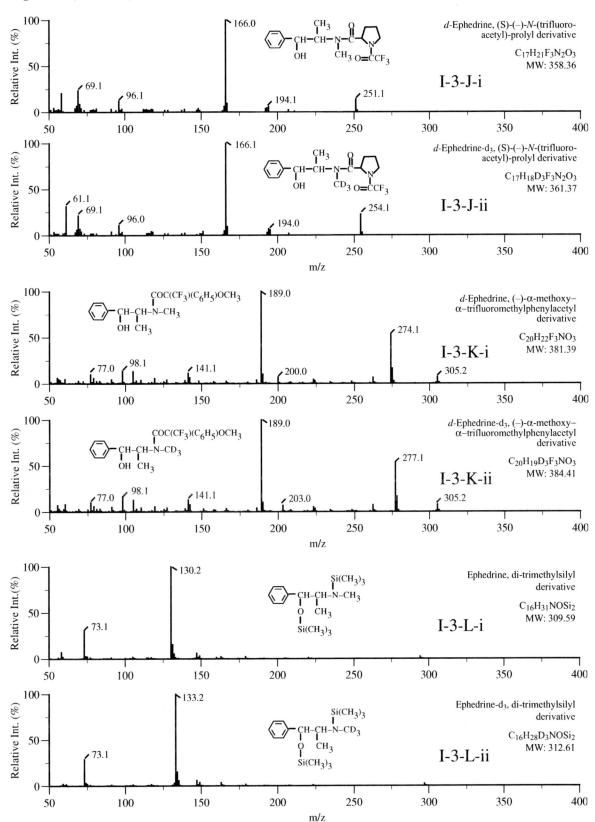

Figure I — Stimulants

Figure I-4. Mass spectra of phenylpropanolamine (PPA) and its deuterated analog (PPA-d_3): (A) underivatized; (B) acetyl-derivatized; (C) TCA-derivatized; (D) [TFA]$_2$-derivatized; (E) [PFP]$_2$-derivatized; (F) [HFB]$_2$-derivatized; (G) 4-CB-derivatized; (H) PFB-derivatized; (I,J) l-TPC-derivatized; (K,L) l-MTPA-derivatized; (M) [TMS]$_2$-derivatized; (N) t-BDMS-derivatized; (O) [t-BDMS]$_2$-derivatized; (P) TFA/[t-BDMS]$_2$-derivatized; (Q) PFP/[t-BDMS]$_2$-derivatized; (R) HFB/[t-BDMS]$_2$-derivatized.

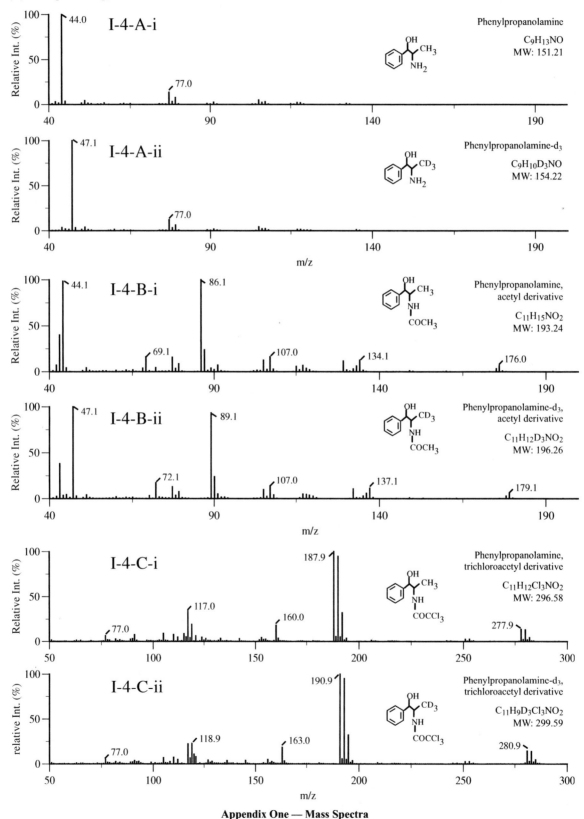

Figure I-4. (Continued)

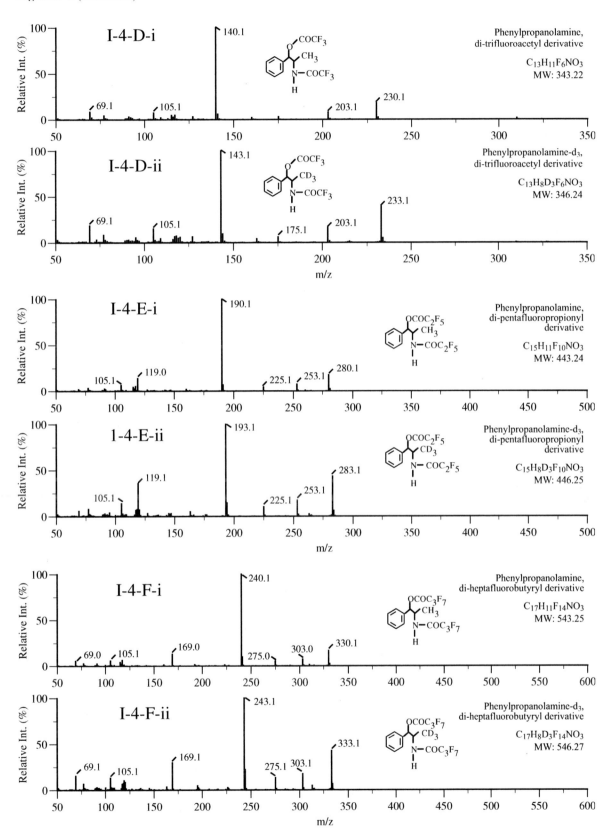

I-4-D-i — 140.1, 69.1, 105.1, 203.1, 230.1 — Phenylpropanolamine, di-trifluoroacetyl derivative — $C_{13}H_{11}F_6NO_3$ — MW: 343.22

I-4-D-ii — 143.1, 69.1, 105.1, 175.1, 203.1, 233.1 — Phenylpropanolamine-d$_3$, di-trifluoroacetyl derivative — $C_{13}H_8D_3F_6NO_3$ — MW: 346.24

I-4-E-i — 190.1, 105.1, 119.0, 225.1, 253.1, 280.1 — Phenylpropanolamine, di-pentafluoropropionyl derivative — $C_{15}H_{11}F_{10}NO_3$ — MW: 443.24

1-4-E-ii — 193.1, 105.1, 119.1, 225.1, 253.1, 283.1 — Phenylpropanolamine-d$_3$, di-pentafluoropropionyl derivative — $C_{15}H_8D_3F_{10}NO_3$ — MW: 446.25

I-4-F-i — 240.1, 69.0, 105.1, 169.0, 275.0, 303.0, 330.1 — Phenylpropanolamine, di-heptafluorobutyryl derivative — $C_{17}H_{11}F_{14}NO_3$ — MW: 543.25

I-4-F-ii — 243.1, 69.1, 105.1, 169.1, 275.1, 303.1, 333.1 — Phenylpropanolamine-d$_3$, di-heptafluorobutyryl derivative — $C_{17}H_8D_3F_{14}NO_3$ — MW: 546.27

Figure I — Stimulants

Figure I-4. (Continued)

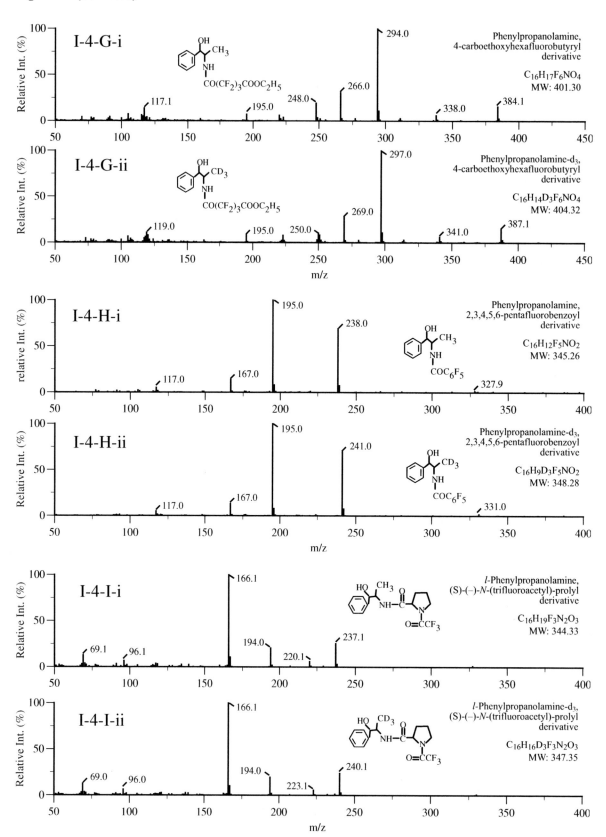

Figure I-4. (Continued)

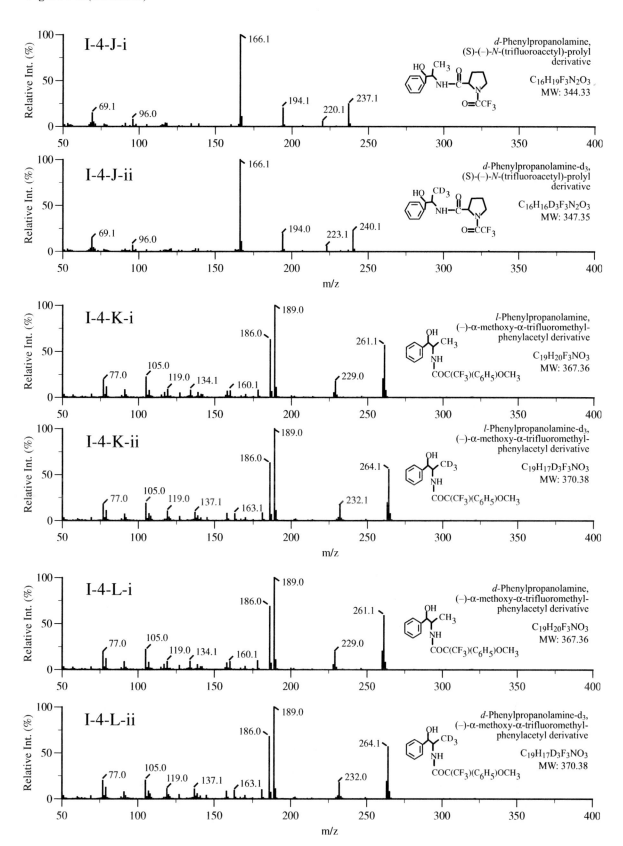

I-4-J-i

d-Phenylpropanolamine,
(S)-(–)-*N*-(trifluoroacetyl)-prolyl
derivative

$C_{16}H_{19}F_3N_2O_3$
MW: 344.33

I-4-J-ii

d-Phenylpropanolamine-d_3,
(S)-(–)-*N*-(trifluoroacetyl)-prolyl
derivative

$C_{16}H_{16}D_3F_3N_2O_3$
MW: 347.35

I-4-K-i

l-Phenylpropanolamine,
(–)-α-methoxy-α-trifluoromethyl-
phenylacetyl derivative

$C_{19}H_{20}F_3NO_3$
MW: 367.36

I-4-K-ii

l-Phenylpropanolamine-d_3,
(–)-α-methoxy-α-trifluoromethyl-
phenylacetyl derivative

$C_{19}H_{17}D_3F_3NO_3$
MW: 370.38

I-4-L-i

d-Phenylpropanolamine,
(–)-α-methoxy-α-trifluoromethyl-
phenylacetyl derivative

$C_{19}H_{20}F_3NO_3$
MW: 367.36

I-4-L-ii

d-Phenylpropanolamine-d_3,
(–)-α-methoxy-α-trifluoromethyl-
phenylacetyl derivative

$C_{19}H_{17}D_3F_3NO_3$
MW: 370.38

Figure I — Stimulants

Figure I-4. (Continued)

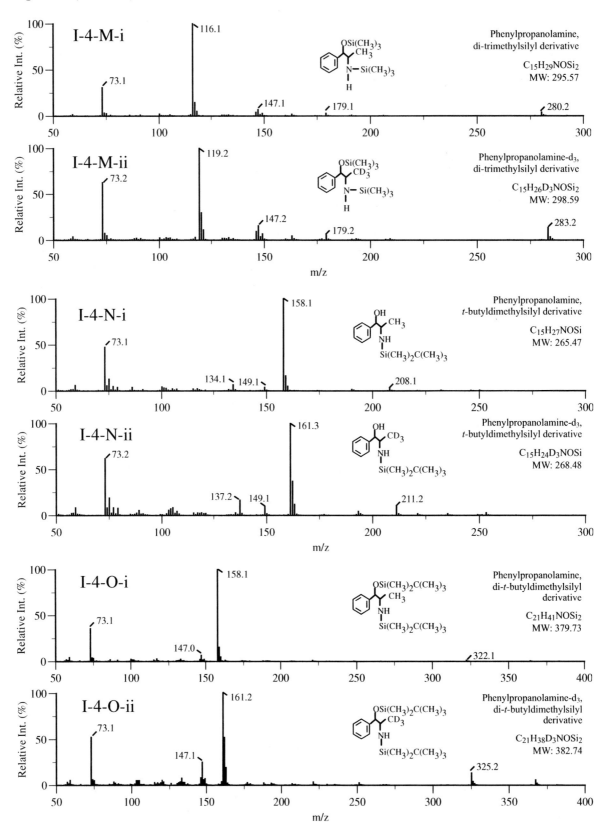

Figure I-4. (Continued)

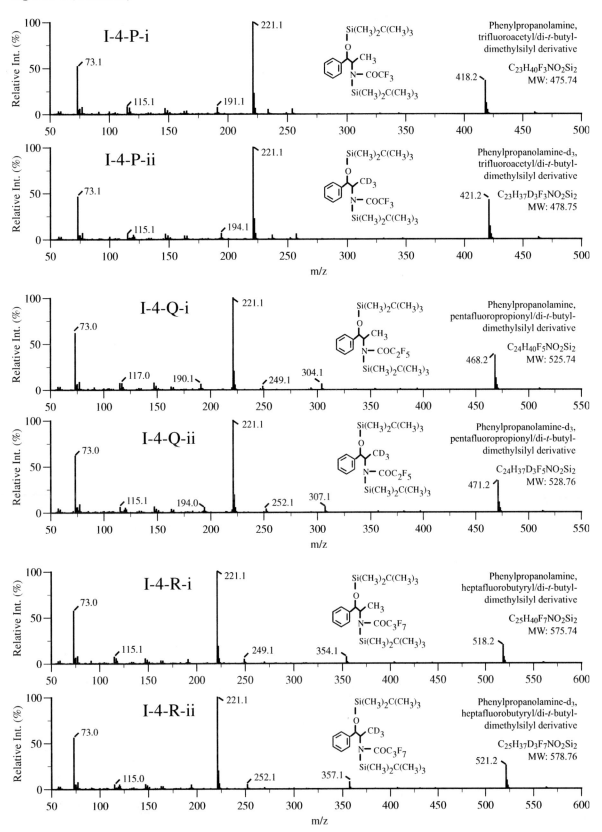

I-4-P-i — Phenylpropanolamine, trifluoroacetyl/di-*t*-butyl-dimethylsilyl derivative — C$_{23}$H$_{40}$F$_3$NO$_2$Si$_2$ — MW: 475.74

I-4-P-ii — Phenylpropanolamine-d$_3$, trifluoroacetyl/di-*t*-butyl-dimethylsilyl derivative — C$_{23}$H$_{37}$D$_3$F$_3$NO$_2$Si$_2$ — MW: 478.75

I-4-Q-i — Phenylpropanolamine, pentafluoropropionyl/di-*t*-butyl-dimethylsilyl derivative — C$_{24}$H$_{40}$F$_5$NO$_2$Si$_2$ — MW: 525.74

I-4-Q-ii — Phenylpropanolamine-d$_3$, pentafluoropropionyl/di-*t*-butyl-dimethylsilyl derivative — C$_{24}$H$_{37}$D$_3$F$_5$NO$_2$Si$_2$ — MW: 528.76

I-4-R-i — Phenylpropanolamine, heptafluorobutyryl/di-*t*-butyl-dimethylsilyl derivative — C$_{25}$H$_{40}$F$_7$NO$_2$Si$_2$ — MW: 575.74

I-4-R-ii — Phenylpropanolamine-d$_3$, heptafluorobutyryl/di-*t*-butyl-dimethylsilyl derivative — C$_{25}$H$_{37}$D$_3$F$_7$NO$_2$Si$_2$ — MW: 578.76

Figure I — Stimulants

Figure I-5. Mass spectra of 3,4-methylenedioxyamphetamine (MDA) and its deuterated analog (MDA-d$_5$): (A) underivatized; (B) acetyl-derivatized; (C) TCA-derivatized; (D) TFA-derivatized; (E) PFP-derivatized; (F) HFB-derivatized; (G) 4-CB-derivatized; (H) PFB-derivatized; (I) propylformyl-derivatized; (J,K) *l*-TPC-derivatized; (L,M) *l*-MTPA-derivatized; (N) TMS-derivatized; (O) TFA/*t*-BDMS-derivatized; (P) PFP/*t*-BDMS-derivatized; (Q) HFB/*t*-BDMS-derivatized

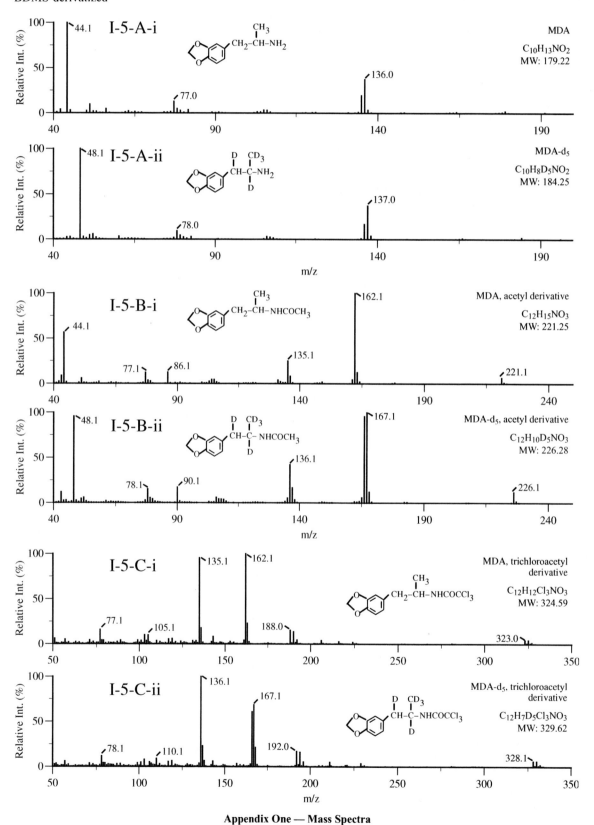

Figure I-5. (Continued)

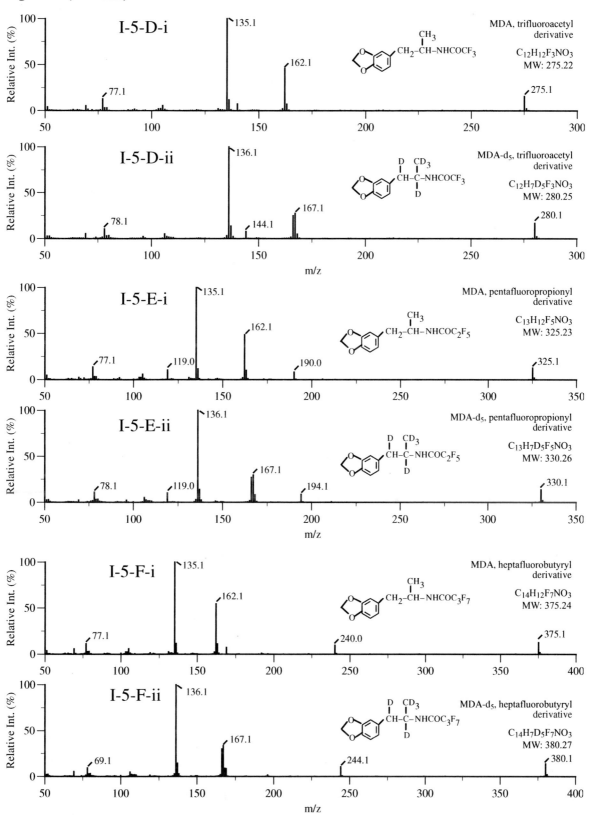

Figure I — Stimulants

Figure I-5. (Continued)

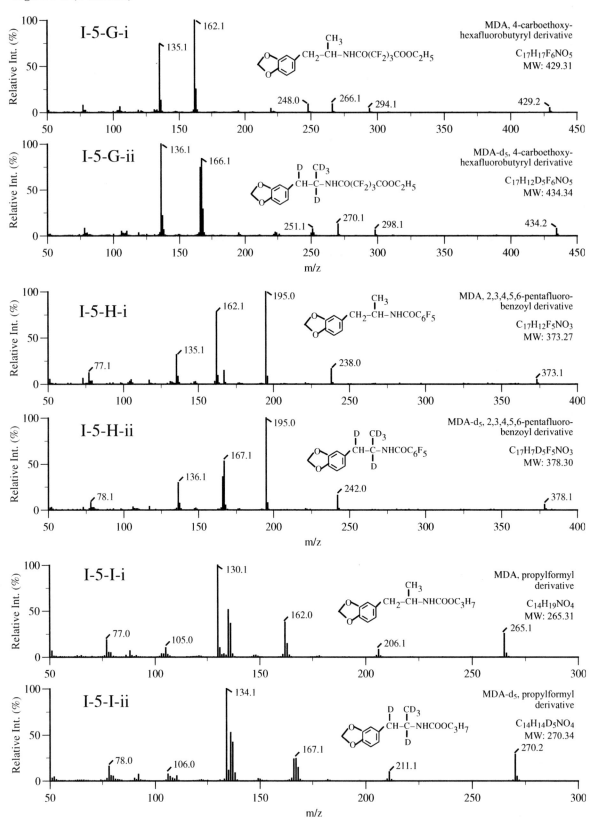

Figure I-5. (Continued)

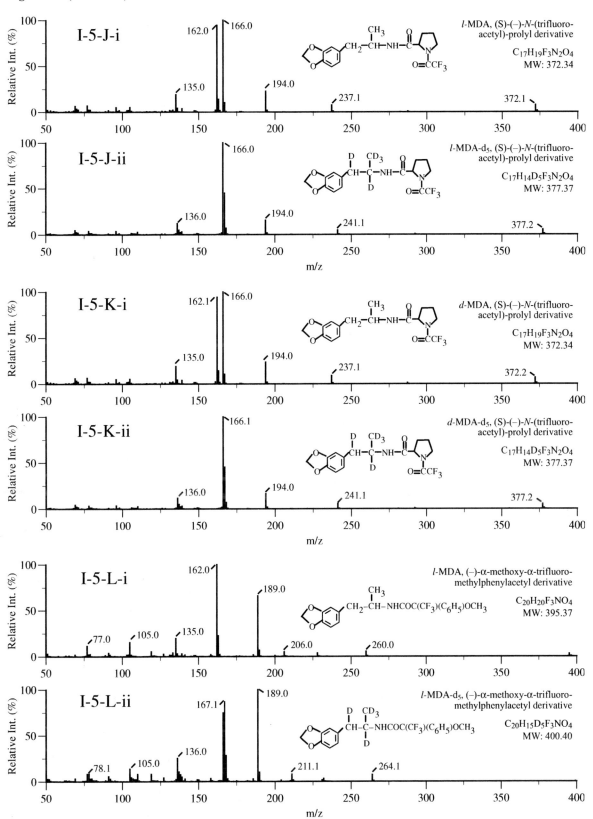

Figure I — Stimulants

Figure I-5. (Continued)

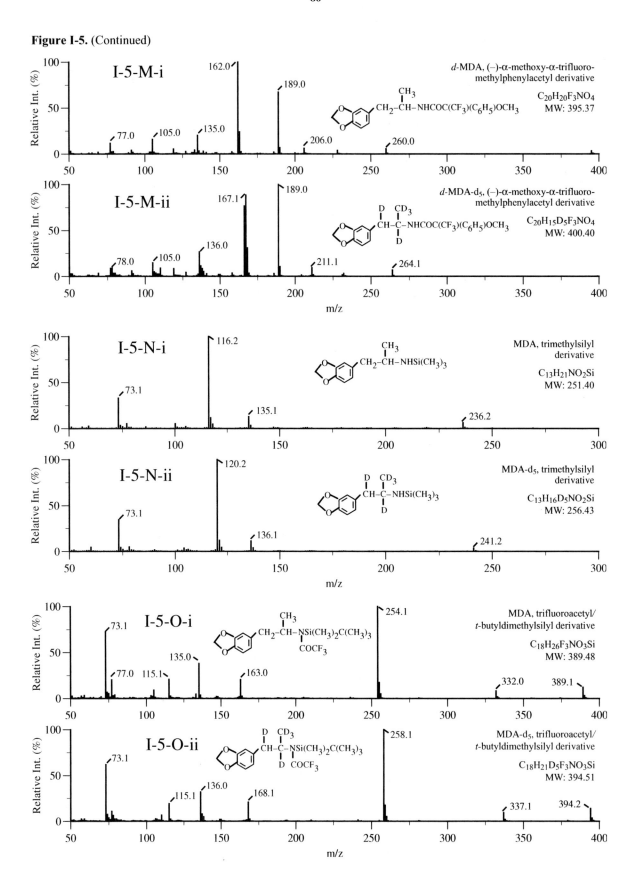

Figure I-5. (Continued)

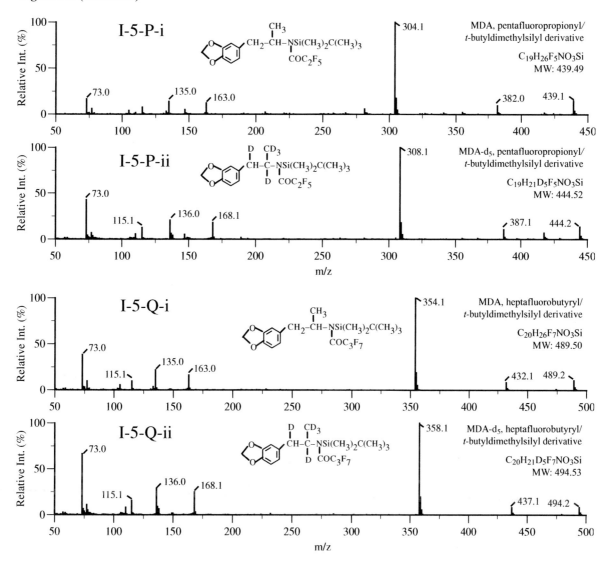

Figure I — Stimulants

Figure I-6. Mass spectra of 3,4-methylenedioxymethamphetamine (MDMA) and its deuterated analog (MDMA-d_5): (A) underivatized; (B) acetyl-derivatized; (C) TCA-derivatized; (D) TFA-derivatized; (E) PFP-derivatized; (F) HFB-derivatized; (G) 4-CB-derivatized; (H) PFB-derivatized; (I) propylformyl-derivatized; (J,K) *l*-TPC-derivatized; (L,M) *l*-MTPA-derivatized; (N) TMS-derivatized.

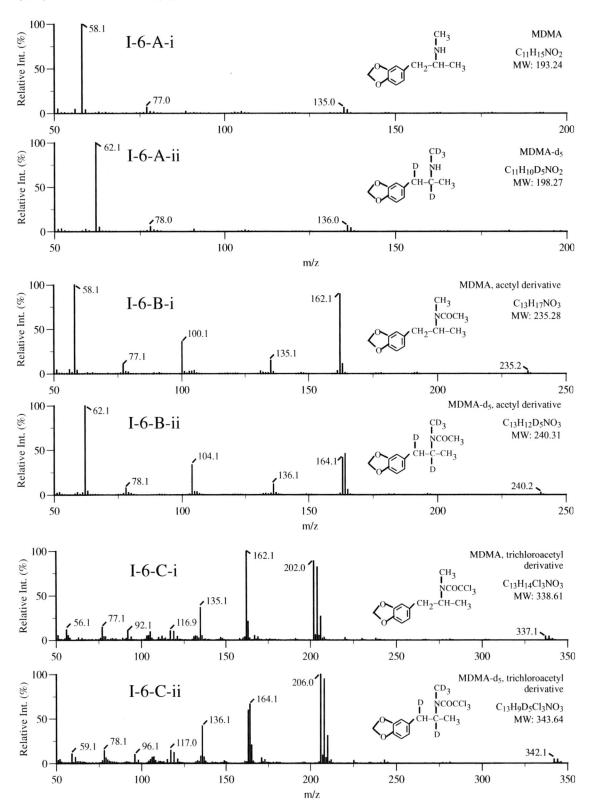

89

Figure I-6. (Continued)

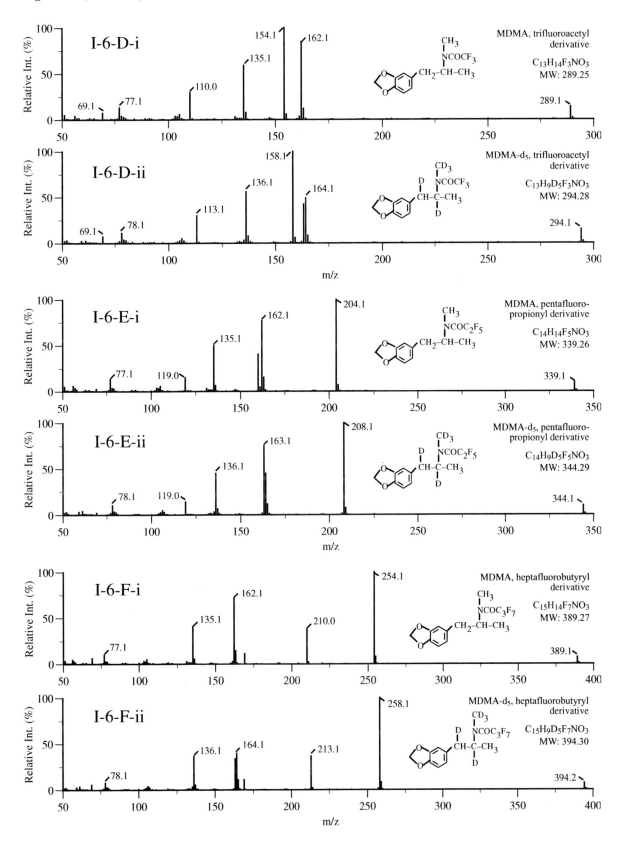

Figure 1 — Stimulants

Figure I-6. (Continued)

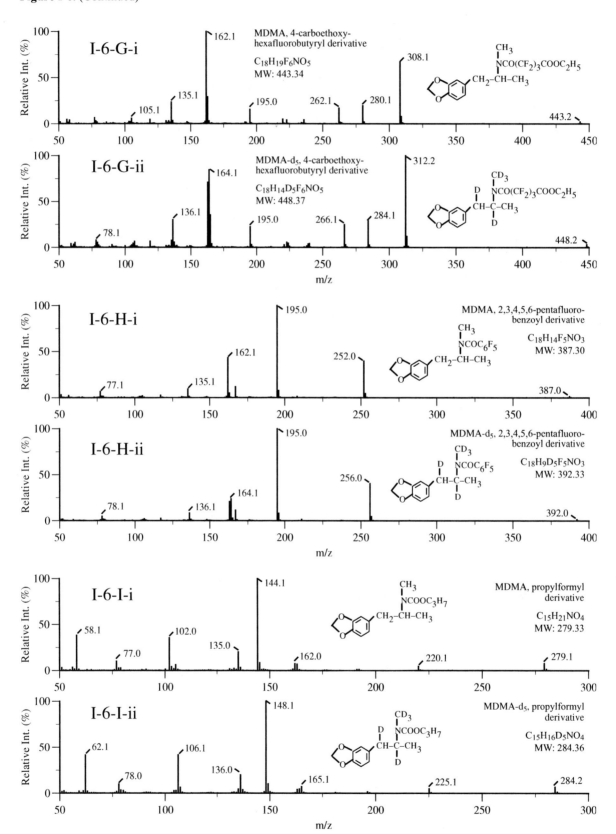

Figure I-6. (Continued)

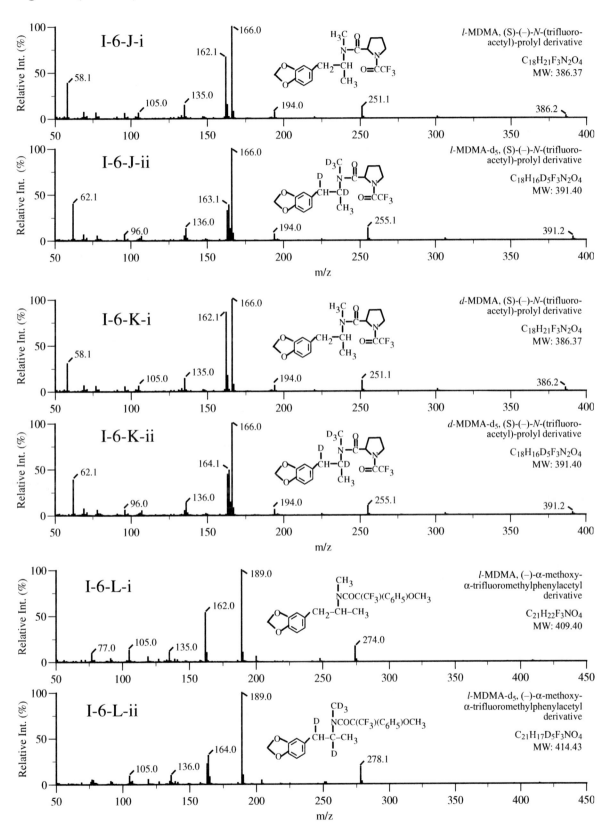

Figure 1 — Stimulants

Figure I-6. (Continued)

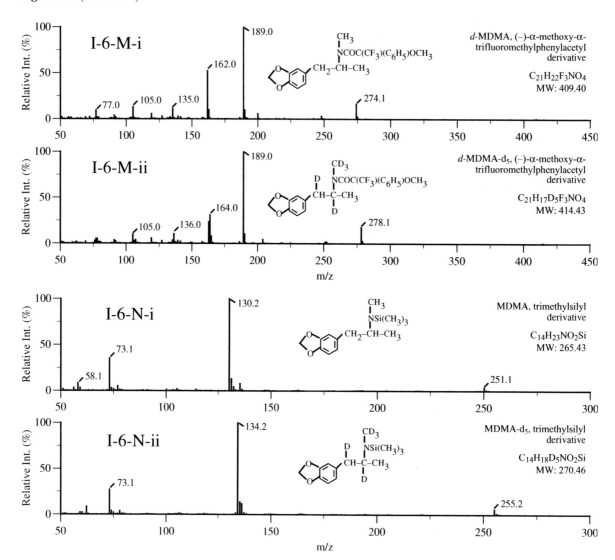

Figure I-7. Mass spectra of 3,4-methylenedioxyethylamphetamine (MDEA) and its deuterated analogs (MDEA-d₅, -d₆); (A) underivatized; (B) acetyl-derivatized; (C) TCA-derivatized; (D) TFA-derivatized; (E) PFP-derivatized; (F) HFB-derivatized; (G) 4-CB-derivatized; (H) PFB-derivatized; (I) propylformyl-derivatized; (J,K) *l*-TPC-derivatized; (L,M) *l*-MTPA-derivatized; (N) TMS-derivatized.

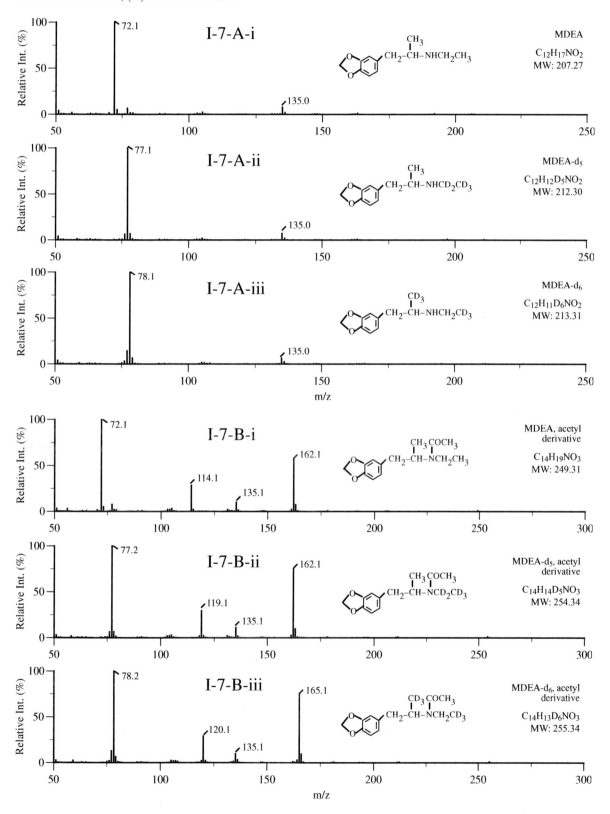

Figure 1 — Stimulants

94

Figure I-7. (Continued)

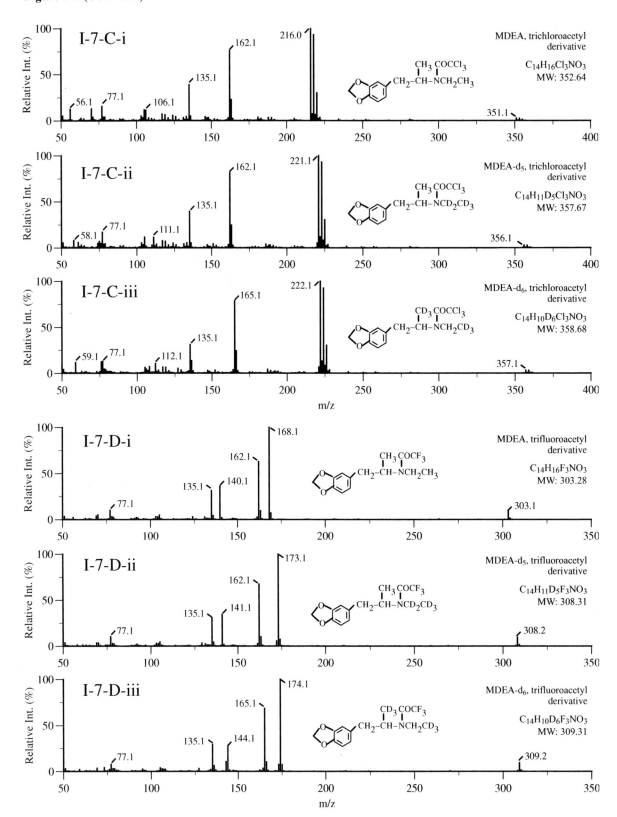

Appendix One — Mass Spectra

Figure I-7. (Continued)

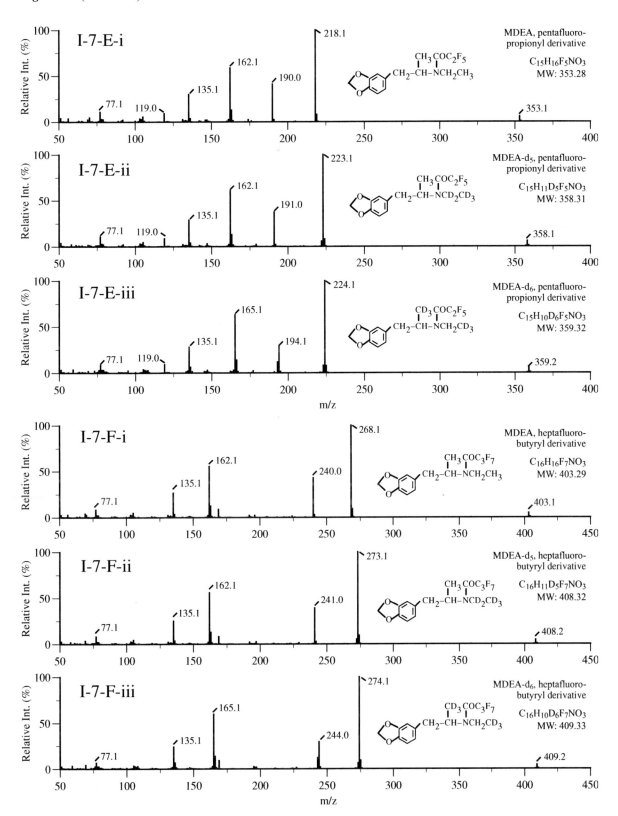

Figure 1 — Stimulants

Figure I-7. (Continued)

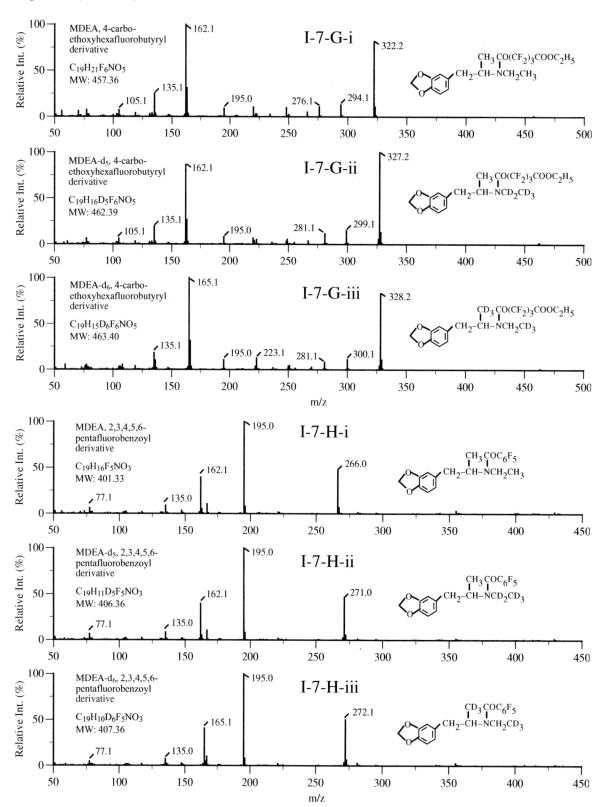

Figure I-7. (Continued)

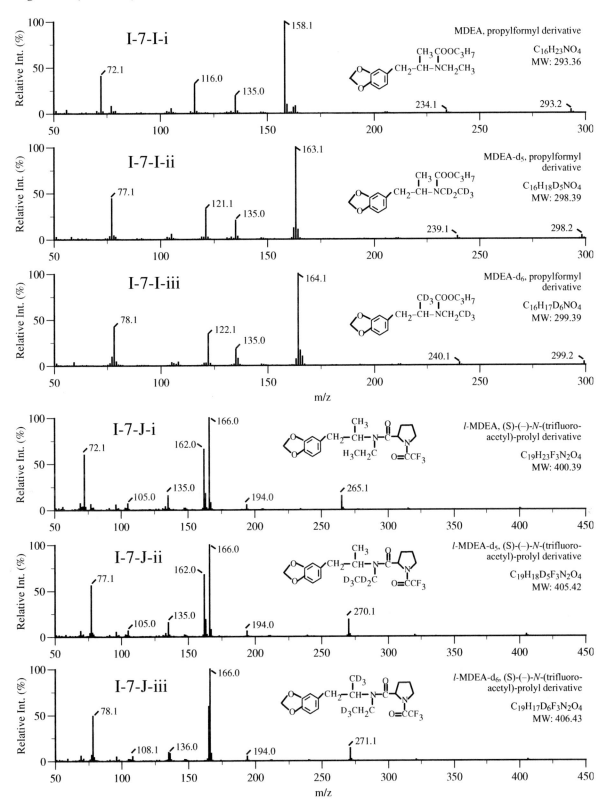

Figure 1 — Stimulants

Figure I-7. (Continued)

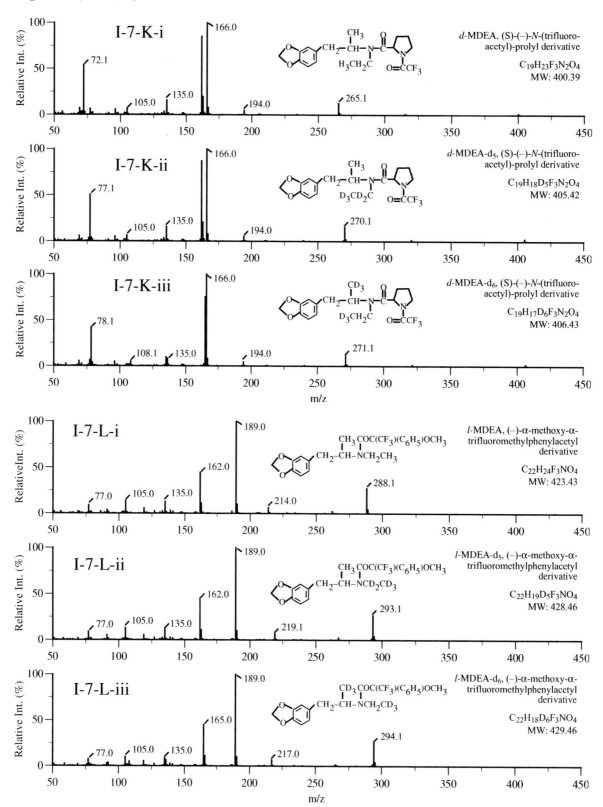

Figure I-7. (Continued)

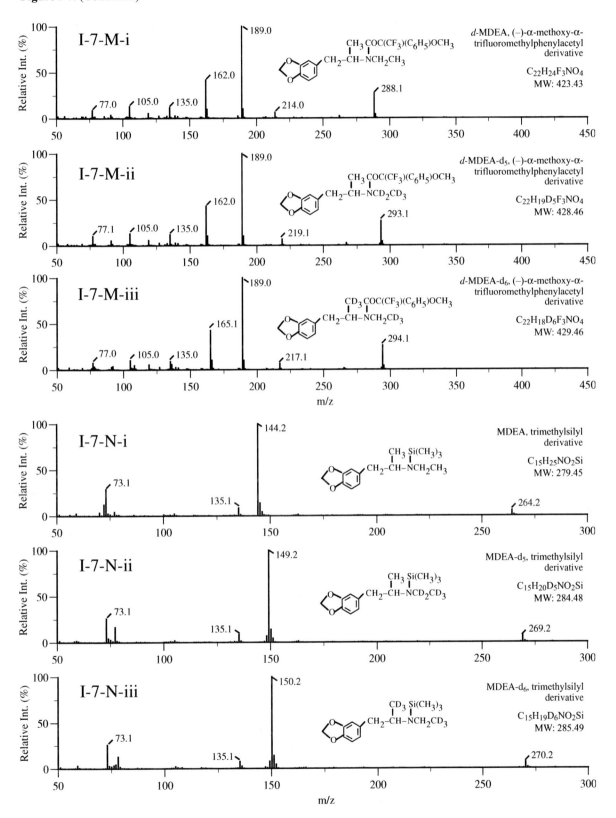

I-7-M-i

189.0
162.0
77.0 105.0 135.0
214.0
288.1

d-MDEA, (–)-α-methoxy-α-
trifluoromethylphenylacetyl
derivative

C$_{22}$H$_{24}$F$_3$NO$_4$
MW: 423.43

I-7-M-ii

189.0
162.0
77.1 105.0 135.0
219.1
293.1

d-MDEA-d$_5$, (–)-α-methoxy-α-
trifluoromethylphenylacetyl
derivative

C$_{22}$H$_{19}$D$_5$F$_3$NO$_4$
MW: 428.46

I-7-M-iii

189.0
165.1
77.0 105.0 135.0
217.1
294.1

d-MDEA-d$_6$, (–)-α-methoxy-α-
trifluoromethylphenylacetyl
derivative

C$_{22}$H$_{18}$D$_6$F$_3$NO$_4$
MW: 429.46

m/z

I-7-N-i

144.2
73.1
135.1
264.2

MDEA, trimethylsilyl
derivative

C$_{15}$H$_{25}$NO$_2$Si
MW: 279.45

I-7-N-ii

149.2
73.1
135.1
269.2

MDEA-d$_5$, trimethylsilyl
derivative

C$_{15}$H$_{20}$D$_5$NO$_2$Si
MW: 284.48

I-7-N-iii

150.2
73.1
135.1
270.2

MDEA-d$_6$, trimethylsilyl
derivative

C$_{15}$H$_{19}$D$_6$NO$_2$Si
MW: 285.49

m/z

Figure 1 — Stimulants

Figure I-8. Mass spectra of *N*-methyl-1-(3,4-methylenedioxyphenyl)-2-butanamine (MBDB), and its deuterated analog (MBDB-d₅): (A) underivatized; (B) acetyl-derivatized; (C) TCA-derivatized; (D) TFA-derivatized; (E) PFP-derivatized; (F) HFB-derivatized; (G) 4-CB-derivatized; (H) PFB-derivatized; (I) propylformyl-derivatized; (J,K) *l*-TPC-derivatized; (L,M) *l*-MTPA-derivatized; (N) TMS-derivatized.

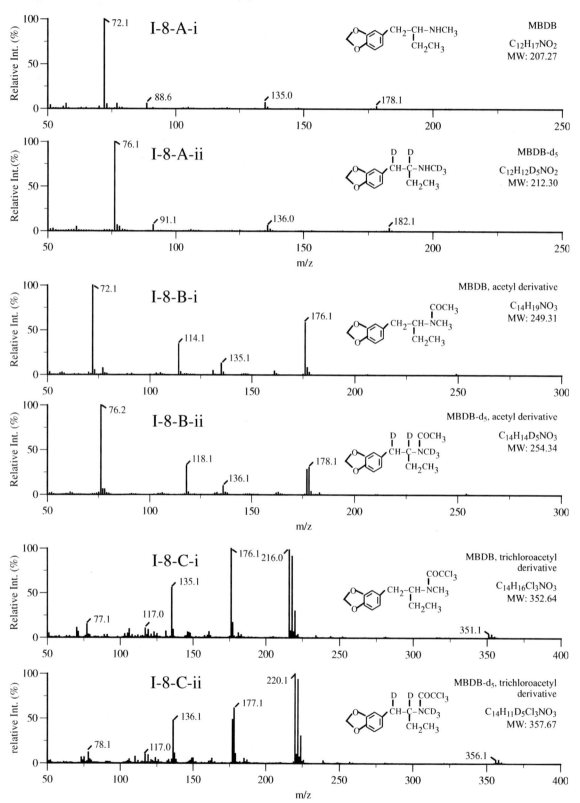

Figure I-8. (Continued)

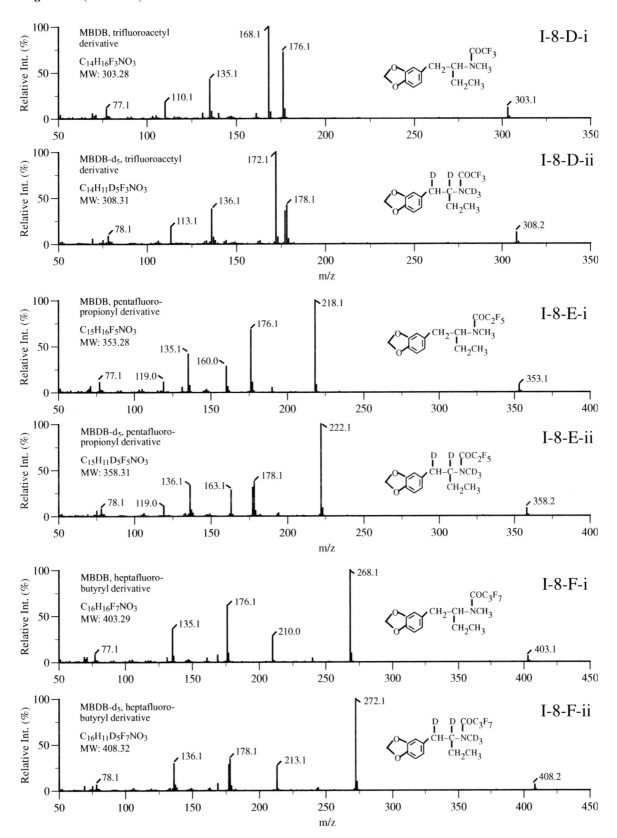

I-8-D-i

MBDB, trifluoroacetyl derivative

$C_{14}H_{16}F_3NO_3$
MW: 303.28

I-8-D-ii

MBDB-d5, trifluoroacetyl derivative

$C_{14}H_{11}D_5F_3NO_3$
MW: 308.31

I-8-E-i

MBDB, pentafluoro-propionyl derivative

$C_{15}H_{16}F_5NO_3$
MW: 353.28

I-8-E-ii

MBDB-d5, pentafluoro-propionyl derivative

$C_{15}H_{11}D_5F_5NO_3$
MW: 358.31

I-8-F-i

MBDB, heptafluoro-butyryl derivative

$C_{16}H_{16}F_7NO_3$
MW: 403.29

I-8-F-ii

MBDB-d5, heptafluoro-butyryl derivative

$C_{16}H_{11}D_5F_7NO_3$
MW: 408.32

Figure 1 — Stimulants

102

Figure I-8. (Continued)

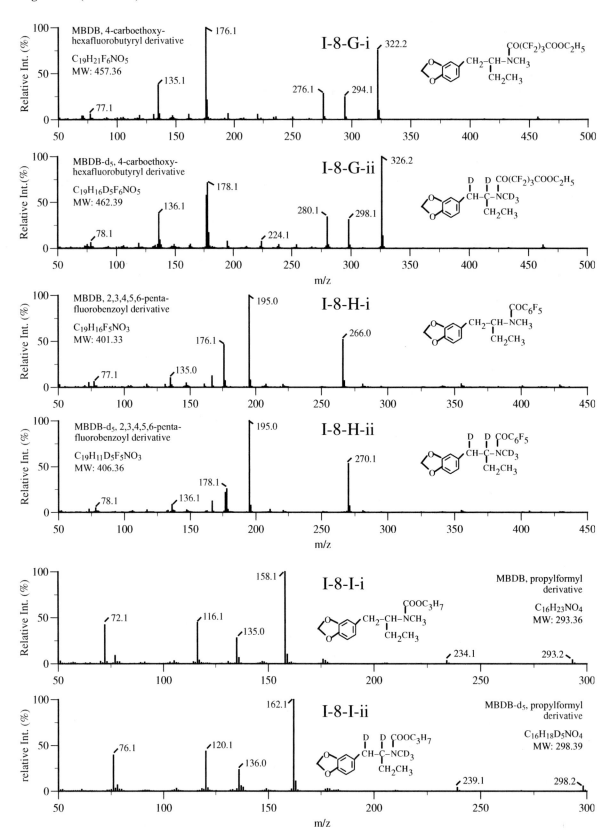

Appendix One — Mass Spectra

Figure I-8. (Continued)

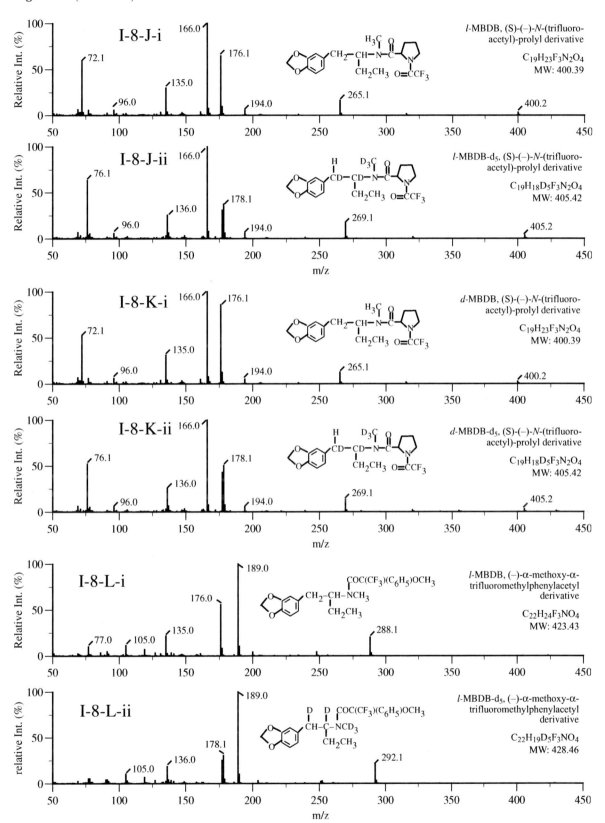

Figure 1 — Stimulants

Figure I-8. (Continued)

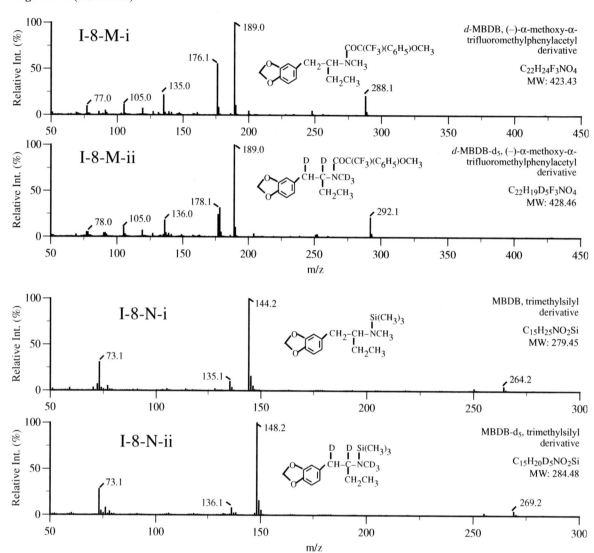

I-8-M-i

189.0

176.1

135.0

77.0 105.0

288.1

d-MBDB, (–)-α-methoxy-α-
trifluoromethylphenylacetyl
derivative

$C_{22}H_{24}F_3NO_4$
MW: 423.43

I-8-M-ii

189.0

178.1

136.0

78.0 105.0

292.1

d-MBDB-d$_5$, (–)-α-methoxy-α-
trifluoromethylphenylacetyl
derivative

$C_{22}H_{19}D_5F_3NO_4$
MW: 428.46

m/z

I-8-N-i

144.2

73.1

135.1

264.2

MBDB, trimethylsilyl
derivative

$C_{15}H_{25}NO_2Si$
MW: 279.45

I-8-N-ii

148.2

73.1

136.1

269.2

MBDB-d$_5$, trimethylsilyl
derivative

$C_{15}H_{20}D_5NO_2Si$
MW: 284.48

m/z

Figure I-9. Mass spectra of selegiline and its deuterated analog (selegiline-d$_8$).

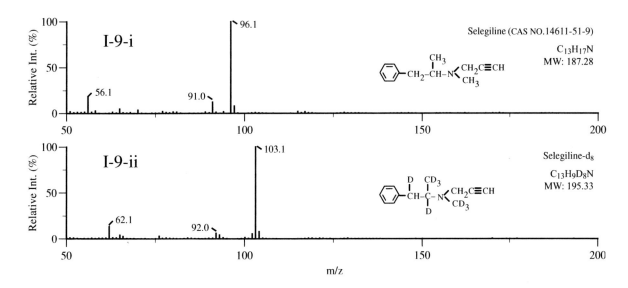

Figure I — Stimulants

Figure I-10. Mass spectra of *N*-desmethylselegiline and its deuterated analogs (*N*-desmethylselegiline-d₁₁): (A) underivatized; (B) acetyl-derivatized; (C) TCA-derivatized; (D) TFA-derivatized; (E) PFP-derivatized; (F) HFB-derivatized; (G) 4-CB-derivatized; (H) TMS-derivatized.

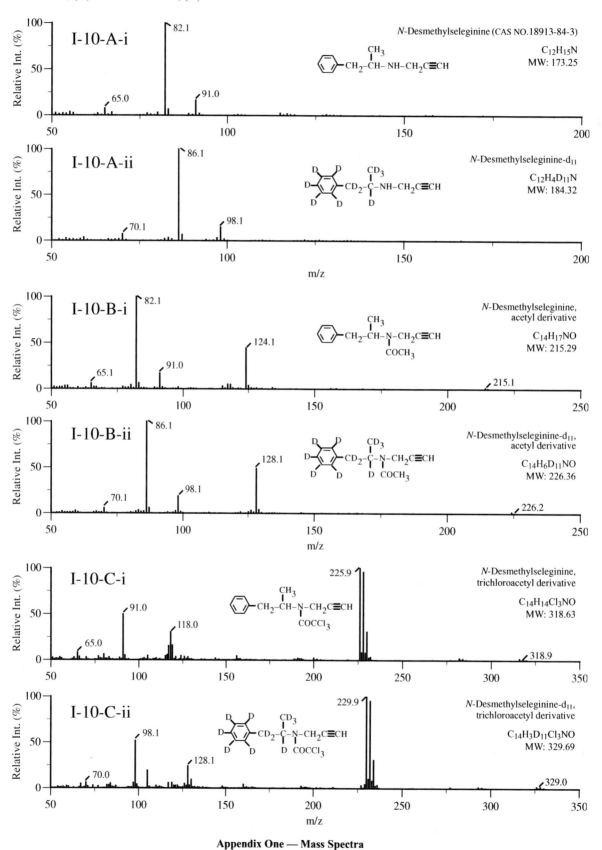

Figure I-10. (Continued)

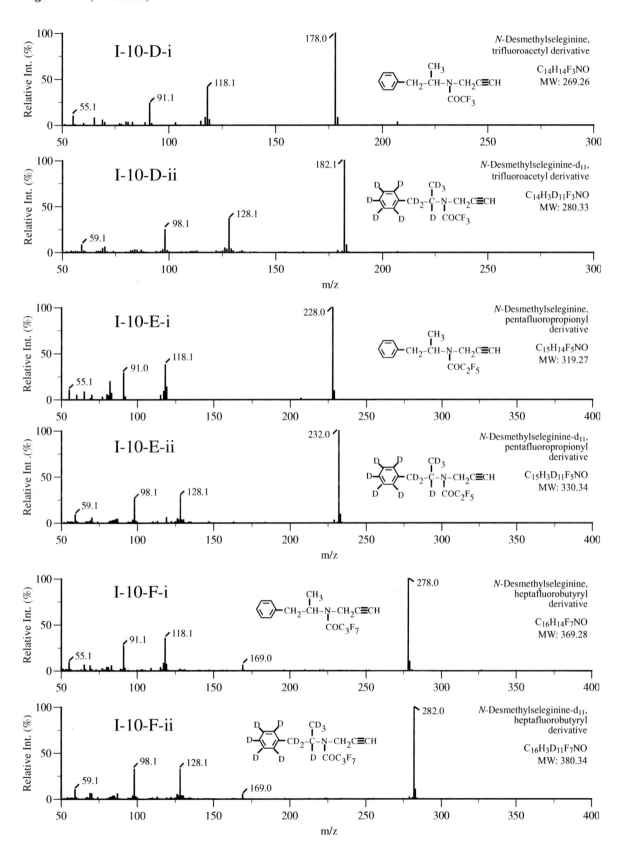

Figure I — Stimulants

Figure I-10. (Continued)

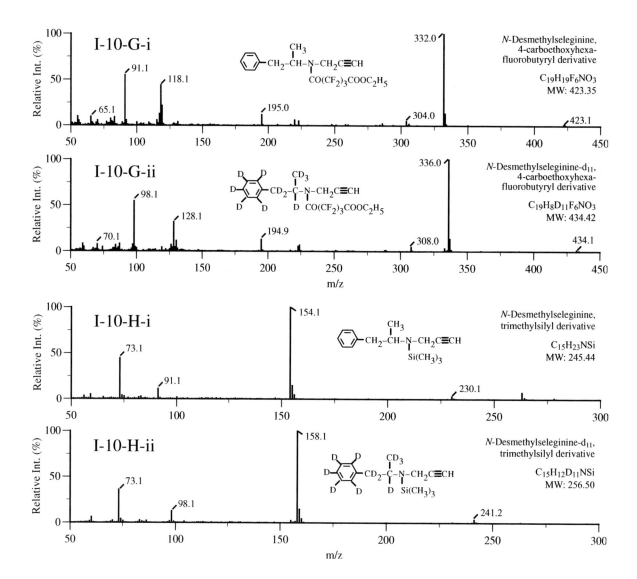

Figure I-11. Mass spectra of fenfluramine and its deuterated analogs (fenfluramine-d₁₀): (A) underivatized; (B) acetyl-derivatized; (C) TCA-derivatized; (D) TFA-derivatized; (E) PFP-derivatized; (F) HFB-derivatized; (G) 4-CB-derivatized.

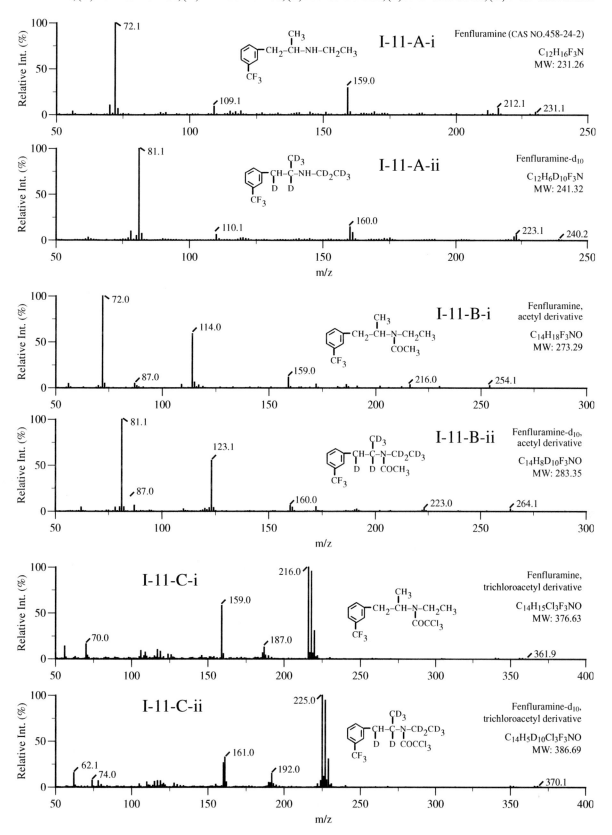

Figure I — Stimulants

Figure I-11. (Continued)

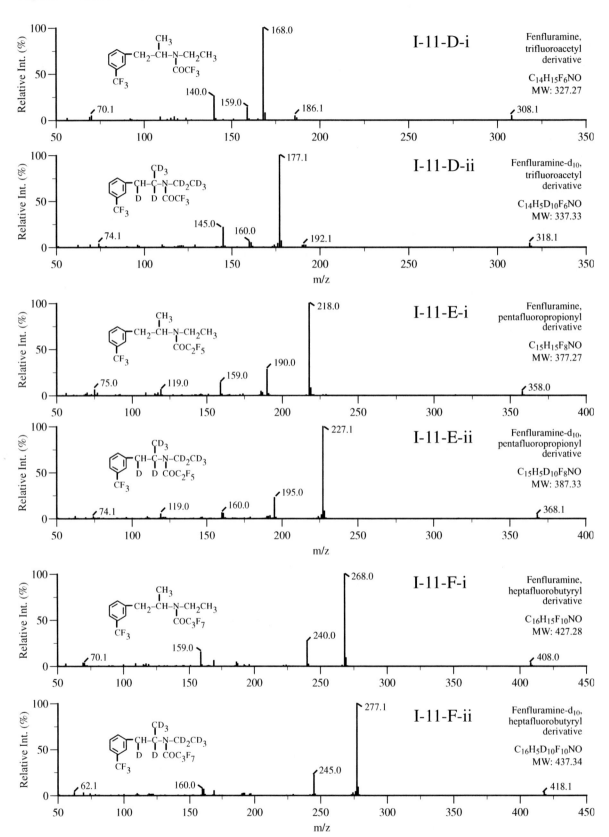

Figure I-11. (Continued)

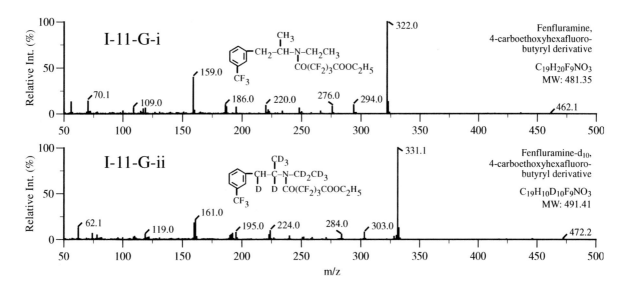

Figure I — Stimulants

Figure I-12. Mass spectra of norcocaine, and its deuterated analog (norcocaine-d₃): (A) underivatized; (B) TFA-derivatized; (C) PFP-derivatized; (D) HFB-derivatized; (E) TMS-derivatized.

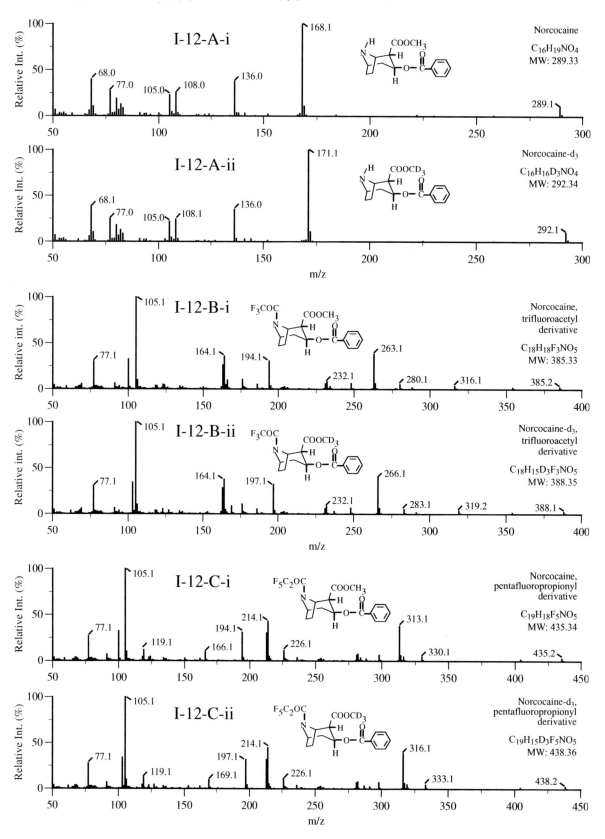

Figure I-12. (Continued)

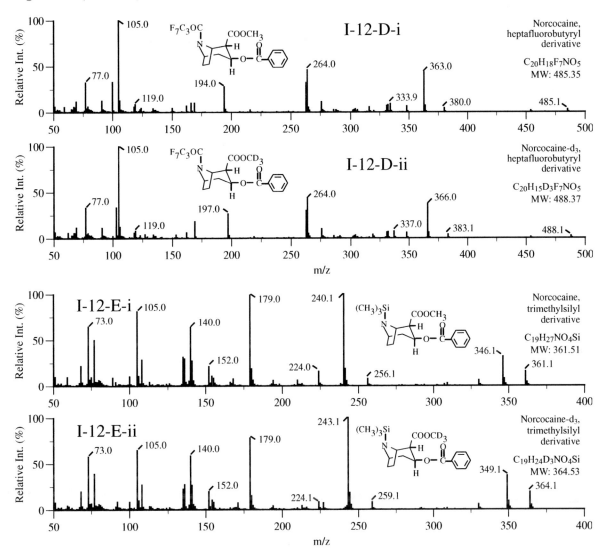

Figure I — Stimulants

114

Figure I-13. Mass spectra of cocaine, and its deuterated analog (cocaine-d₃).

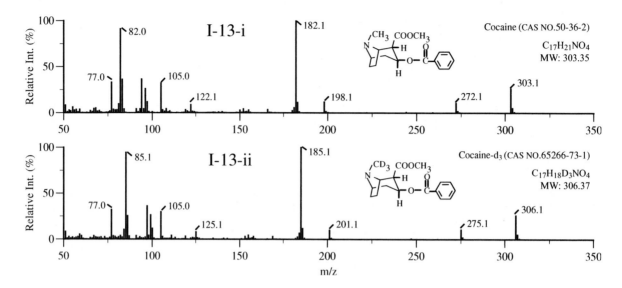

Figure I-14. Mass spectra of cocaethylene, and its deuterated analog (cocaethylene-d$_3$,-d$_8$).

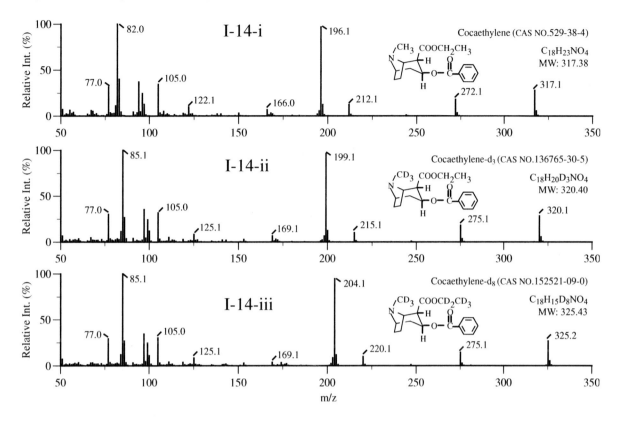

Figure I — Stimulants

Figure I-15. Mass spectra of ecgonine methyl ester, and its deuterated analog (ecgonine methyl ester-d₃): (A) underivatized; (B) TFA-derivatized; (C) PFP-derivatized; (D) HFB-derivatized; (E) TMS-derivatized; (F) *t*-BDMS-derivatized.

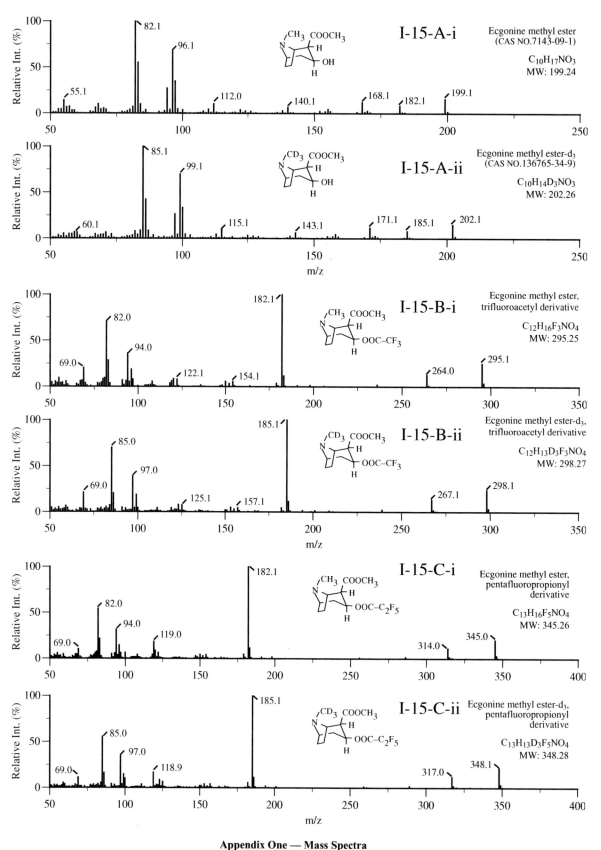

Figure I-15. (Continued)

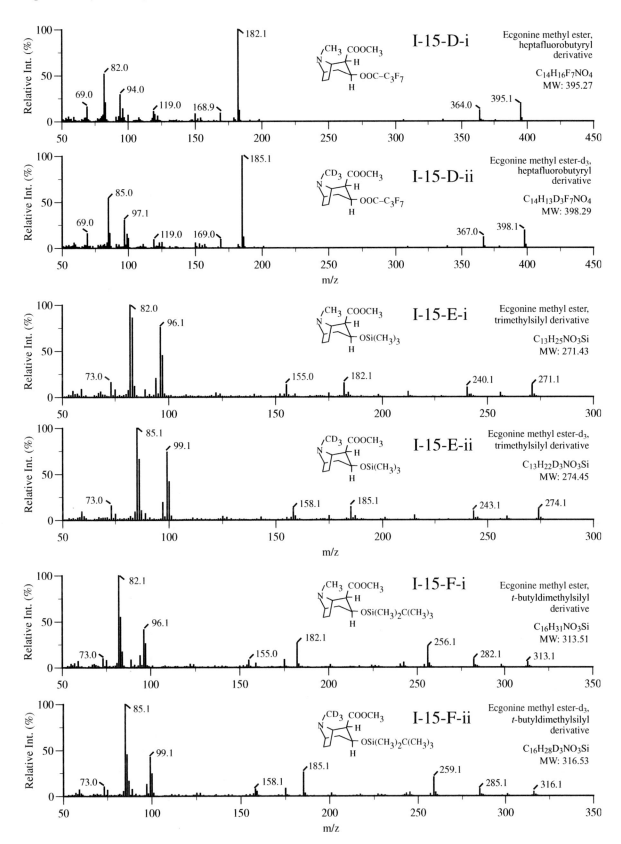

I-15-D-i
Ecgonine methyl ester, heptafluorobutyryl derivative
$C_{14}H_{16}F_7NO_4$
MW: 395.27

I-15-D-ii
Ecgonine methyl ester-d_3, heptafluorobutyryl derivative
$C_{14}H_{13}D_3F_7NO_4$
MW: 398.29

I-15-E-i
Ecgonine methyl ester, trimethylsilyl derivative
$C_{13}H_{25}NO_3Si$
MW: 271.43

I-15-E-ii
Ecgonine methyl ester-d_3, trimethylsilyl derivative
$C_{13}H_{22}D_3NO_3Si$
MW: 274.45

I-15-F-i
Ecgonine methyl ester, t-butyldimethylsilyl derivative
$C_{16}H_{31}NO_3Si$
MW: 313.51

I-15-F-ii
Ecgonine methyl ester-d_3, t-butyldimethylsilyl derivative
$C_{16}H_{28}D_3NO_3Si$
MW: 316.53

Figure I — Stimulants

Figure I-16. Mass spectra of benzoylecgonine and its deuterated analog (benzoylecgonine-d₃,-d₈): (A) methyl-derivatized; (B) ethyl-derivatized; (C) propyl-derivatized; (D) butyl-derivatized; (E) PFPoxy-derivatized; (F) HFPoxy-derivatized; (G) TMS-derivatized; (H) t-BDMS-derivatized.

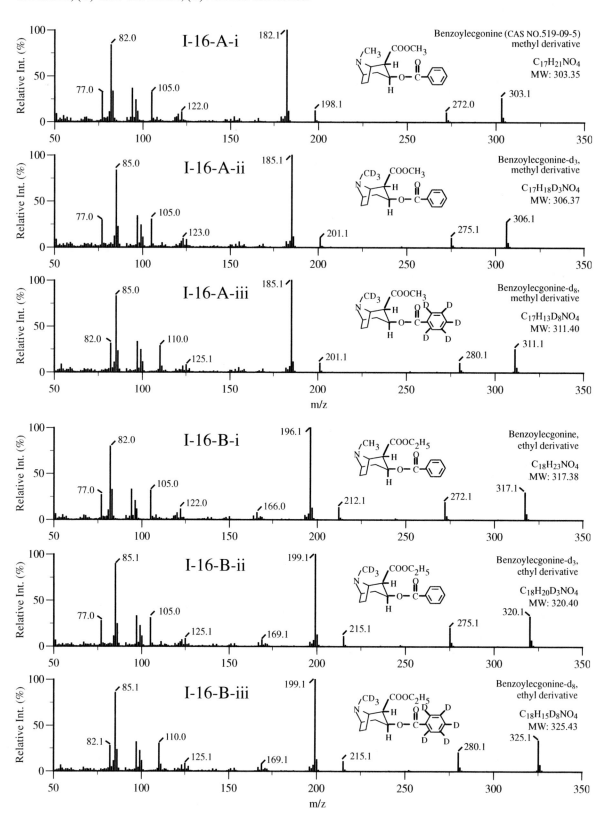

Figure I-16. (Continued)

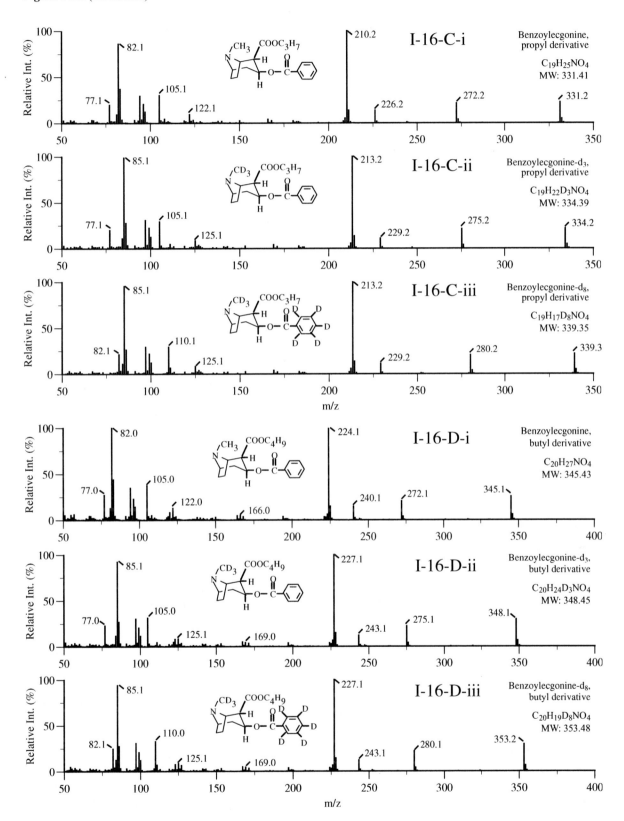

I-16-C-i — Benzoylecgonine, propyl derivative — $C_{19}H_{25}NO_4$ — MW: 331.41

I-16-C-ii — Benzoylecgonine-d_3, propyl derivative — $C_{19}H_{22}D_3NO_4$ — MW: 334.39

I-16-C-iii — Benzoylecgonine-d_8, propyl derivative — $C_{19}H_{17}D_8NO_4$ — MW: 339.35

I-16-D-i — Benzoylecgonine, butyl derivative — $C_{20}H_{27}NO_4$ — MW: 345.43

I-16-D-ii — Benzoylecgonine-d_3, butyl derivative — $C_{20}H_{24}D_3NO_4$ — MW: 348.45

I-16-D-iii — Benzoylecgonine-d_8, butyl derivative — $C_{20}H_{19}D_8NO_4$ — MW: 353.48

Figure I — Stimulants

Figure I-16. (Continued)

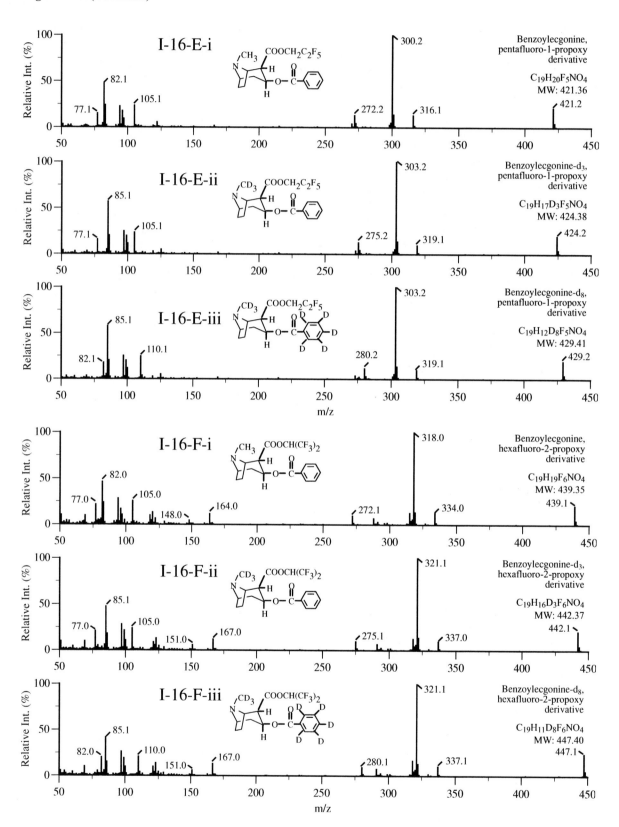

121

Figure I-16. (Continued)

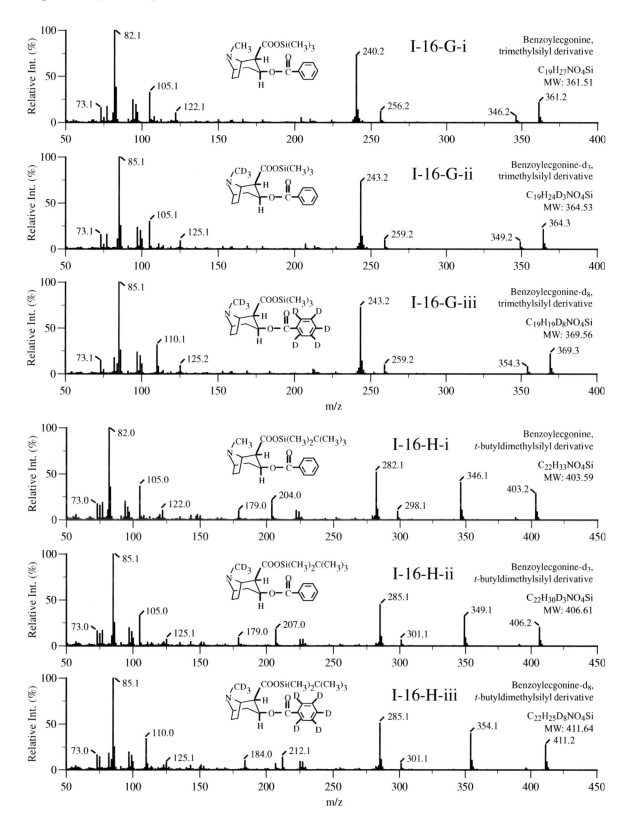

Figure I — Stimulants

Figure I-17. Mass spectra of ecgonine and its deuterated analogs (ecgonine-d₃); (A) [TMS]₂-underivatized; (B) [*t*-BDMS]₂-underivatized; (C) HFPoxy/TFA-derivatized; (D) PFPoxy/PFP-derivatized; (E) HFPoxy/HFB-derivatized.

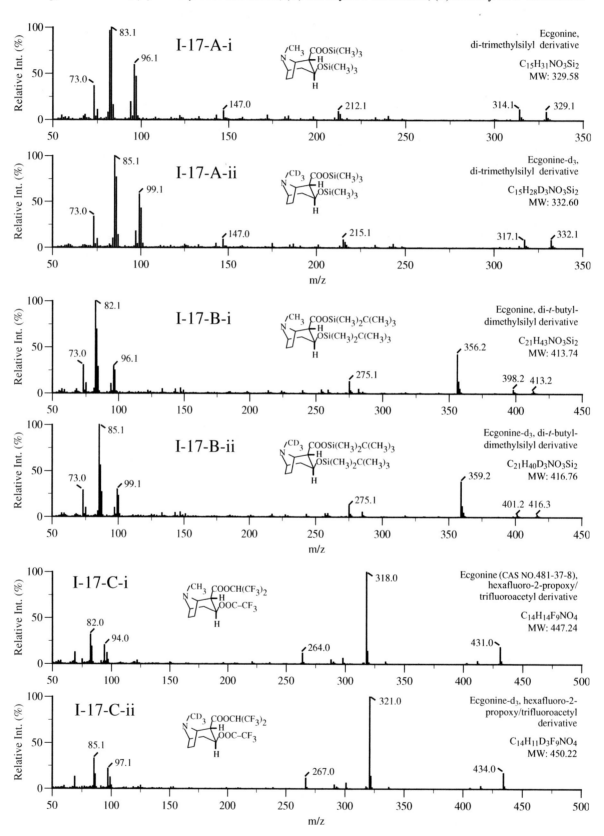

Figure I-17. (Continued)

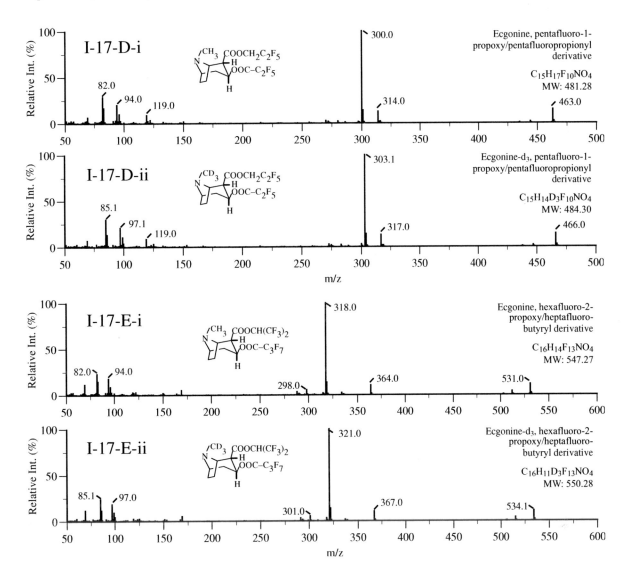

I-17-D-i — Ecgonine, pentafluoro-1-propoxy/pentafluoropropionyl derivative — $C_{15}H_{17}F_{10}NO_4$ — MW: 481.28

I-17-D-ii — Ecgonine-d_3, pentafluoro-1-propoxy/pentafluoropropionyl derivative — $C_{15}H_{14}D_3F_{10}NO_4$ — MW: 484.30

I-17-E-i — Ecgonine, hexafluoro-2-propoxy/heptafluoro-butyryl derivative — $C_{16}H_{14}F_{13}NO_4$ — MW: 547.27

I-17-E-ii — Ecgonine-d_3, hexafluoro-2-propoxy/heptafluoro-butyryl derivative — $C_{16}H_{11}D_3F_{13}NO_4$ — MW: 550.28

Figure I — Stimulants

Figure I-18. Mass spectra of anhydroecgonine methyl ester, and its deuterated analog (anhydroecgoninemethyl ester-d₃).

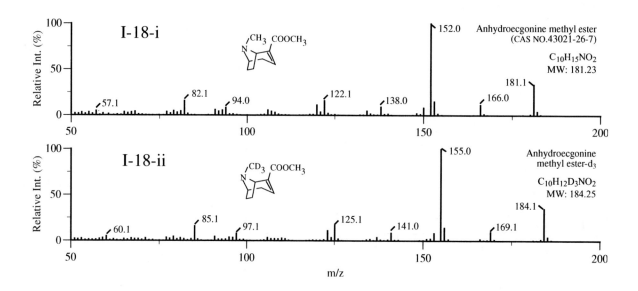

Figure I-19. Mass spectra of caffeine, and its deuterated analog (caffeine-$^{13}C_3$).

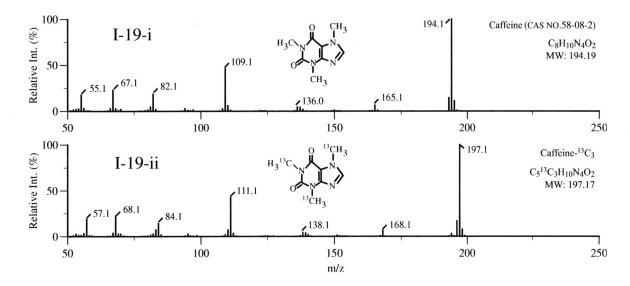

Figure I — Stimulants

Figure I-20. Mass spectra of methylphenidate, and its deuterated analog (methylphenidate-d₃): (A) underivatized; (B) TFA-derivatized; (C) PFP-derivatized; (D) HFB-derivatized; (E) 4-CB-derivatized; (F) TMS-derivatized.

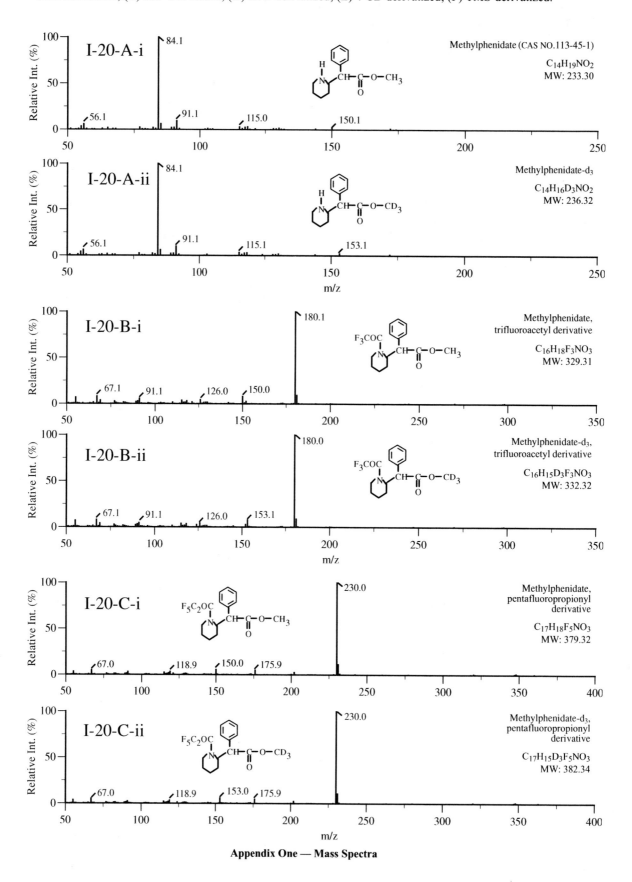

Figure I-20. (Continued)

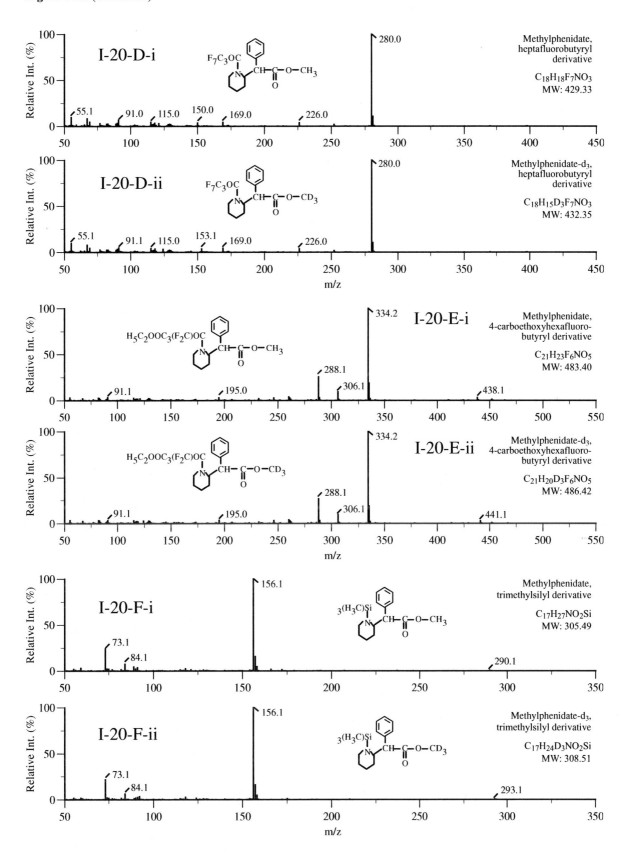

Figure I — Stimulants

Figure I-21. Mass spectra of ritalinic acid, and its deuterated analog (ritalinic acid-d₅): (A) 4-CB-derivatized; (B) [TMS]₂-derivatized; (C) *t*-BDMS-derivatized.

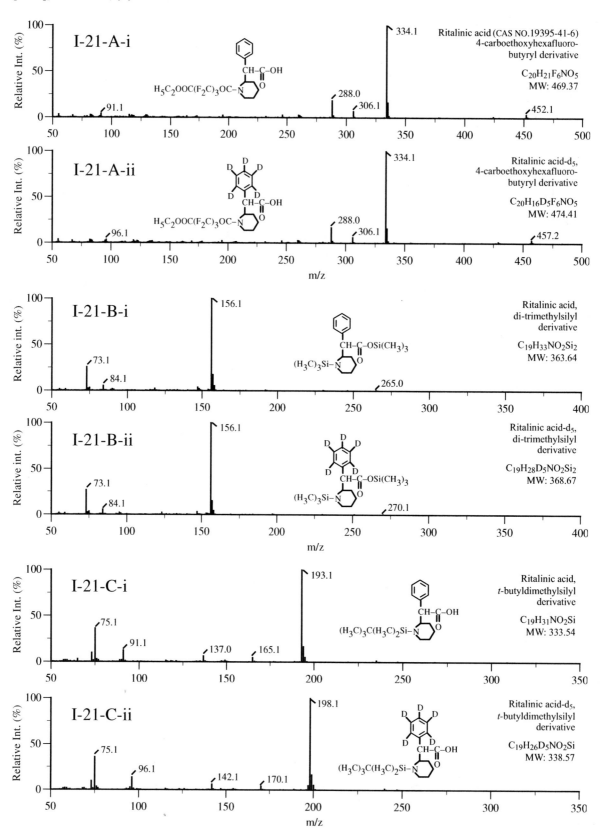

Summary of Drugs, Isotopic Analogs, and Chemical Derivatization Groups Included in Figure II (Opioids)

Compound	Isotopic analog	Chemical derivatization group (no. of spectra)	Figure #
Heroin	d_3, d_9	None (3)	II-1
6-Acetylmorphine	d_3, d_6	None, acetyl, TFA, propionyl, PFP, HFB, TMS, t-BDMS (24)	II-2
Morphine	d_3, d_6	Ethyl, propyl, butyl, [acetyl]$_2$, [TFA]$_2$, propionyl, [propionyl]$_2$, [PFP]$_2$, [HFB]$_2$, [TMS]$_2$, t-BDMS, [t-BDMS]$_2$, ethyl/acetyl, ethyl/TMS, propyl/TMS, propyl/t-BDMS, butyl/TMS, butyl/t-BDMS, acetyl/TMS, acetyl/t-BDMS, propionyl/TMS (63)	II-3
Hydromorphone	d_3, d_6	Acetyl, [acetyl]$_2$, [TFA]$_2$, propionyl, PFP, [PFP]$_2$, HFB, [HFB]$_2$, TMS, [TMS]$_2$, t-BDMS, [t-BDMS]$_2$, MA/ethyl, MA/acetyl, MA/propionyl, MA/TMS, MA/t-BDMS, HA/[TMS]$_2$ (54)	II-4
Oxymorphone	d_3	[acetyl]$_2$, [acetyl]$_3$, [TFA]$_2$, propionyl, [propionyl]$_2$, [propionyl]$_3$, [PFP]$_2$, [HFB]$_2$, [TMS]$_2$, [TMS]$_3$, t-BDMS, MA/ethyl, MA/acetyl, MA/[acetyl]$_2$, MA/propionyl, MA/[HFB]$_2$, MA/[TMS]$_2$, MA/[t-BDMS]$_2$, MA/ethyl/propionyl, MA/ethyl/TMS, MA/ethyl/t-BDMS, MA/acetyl/TMS, MA/propionyl/TMS, HA/[TMS]$_3$, HA/[ethyl]$_2$/propionyl, HA/[ethyl]$_2$/TMS (52)	II-5
6-Acetylcodeine	d_3	None (2)	II-6
Codeine	d_3, d_6, $^{13}C_1d_3$	None, acetyl, TFA, propionyl, PFP, HFB, TMS, t-BDMS (32)	II-7
Hydrocodone	d_3, d_6	None, ethyl, acetyl, TMS, t-BDMS, MA, HA/TMS (21)	II-8
Dihydrocodeine	d_3, d_6	None, acetyl, TFA, propionyl, PFP, HFB, TMS, t-BDMS (24)	II-9
Oxycodone	d_3, d_6	None, acetyl, [acetyl]$_2$, propionyl, TMS, [TMS]$_2$, t-BDMS, [t-BDMS]$_2$, MA, MA/propionyl, MA/TMS, HA/[propionyl]$_2$, HA/[TMS]$_2$, HA/ethyl/propionyl (42)	II-10
Noroxycodone	d_3	None, [acetyl]$_2$, [TFA]$_3$, propionyl, [PFP]$_2$, [HFB]$_2$, [TMS]$_2$, [TMS]$_3$, MA/ethyl, MA/acetyl, MA/[TFA]$_2$, MA/propionyl, MA/PFP, MA/[HFB]$_2$, MA/[TMS]$_2$, MA/t-BDMS, MA/ethyl/propionyl, MA/ethyl/TMS, MA/ethyl/t-BDMS, MA/acetyl/TMS, MA/propionyl/TMS, HA/[ethyl]$_2$/TMS (44)	II-11
Buprenorphine	d_4	Methyl, ethyl, acetyl, MBTFA, PFP, HFB, TMS, [TMS]$_2$, t-BDMS (18)	II-12
Norbuprenorphine	d_3	[Methyl]$_2$, [ethyl]$_2$, [acetyl]$_2$, [MBTFA]$_2$, [PFP]$_2$, [HFB]$_2$, [TMS]$_2$, [TMS]$_3$, t-BDMS (18)	II-13
Fentanyl	d_5	None (2)	II-14
Norfentanyl	d_5	None, acetyl, TCA, TFA, PFP, HFB, 4-CB, TMS, t-BDMS (18)	II-15
Methadone	d_3, d_9	None (3)	II-16
EDDP	d_3	None (2)	II-17
Propoxyphene	d_5, d_7, d_{11}	None (4)	II-18
Norpropoxyphene	d_5	None (2)	II-19
Meperidine	d_4	None (2)	II-20
Normeperidine	d_4	None, ethyl, propyl, butyl, acetyl, TCA, TFA, PFP, HFB, 4-CB, TMS, t-BDMS (24)	II-21

Total no. of mass spectra: 454

Figure II — Opioids

Appendix One — Figure II
Mass Spectra of Commonly Abused Drugs and Their Isotopically Labeled Analogs
in Various Derivatization Forms — Opioids

Figure II — Opioids

133

Figure II-1. Mass spectra of heroin and its deuterated analogs (heroin-d₃, -d₉).

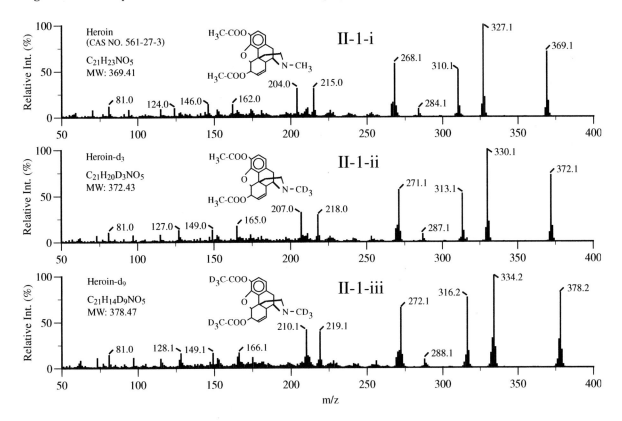

Figure II — Opioids

134

Figure II-2. Mass spectra of 6-acetylmorphine and its deuterated analogs (6-acetylmorphine-d₃, -d₆): (A) underivatized; (B) acetyl-derivatized; (C) TFA-derivatized; (D) propionyl-derivatized; (E) PFP-derivatized; (F) HFB-derivatized; (G) TMS-derivatized; (H) *t*-BDMS-derivatized.

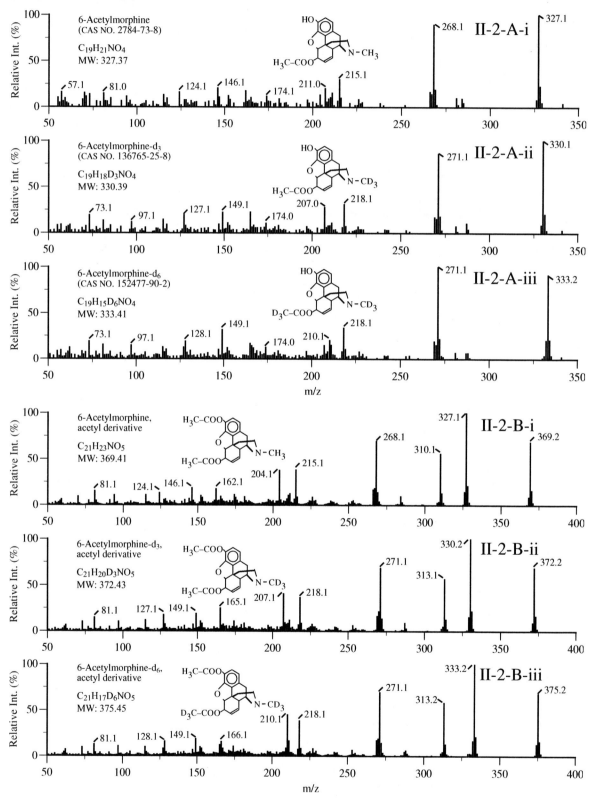

Appendix One — Mass Spectra

Figure II-2. (Continued)

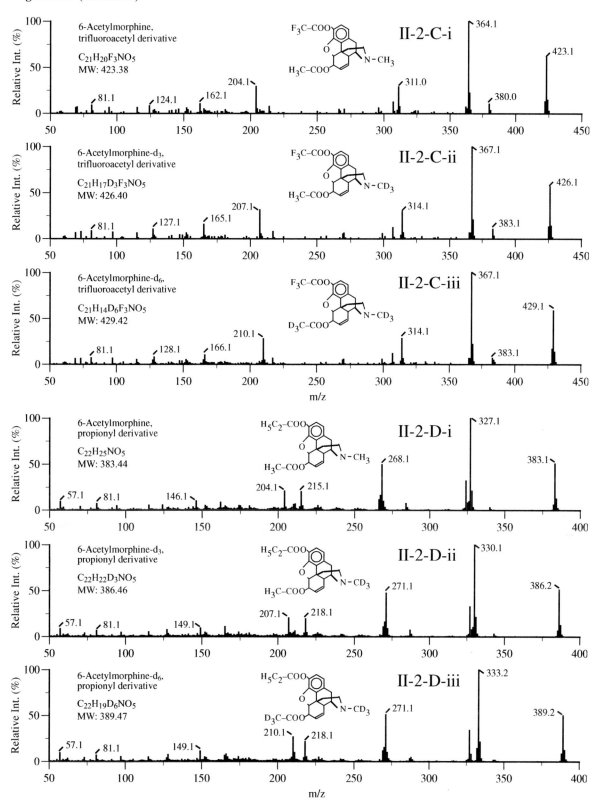

Figure II — Opioids

Figure II-2. (Continued)

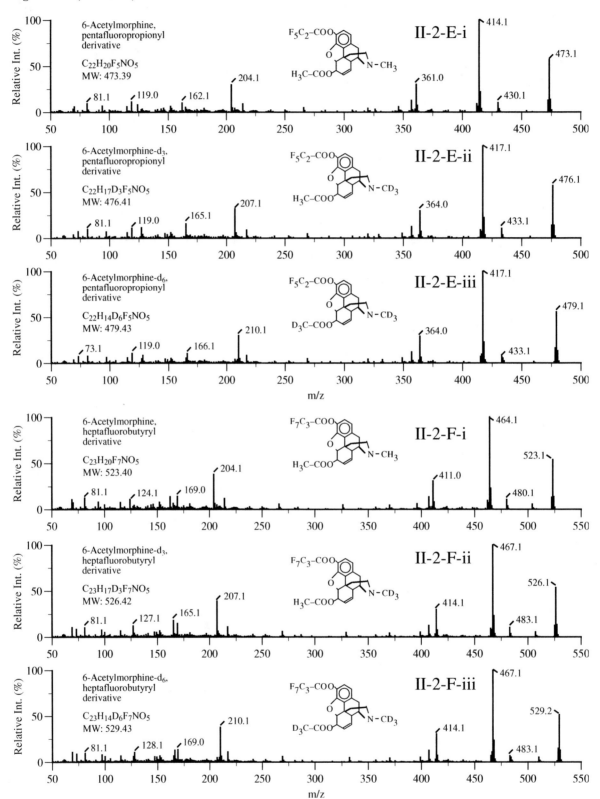

6-Acetylmorphine, pentafluoropropionyl derivative

$C_{22}H_{20}F_5NO_5$
MW: 473.39

II-2-E-i

81.1, 119.0, 162.1, 204.1, 361.0, 414.1, 430.1, 473.1

6-Acetylmorphine-d_3, pentafluoropropionyl derivative

$C_{22}H_{17}D_3F_5NO_5$
MW: 476.41

II-2-E-ii

81.1, 119.0, 165.1, 207.1, 364.0, 417.1, 433.1, 476.1

6-Acetylmorphine-d_6, pentafluoropropionyl derivative

$C_{22}H_{14}D_6F_5NO_5$
MW: 479.43

II-2-E-iii

73.1, 119.0, 166.1, 210.1, 364.0, 417.1, 433.1, 479.1

6-Acetylmorphine, heptafluorobutyryl derivative

$C_{23}H_{20}F_7NO_5$
MW: 523.40

II-2-F-i

81.1, 124.1, 169.0, 204.1, 411.0, 464.1, 480.1, 523.1

6-Acetylmorphine-d_3, heptafluorobutyryl derivative

$C_{23}H_{17}D_3F_7NO_5$
MW: 526.42

II-2-F-ii

81.1, 127.1, 165.1, 207.1, 414.1, 467.1, 483.1, 526.1

6-Acetylmorphine-d_6, heptafluorobutyryl derivative

$C_{23}H_{14}D_6F_7NO_5$
MW: 529.43

II-2-F-iii

81.1, 128.1, 169.0, 210.1, 414.1, 467.1, 483.1, 529.2

Appendix One — Mass Spectra

Figure II-2. (Continued)

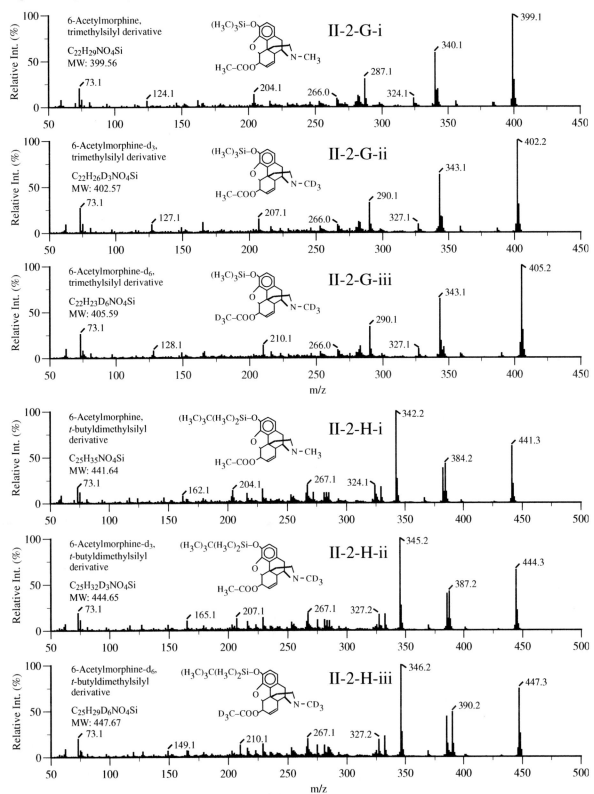

Figure II — Opioids

Figure II-3. Mass spectra of morphine and its deuterated analogs (morphine-d₃, -d₆): (A) ethyl-derivatized; (B) propyl-derivatized; (C) butyl-derivatized; (D) [acetyl]₂-derivatized; (E) [TFA]₂-derivatized; (F) propionyl-derivatized; (G) [propionyl]₂-derivatized; (H) [PFP]₂-derivatized; (I) [HFB]₂-derivatized; (J) [TMS]₂-derivatized; (K) *t*-BDMS-derivatized; (L) [*t*-BDMS]₂-derivatized; (M) ethyl/acetyl-derivatized; (N) ethyl/TMS-derivatized; (O) propyl/TMS-derivatized; (P) propyl/*t*-BDMS-derivatized; (Q) butyl/TMS-derivatized; (R) butyl/*t*-BDMS-derivatized; (S) acetyl/TMS-derivatized; (T) acetyl/*t*-BDMS-derivatized; (U) propionyl/TMS-derivatized.

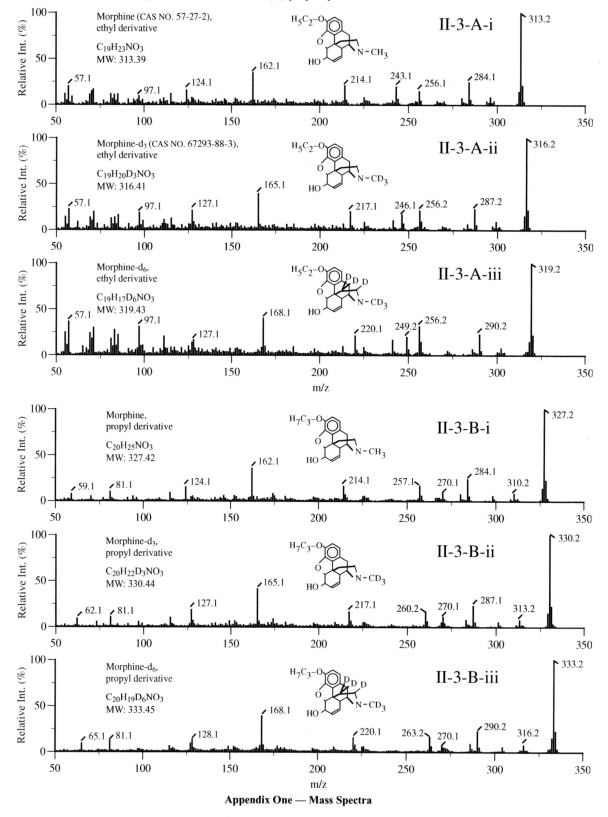

Figure II-3. (Continued)

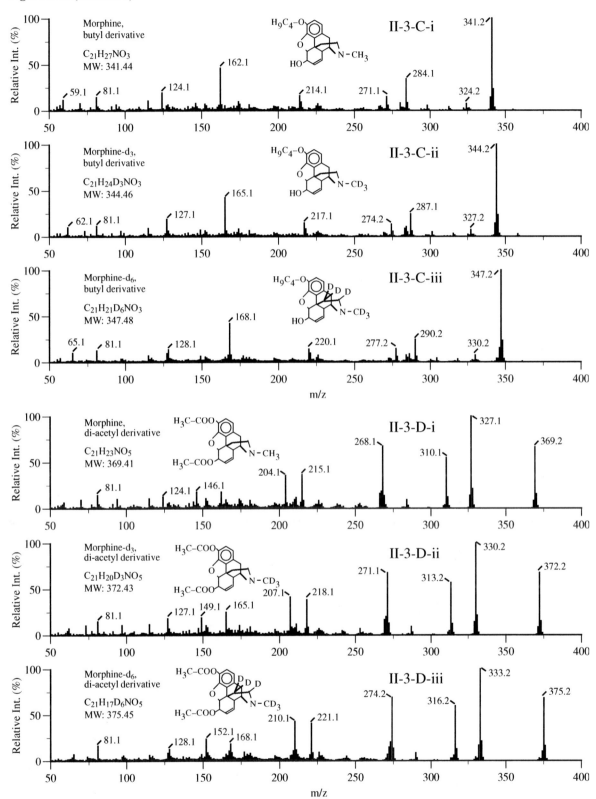

Figure II — Opioids

Figure II-3. (Continued)

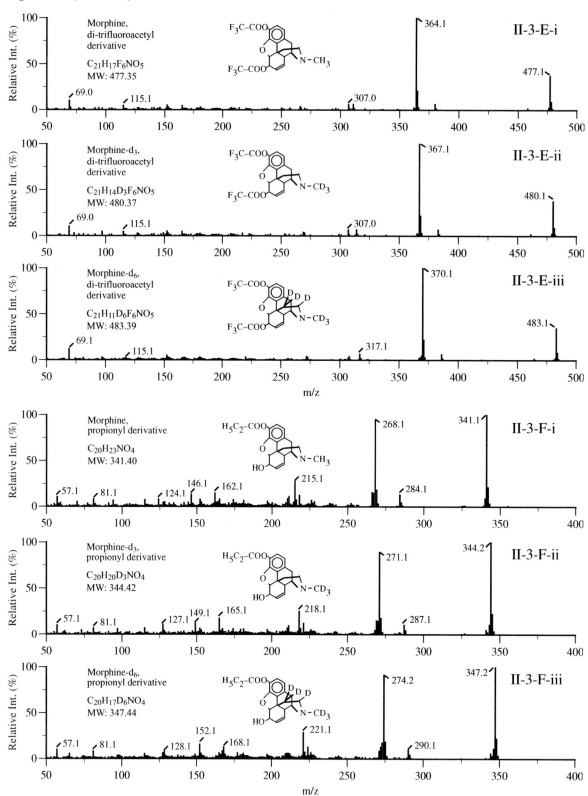

Figure II-3. (Continued)

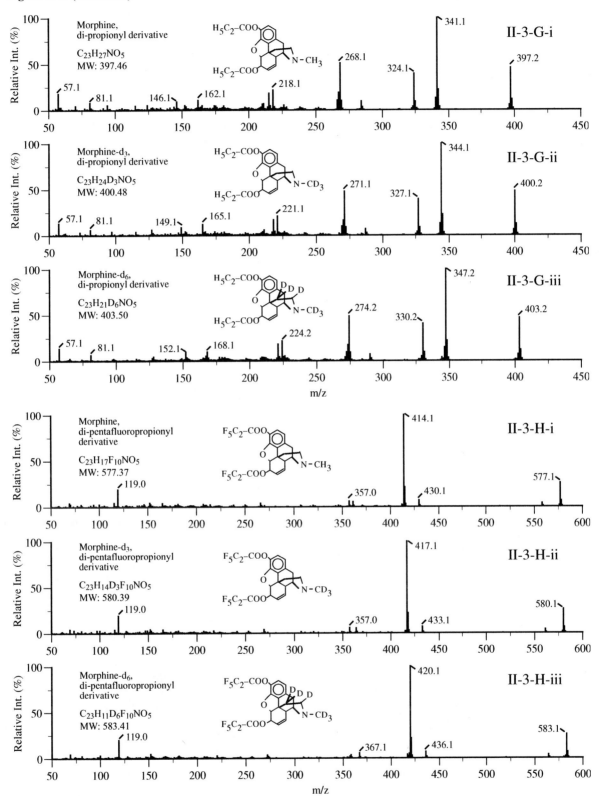

II-3-G-i

Morphine, di-propionyl derivative

$C_{23}H_{27}NO_5$
MW: 397.46

II-3-G-ii

Morphine-d_3, di-propionyl derivative

$C_{23}H_{24}D_3NO_5$
MW: 400.48

II-3-G-iii

Morphine-d_6, di-propionyl derivative

$C_{23}H_{21}D_6NO_5$
MW: 403.50

II-3-H-i

Morphine, di-pentafluoropropionyl derivative

$C_{23}H_{17}F_{10}NO_5$
MW: 577.37

II-3-H-ii

Morphine-d_3, di-pentafluoropropionyl derivative

$C_{23}H_{14}D_3F_{10}NO_5$
MW: 580.39

II-3-H-iii

Morphine-d_6, di-pentafluoropropionyl derivative

$C_{23}H_{11}D_6F_{10}NO_5$
MW: 583.41

Figure II — Opioids

Figure II-3. (Continued)

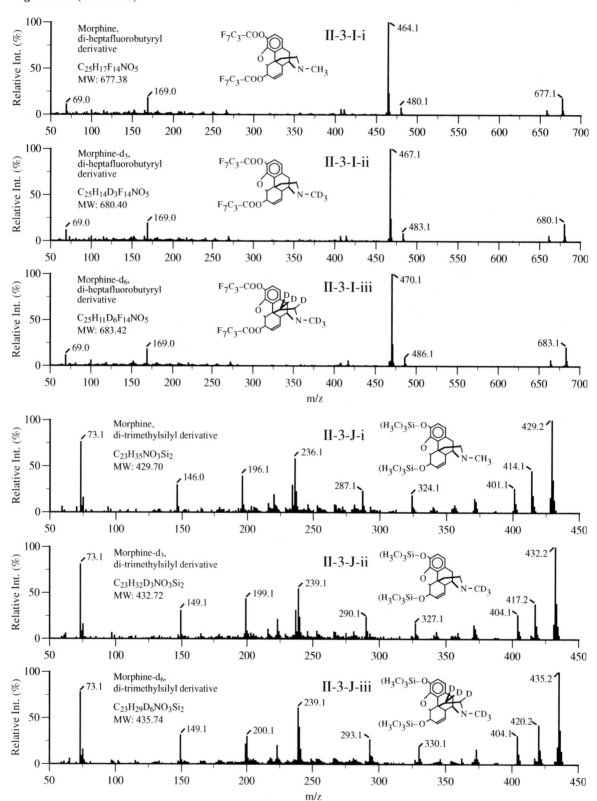

Figure II-3. (Continued)

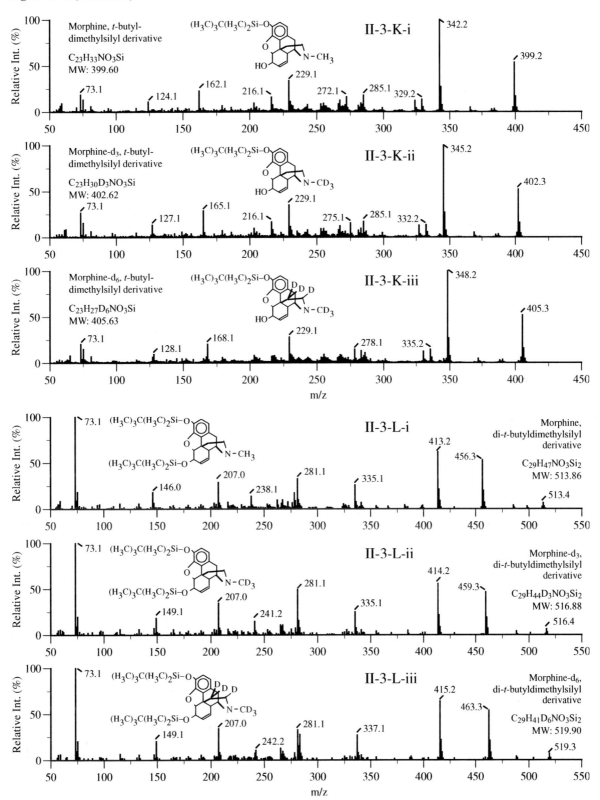

Morphine, *t*-butyl-dimethylsilyl derivative
C₂₃H₃₃NO₃Si
MW: 399.60

II-3-K-i

Morphine-d₃, *t*-butyl-dimethylsilyl derivative
C₂₃H₃₀D₃NO₃Si
MW: 402.62

II-3-K-ii

Morphine-d₆, *t*-butyl-dimethylsilyl derivative
C₂₃H₂₇D₆NO₃Si
MW: 405.63

II-3-K-iii

II-3-L-i
Morphine, di-*t*-butyldimethylsilyl derivative
C₂₉H₄₇NO₃Si₂
MW: 513.86

II-3-L-ii
Morphine-d₃, di-*t*-butyldimethylsilyl derivative
C₂₉H₄₄D₃NO₃Si₂
MW: 516.88

II-3-L-iii
Morphine-d₆, di-*t*-butyldimethylsilyl derivative
C₂₉H₄₁D₆NO₃Si₂
MW: 519.90

Figure II — Opioids

Figure II-3. (Continued)

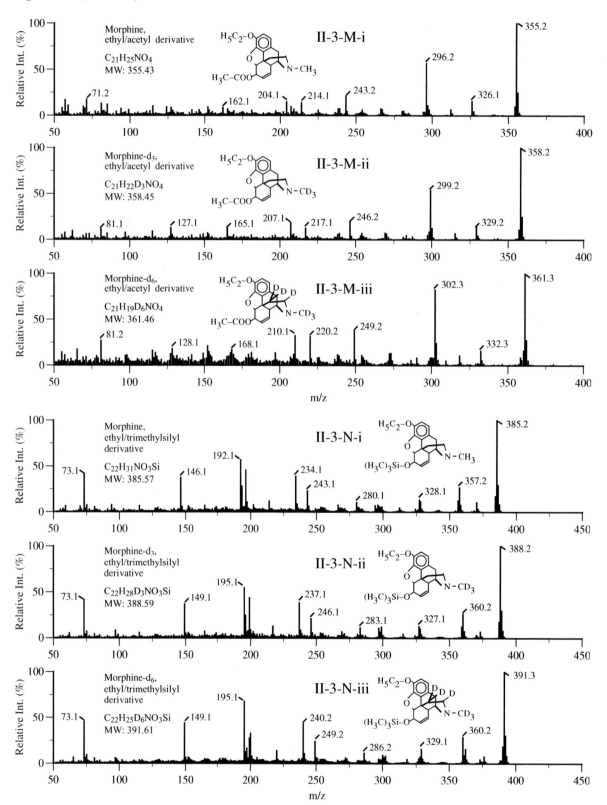

Figure II-3. (Continued)

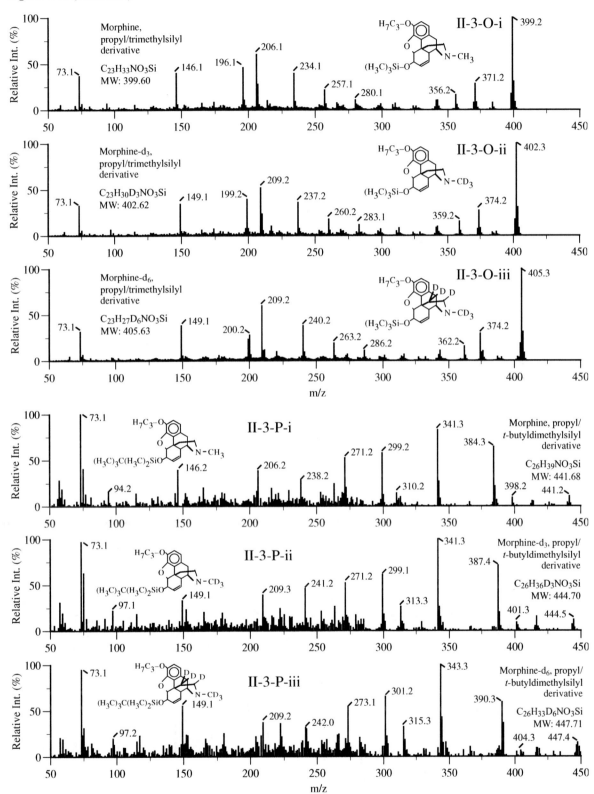

Figure II — Opioids

Figure II-3. (Continued)

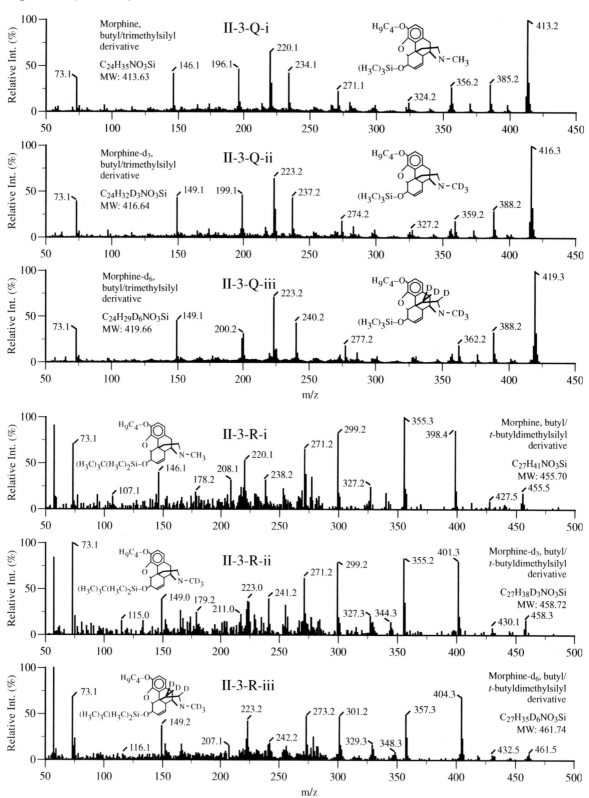

Figure II-3. (Continued)

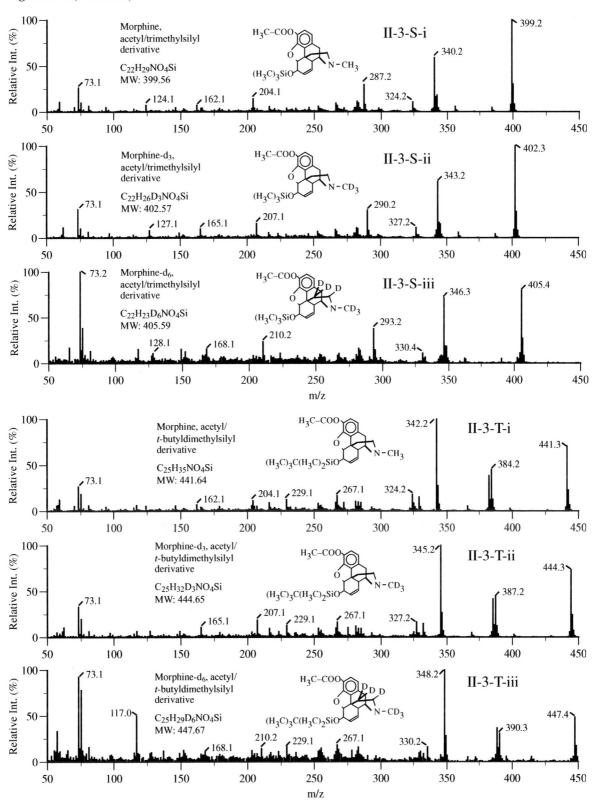

Figure II — Opioids

Figure II-3. (Continued)

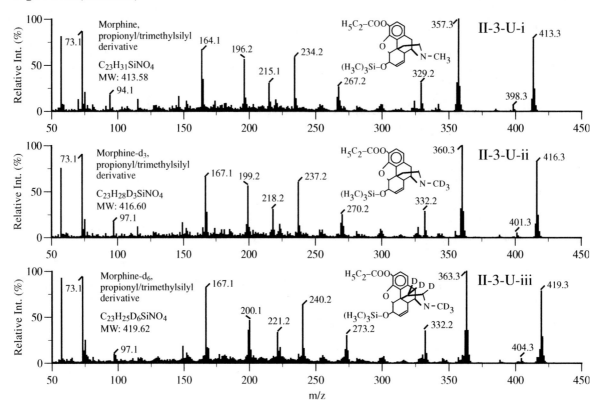

Figure II-4. Mass spectra of hydromorphone and its deuterated analogs (hydromorphone-d$_3$, -d$_6$): (A) acetyl-derivatized; (B) [acetyl]$_2$-derivatized; (C) [TFA]$_2$-derivatized; (D) propionyl-derivatized; (E) PFP-derivatized; (F) [PFP]$_2$-derivatized; (G) HFB-derivatized (H) [HFB]$_2$-derivatized; (I) TMS-derivatized; (J) [TMS]$_2$-derivatized; (K) *t*-BDMS-derivatized; (L) [*t*-BDMS]$_2$-derivatized; (M) MA/ethyl-derivatized; (N) MA/acetyl-derivatized; (O) MA/propionyl-derivatized; (P) MA/TMS-derivatized; (Q) MA/*t*-BDMS-derivatized; (R) HA/[TMS]$_2$-derivatized.

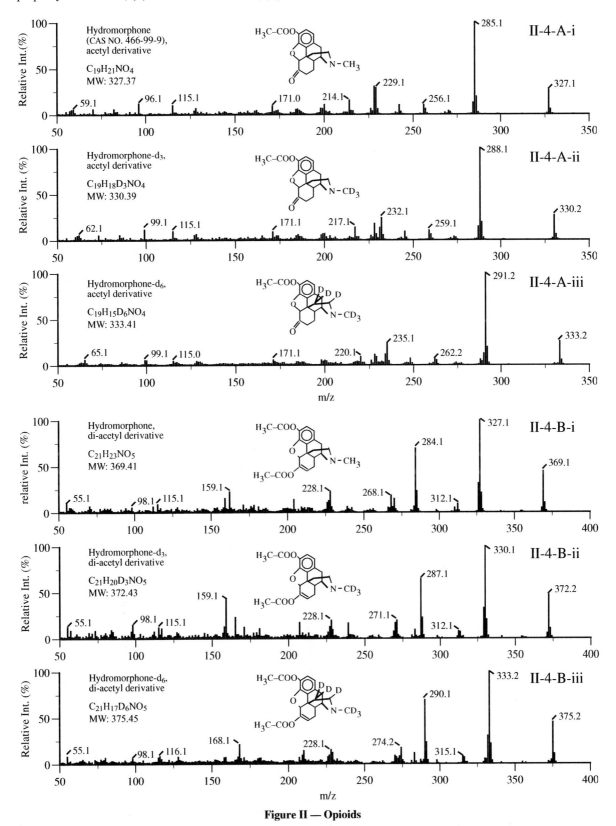

Figure II — Opioids

Figure II-4. (Continued)

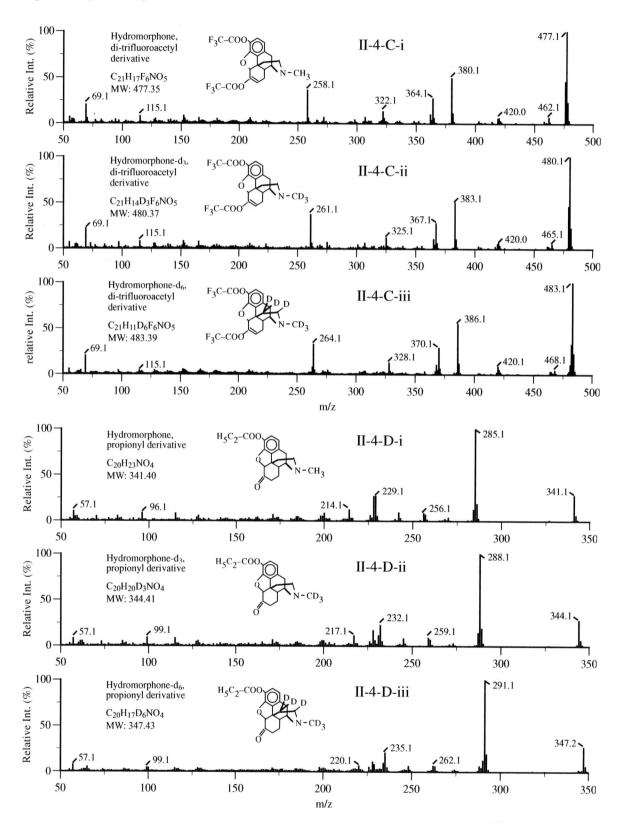

Figure II-4. (Continued)

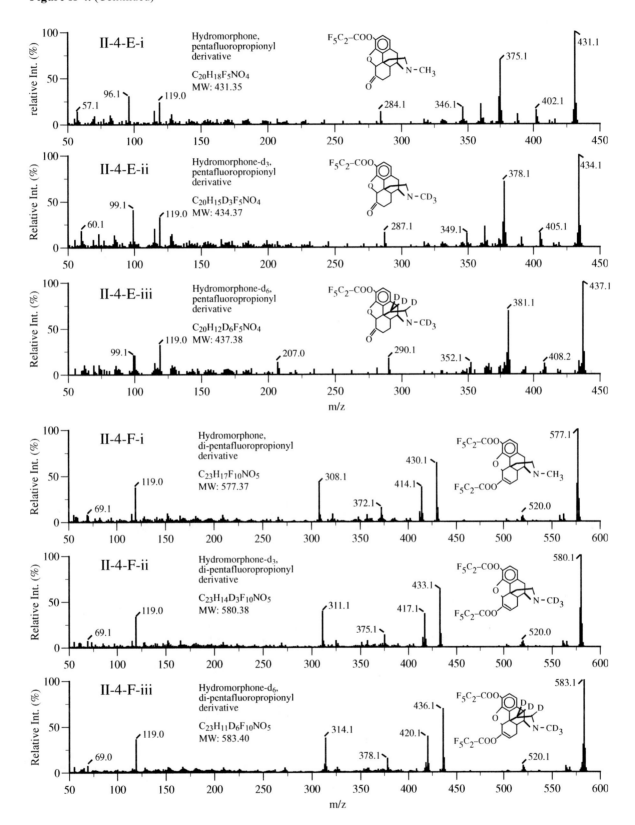

Figure II — Opioids

Figure II-4. (Continued)

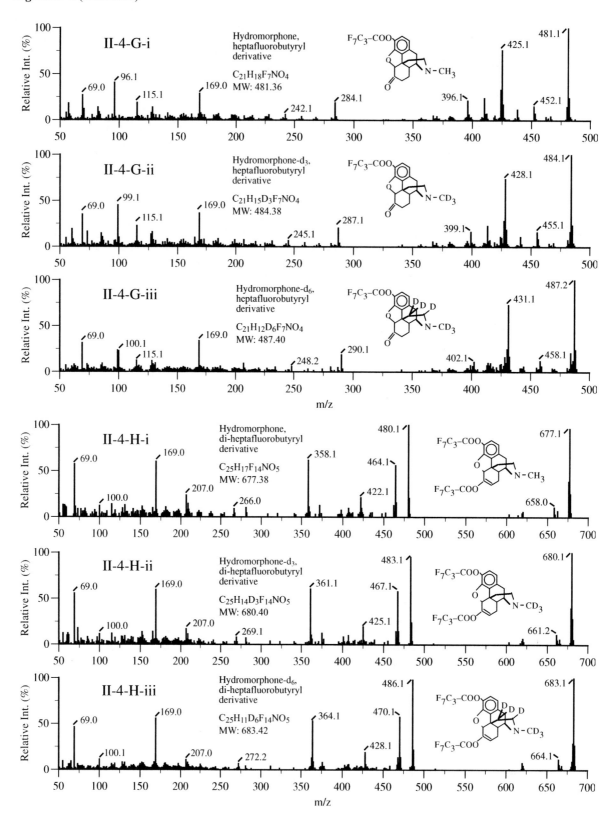

153

Figure II-4. (Continued)

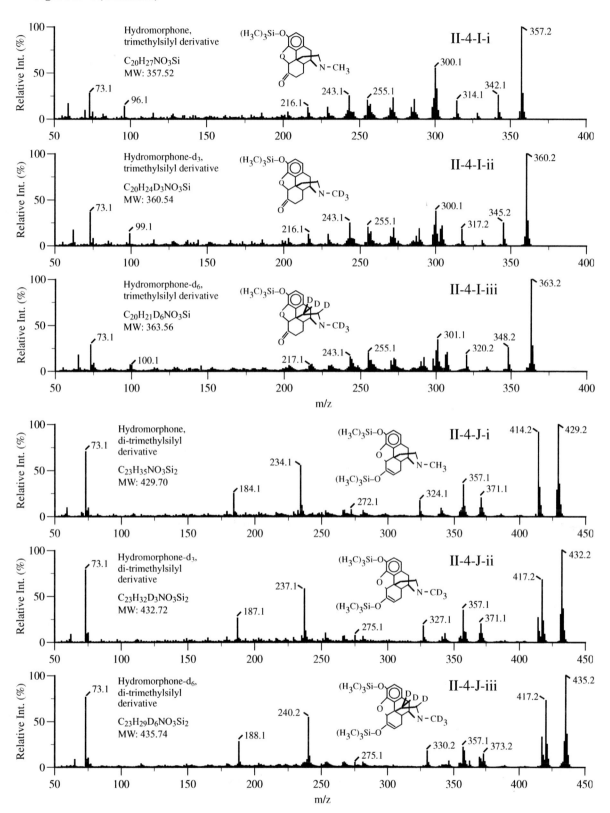

Figure II — Opioids

Figure II-4. (Continued)

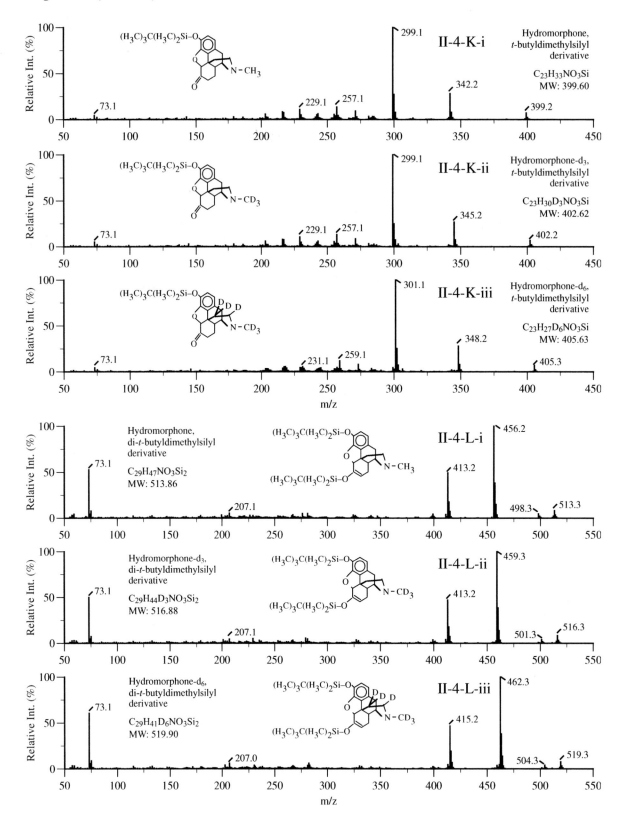

Figure II-4. (Continued)

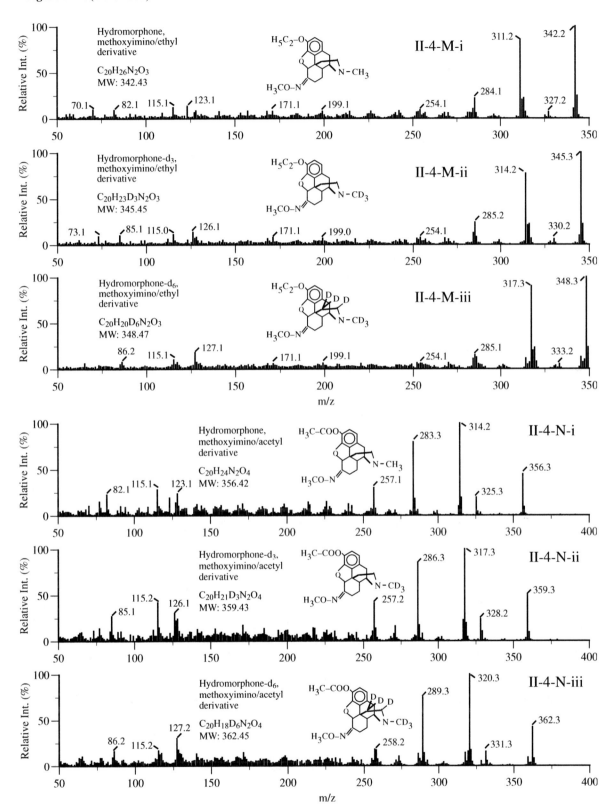

Figure II — Opioids

Figure II-4. (Continued)

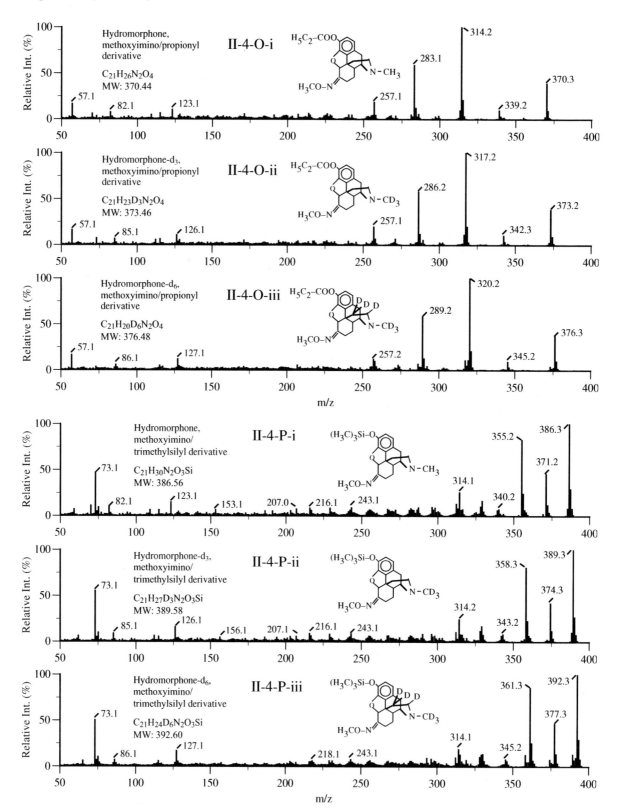

157

Figure II-4. (Continued)

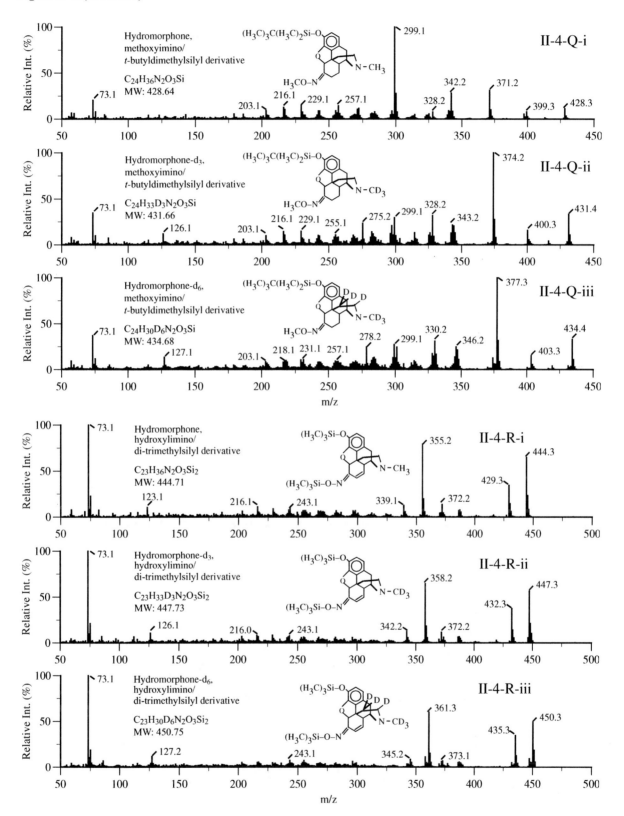

Figure II — Opioids

Figure II-5. Mass spectra of oxymorphone and its deuterated analogs (oxymorphone-d_3): (A) [acetyl]$_2$-derivatized; (B) [acetyl]$_3$-derivatized; (C) [TFA]$_2$-derivatized; (D) propionyl-derivatized; (E) [propionyl]$_2$-derivatized; (F) [propionyl]$_3$-derivatized; (G) [PFP]$_2$-derivatized; (H) [HFB]$_2$-derivatized; (I) [TMS]$_2$-derivatized; (J) [TMS]$_3$-derivatized; (K) t-BDMS-derivatized; (L) MA/ethyl-derivatized; (M) MA/acetyl-derivatized; (N) MA/[acetyl]$_2$-derivatized; (O) MA/propionyl-derivatized; (P) MA/[HFB]$_2$-derivatized; (Q) MA/[TMS]$_2$-derivatized; (R) MA/[t-BDMS]$_2$-derivatized; (S) MA/ethyl/propionyl-derivatized; (T) MA/ethyl/TMS-derivatized; (U) MA/ethyl/t-BDMS-derivatized; (V) MA/acetyl/TMS-derivatized; (W) MA/propionyl/TMS; (X) HA/[TMS]$_3$-derivatized; (Y) HA/[ethyl]$_2$/propionyl-derivatized; (Z) HA/[ethyl]$_2$/TMS-derivatized.

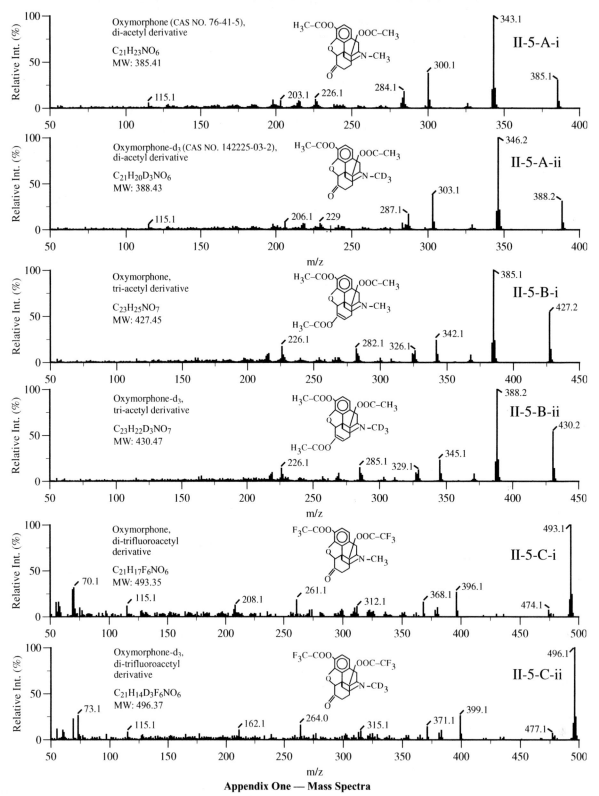

Appendix One — Mass Spectra

Figure II-5. (Continued)

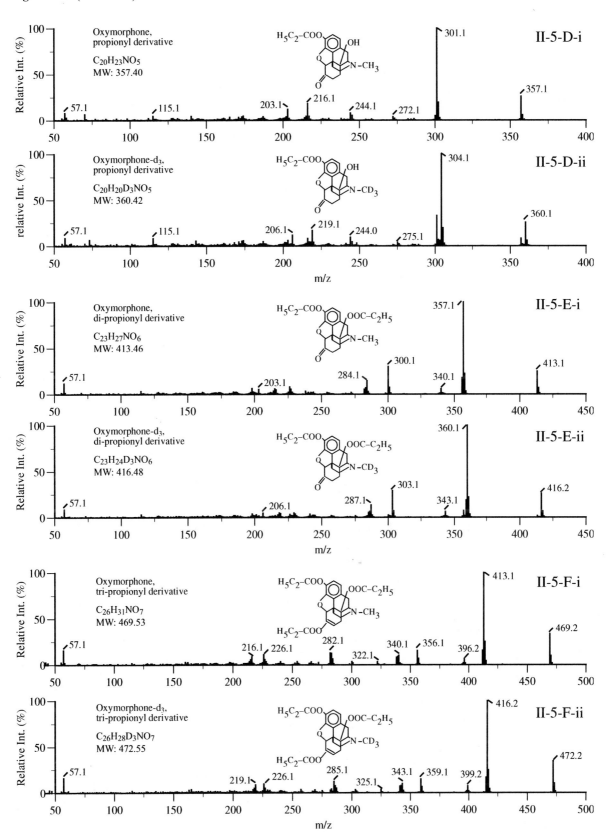

Figure II — Opioids

Figure II-5. (Continued)

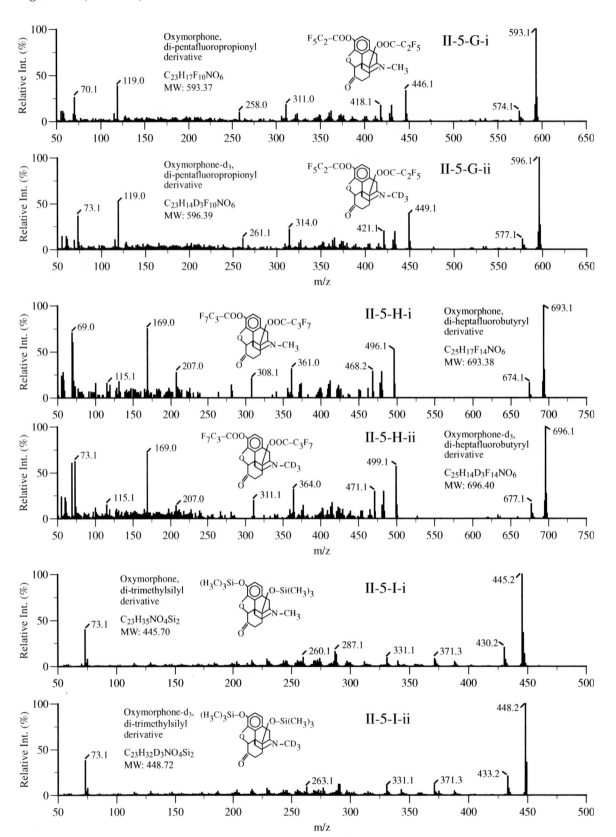

Figure II-5. (Continued)

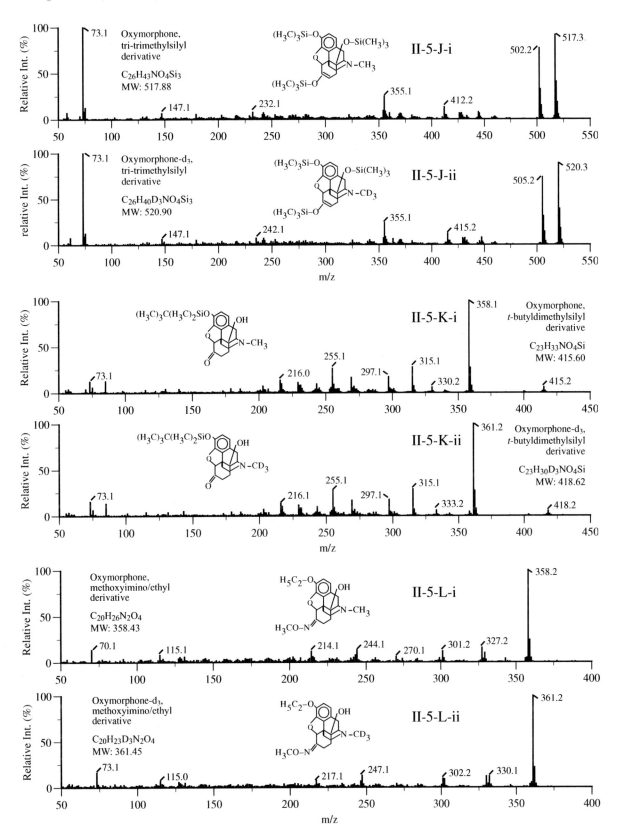

Figure II — Opioids

Figure II-5. (Continued)

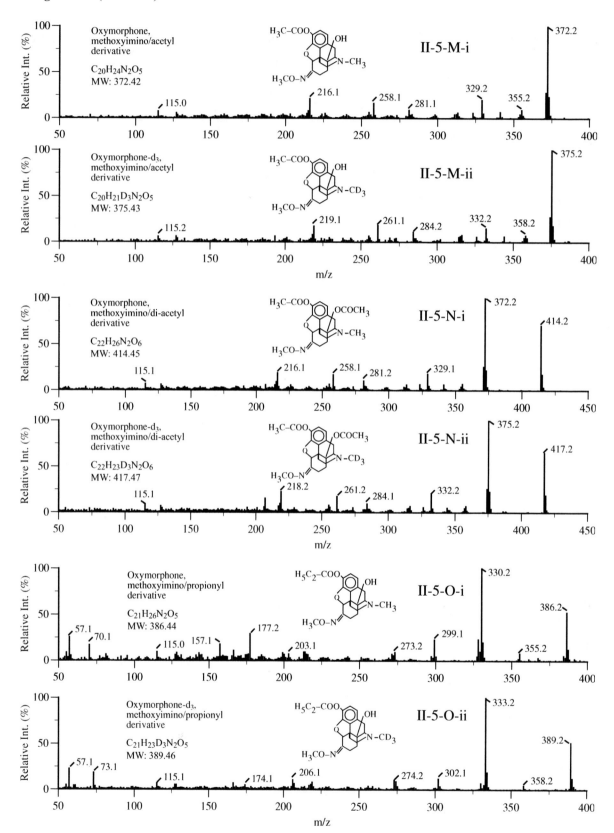

Figure II-5. (Continued)

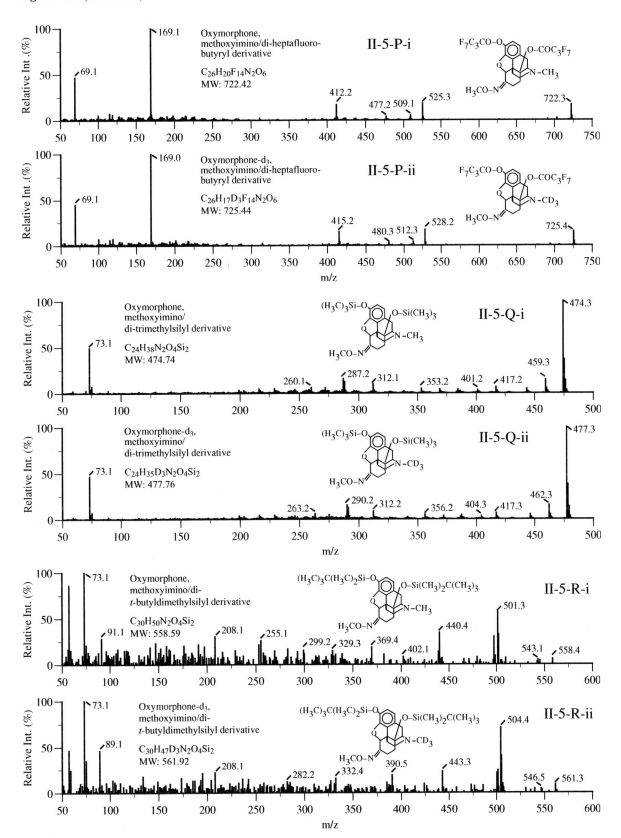

Figure II — Opioids

Figure II-5. (Continued)

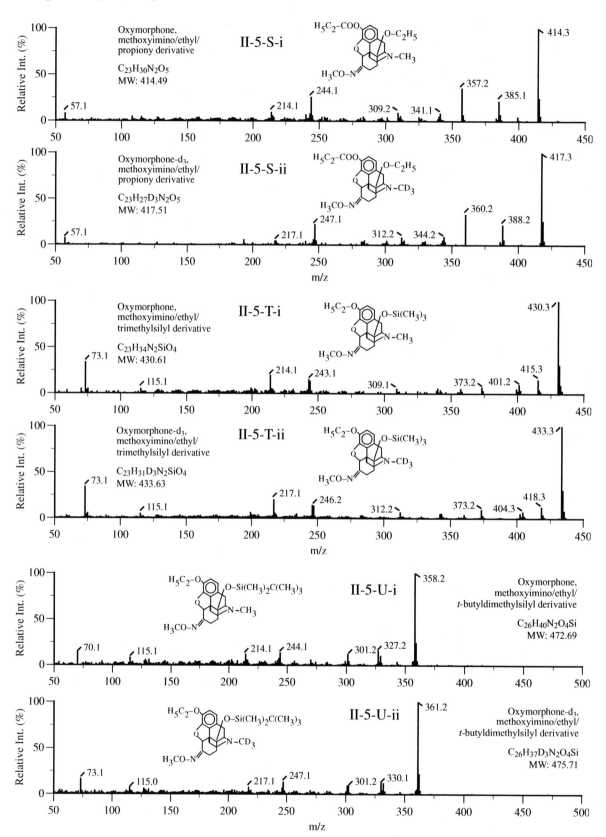

Oxymorphone, methoxyimino/ethyl/propiony derivative

C₂₃H₃₀N₂O₅
MW: 414.49

II-5-S-i

Oxymorphone-d₃, methoxyimino/ethyl/propiony derivative

C₂₃H₂₇D₃N₂O₅
MW: 417.51

II-5-S-ii

Oxymorphone, methoxyimino/ethyl/trimethylsilyl derivative

C₂₃H₃₄N₂SiO₄
MW: 430.61

II-5-T-i

Oxymorphone-d₃, methoxyimino/ethyl/trimethylsilyl derivative

C₂₃H₃₁D₃N₂SiO₄
MW: 433.63

II-5-T-ii

II-5-U-i

Oxymorphone, methoxyimino/ethyl/t-butyldimethylsilyl derivative

C₂₆H₄₀N₂O₄Si
MW: 472.69

II-5-U-ii

Oxymorphone-d₃, methoxyimino/ethyl/t-butyldimethylsilyl derivative

C₂₆H₃₇D₃N₂O₄Si
MW: 475.71

Appendix One — Mass Spectra

Figure II-5. (Continued)

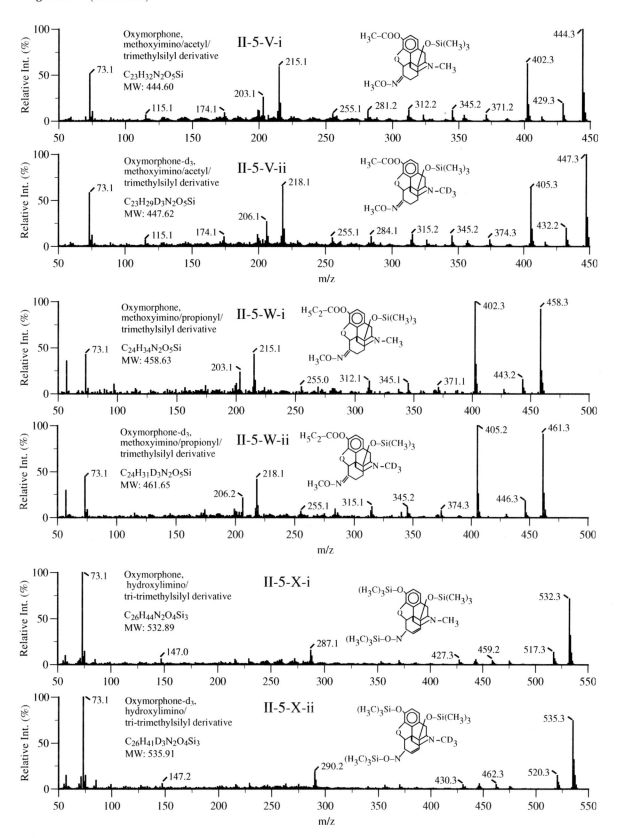

Oxymorphone, methoxyimino/acetyl/trimethylsilyl derivative
II-5-V-i
C₂₃H₃₂N₂O₅Si
MW: 444.60

Oxymorphone-d₃, methoxyimino/acetyl/trimethylsilyl derivative
II-5-V-ii
C₂₃H₂₉D₃N₂O₅Si
MW: 447.62

Oxymorphone, methoxyimino/propionyl/trimethylsilyl derivative
II-5-W-i
C₂₄H₃₄N₂O₅Si
MW: 458.63

Oxymorphone-d₃, methoxyimino/propionyl/trimethylsilyl derivative
II-5-W-ii
C₂₄H₃₁D₃N₂O₅Si
MW: 461.65

Oxymorphone, hydroxyimino/tri-trimethylsilyl derivative
II-5-X-i
C₂₆H₄₄N₂O₄Si₃
MW: 532.89

Oxymorphone-d₃, hydroxyimino/tri-trimethylsilyl derivative
II-5-X-ii
C₂₆H₄₁D₃N₂O₄Si₃
MW: 535.91

Figure II — Opioids

Figure II-5. (Continued)

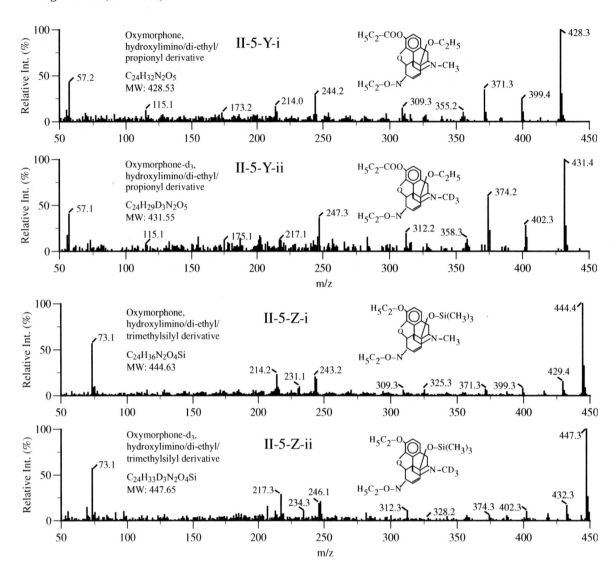

Figure II-6. Mass spectra of 6-acetylcodeine and its deuterated analogs (6-acetylcodeine-d₃).

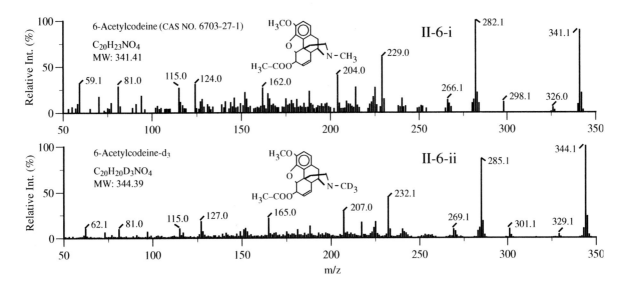

Figure II — Opioids

Figure II-7. Mass spectra of codeine and its deuterated analogs (codeine-d_3, -d_6, $^{13}C_1$-d_3): (A) underivatized; (B) acetyl-derivatized; (C) TFA-derivatized; (D) propionyl-derivatized; (E) PFP-derivatized; (F) HFB-derivatized; (G) TMS-derivatized; (H) t-BDMS-derivatized.

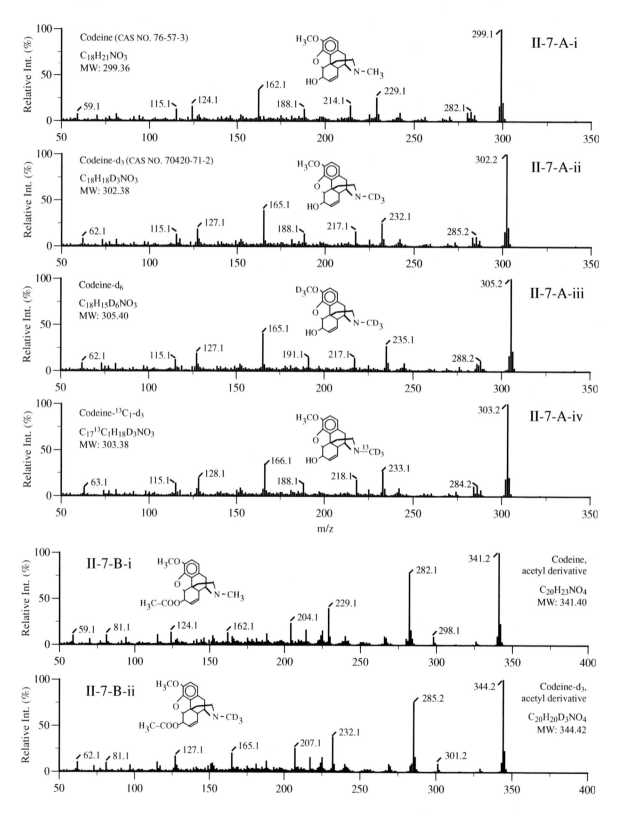

Figure II-7. (Continued)

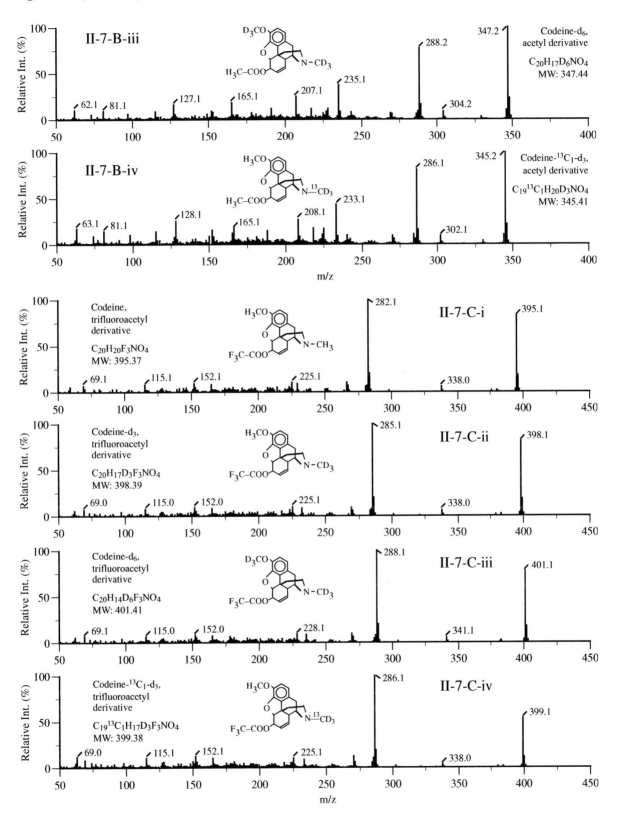

Figure II — Opioids

Figure II-7. (Continued)

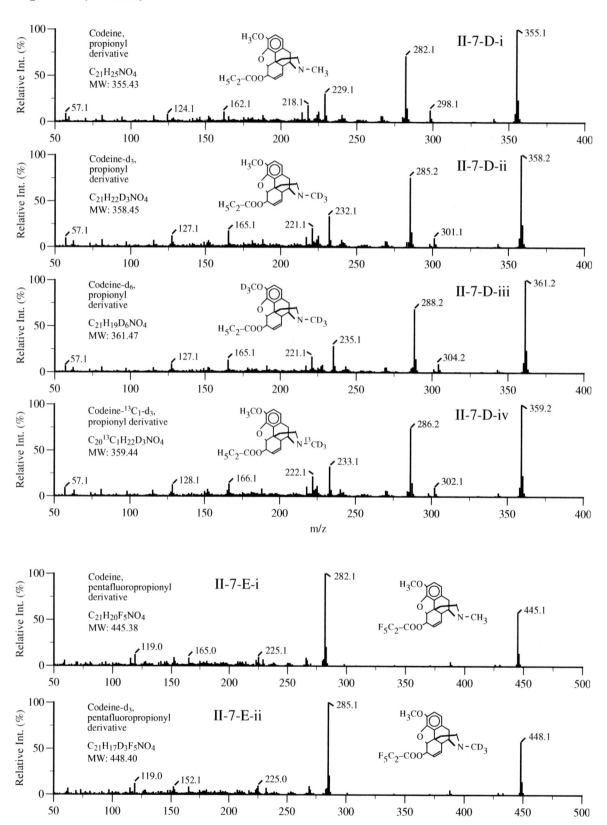

Figure II-7. (Continued)

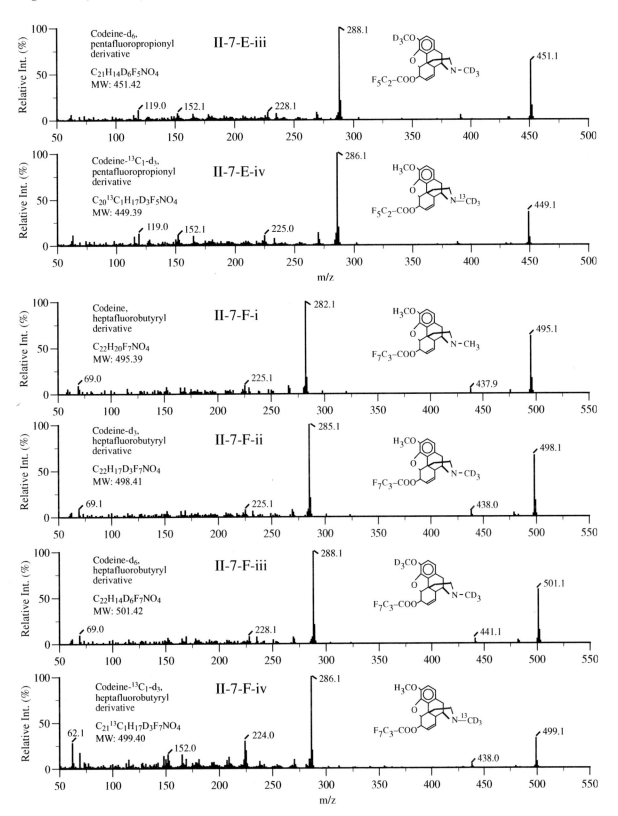

Figure II — Opioids

Figure II-7. (Continued)

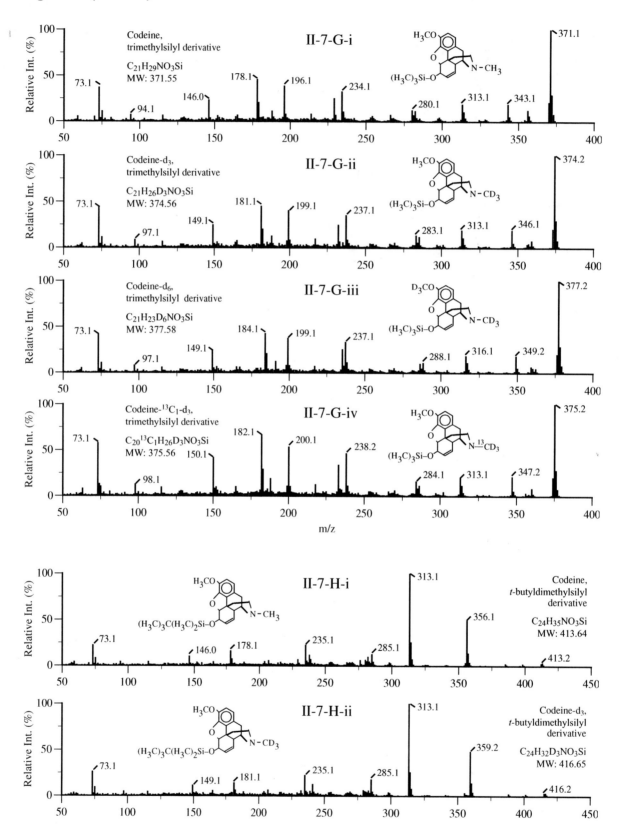

Figure II-7. (Continued)

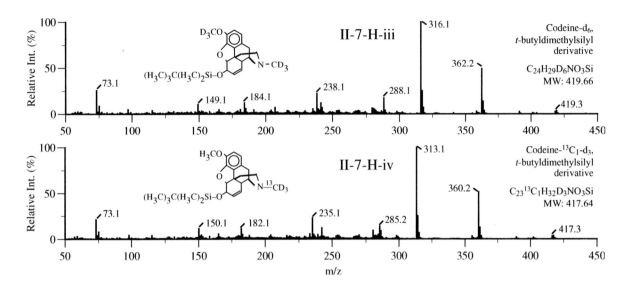

II-7-H-iii
Codeine-d6, t-butyldimethylsilyl derivative
C24H29D6NO3Si
MW: 419.66
Relative Int. (%)
316.1
362.2
73.1
149.1
184.1
238.1
288.1
419.3

II-7-H-iv
Codeine-13C1-d3, t-butyldimethylsilyl derivative
C23 13C1H32D3NO3Si
MW: 417.64
Relative Int. (%)
313.1
360.2
73.1
150.1
182.1
235.1
285.2
417.3
m/z

174

Figure II-8. Mass spectra of hydrocodone and its deuterated analogs (hydrocodone-d_3, -d_6): (A) underivatized; (B) ethyl-derivatized; (C) acetyl-derivatized; (D) TMS-derivatized; (E) *t*-BDMS-derivatized; (F) MA-derivatized; (G) HA/TMS-derivatized.

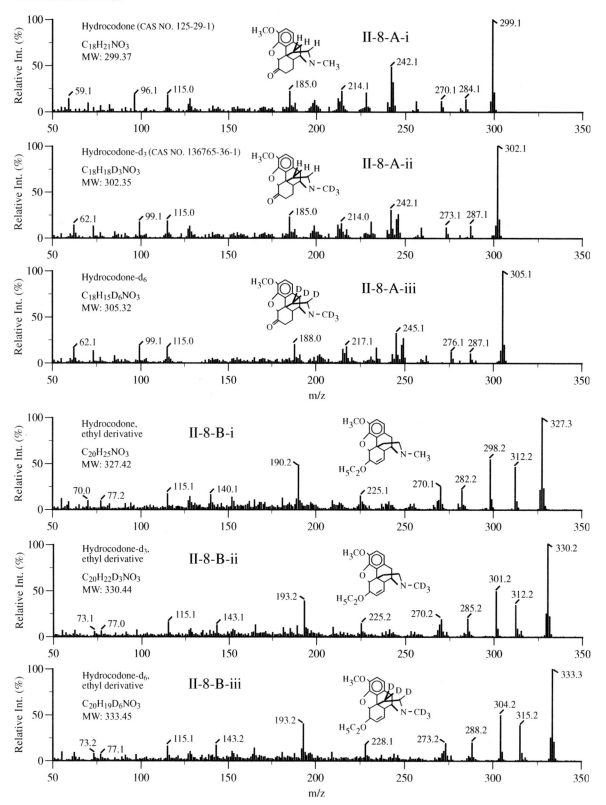

Figure II-8. (Continued)

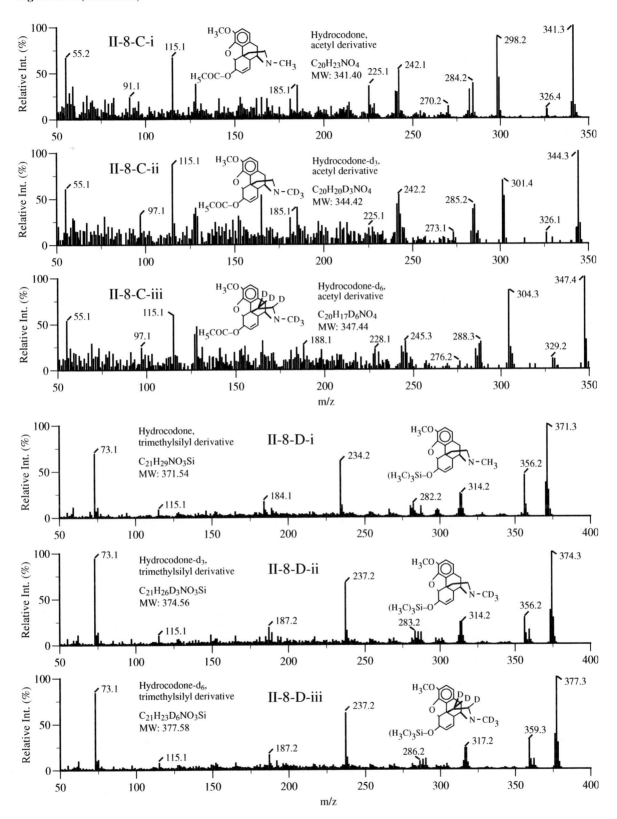

Figure II — Opioids

Figure II-8. (Continued)

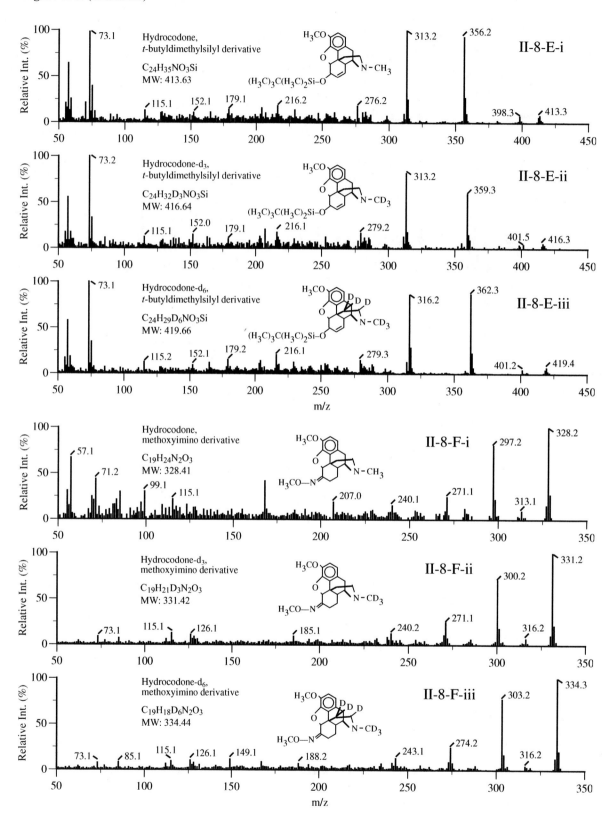

Figure II-8. (Continued)

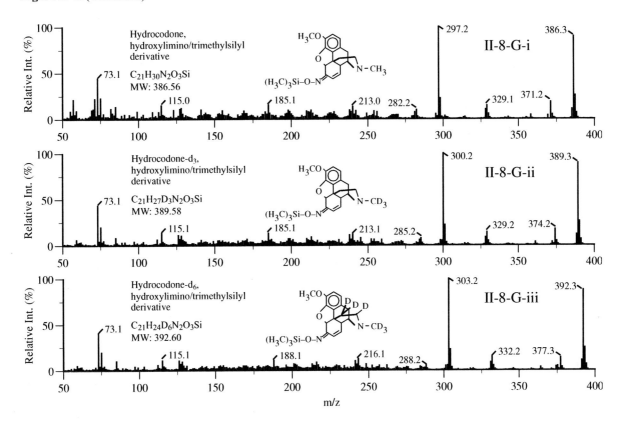

Figure II — Opioids

Figure II-9. Mass spectra of dihydrocodeine and its deuterated analogs (dihydrocodeine-d₃, -d₆): (A) underivatized-derivatized; (B) acetyl-derivatized; (C) TFA-derivatized; (D) propionyl-derivatized; (E) PFP-derivatized; (F) HFB-derivatized; (G) TMS-derivatized; (H) *t*-BDMS-derivatized.

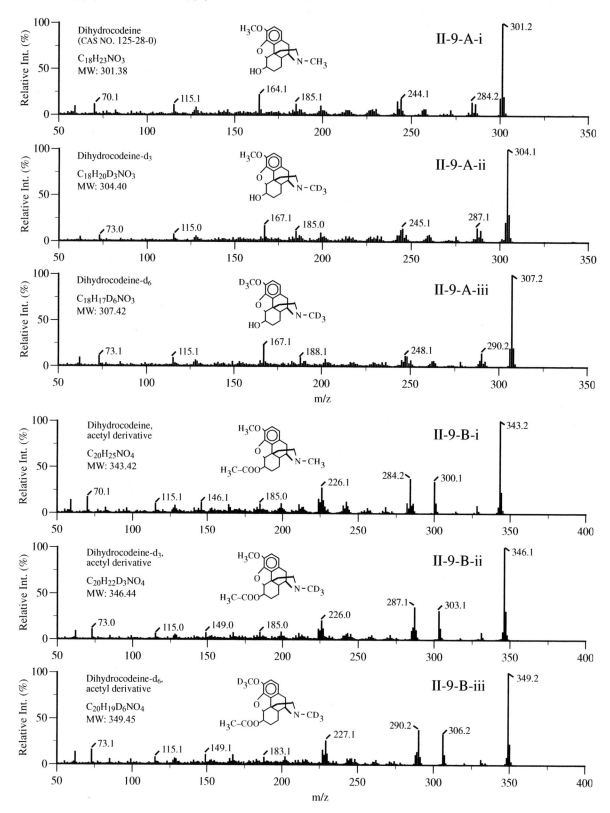

Figure II-9. (Continued)

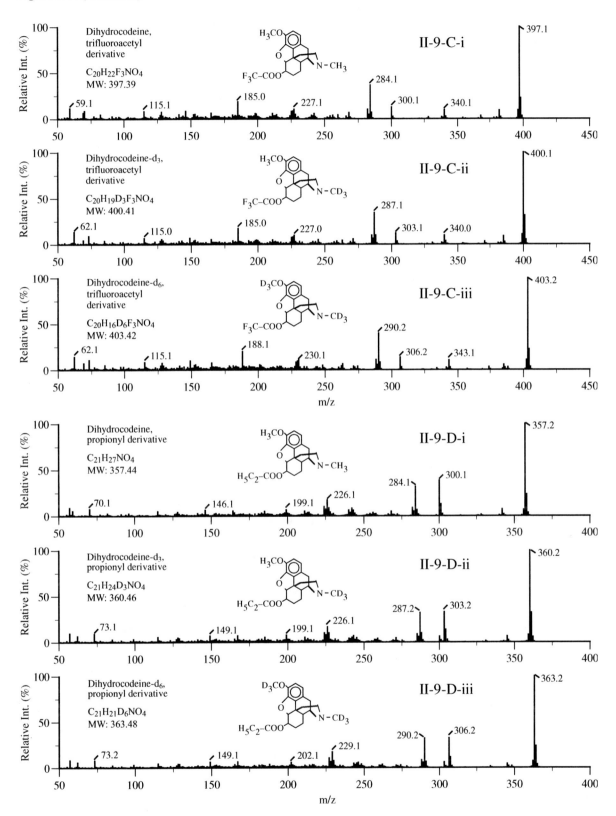

Figure II — Opioids

Figure II-9. (Continued)

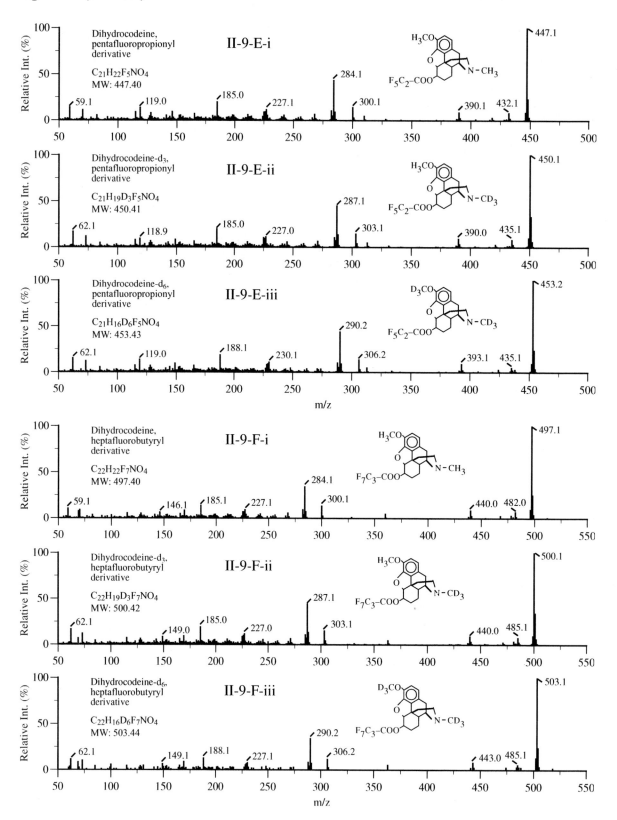

II-9-E-i

Dihydrocodeine, pentafluoropropionyl derivative

$C_{21}H_{22}F_5NO_4$
MW: 447.40

II-9-E-ii

Dihydrocodeine-d_3, pentafluoropropionyl derivative

$C_{21}H_{19}D_3F_5NO_4$
MW: 450.41

II-9-E-iii

Dihydrocodeine-d_6, pentafluoropropionyl derivative

$C_{21}H_{16}D_6F_5NO_4$
MW: 453.43

II-9-F-i

Dihydrocodeine, heptafluorobutyryl derivative

$C_{22}H_{22}F_7NO_4$
MW: 497.40

II-9-F-ii

Dihydrocodeine-d_3, heptafluorobutyryl derivative

$C_{22}H_{19}D_3F_7NO_4$
MW: 500.42

II-9-F-iii

Dihydrocodeine-d_6, heptafluorobutyryl derivative

$C_{22}H_{16}D_6F_7NO_4$
MW: 503.44

Figure II-9. (Continued)

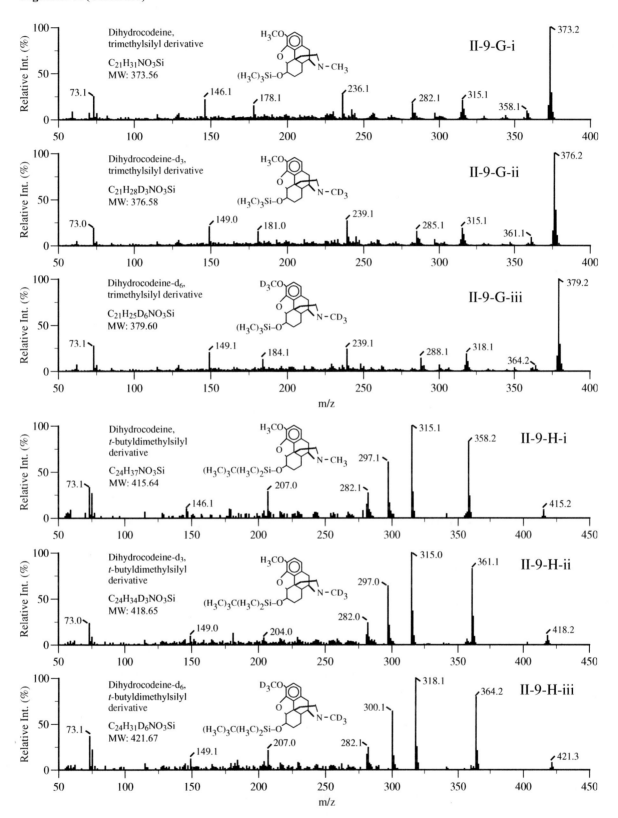

Dihydrocodeine,
trimethylsilyl derivative

$C_{21}H_{31}NO_3Si$
MW: 373.56

II-9-G-i

Dihydrocodeine-d_3,
trimethylsilyl derivative

$C_{21}H_{28}D_3NO_3Si$
MW: 376.58

II-9-G-ii

Dihydrocodeine-d_6,
trimethylsilyl derivative

$C_{21}H_{25}D_6NO_3Si$
MW: 379.60

II-9-G-iii

Dihydrocodeine,
t-butyldimethylsilyl
derivative

$C_{24}H_{37}NO_3Si$
MW: 415.64

II-9-H-i

Dihydrocodeine-d_3,
t-butyldimethylsilyl
derivative

$C_{24}H_{34}D_3NO_3Si$
MW: 418.65

II-9-H-ii

Dihydrocodeine-d_6,
t-butyldimethylsilyl
derivative

$C_{24}H_{31}D_6NO_3Si$
MW: 421.67

II-9-H-iii

Figure II — Opioids

Figure II-10. Mass spectra of oxycodone and its deuterated analogs (oxycodone-d_3, -d_6): (A) underivatized; (B) acetyl-derivatized; (C) [acetyl]$_2$-derivatized; (D) propionyl-derivatized; (E) TMS-derivatized; (F) [TMS]$_2$-derivatized; (G) *t*-BDMS-derivatized; (H) [*t*-BDMS]$_2$-derivatized; (I) MA-derivatized; (J) MA/propionyl-derivatized; (K) MA/TMS-derivatized; (L) HA/[propionyl]$_2$-derivatized; (M) HA/[TMS]$_2$-derivatized; (N) HA/ethyl/propionyl-derivatized.

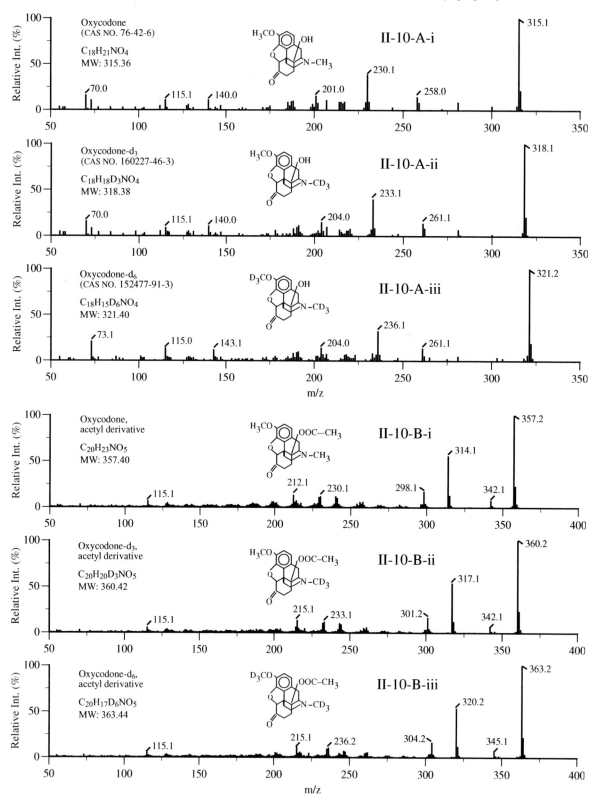

Figure II-10. (Continued)

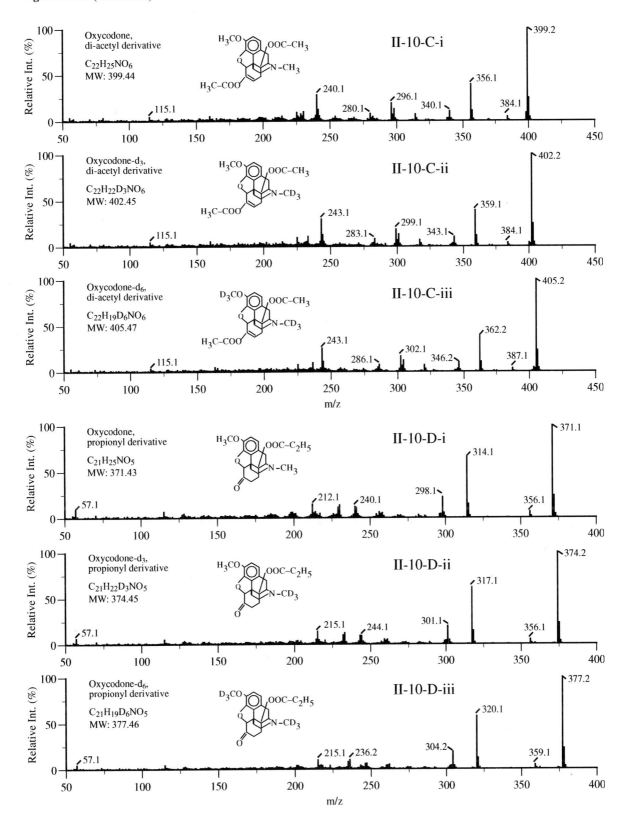

Figure II — Opioids

Figure II-10. (Continued)

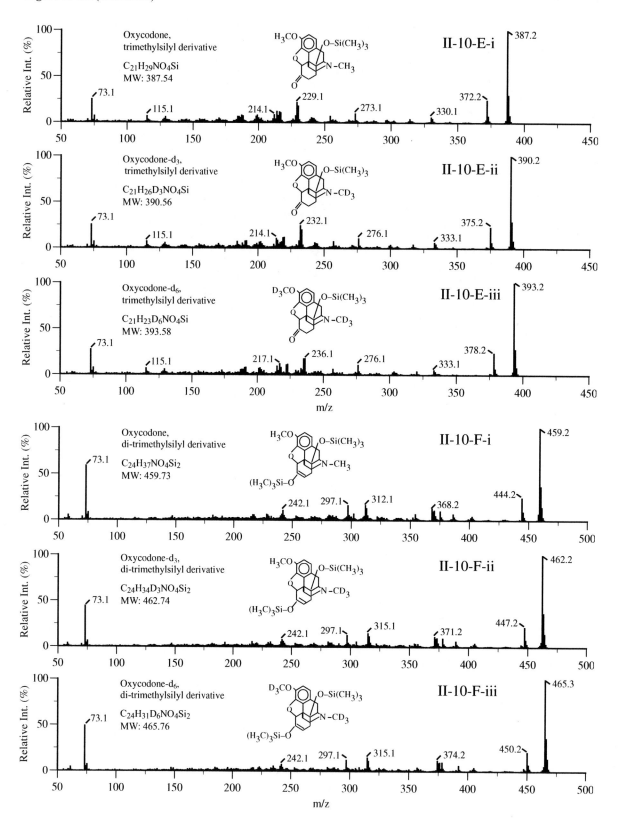

Figure II-10. (Continued)

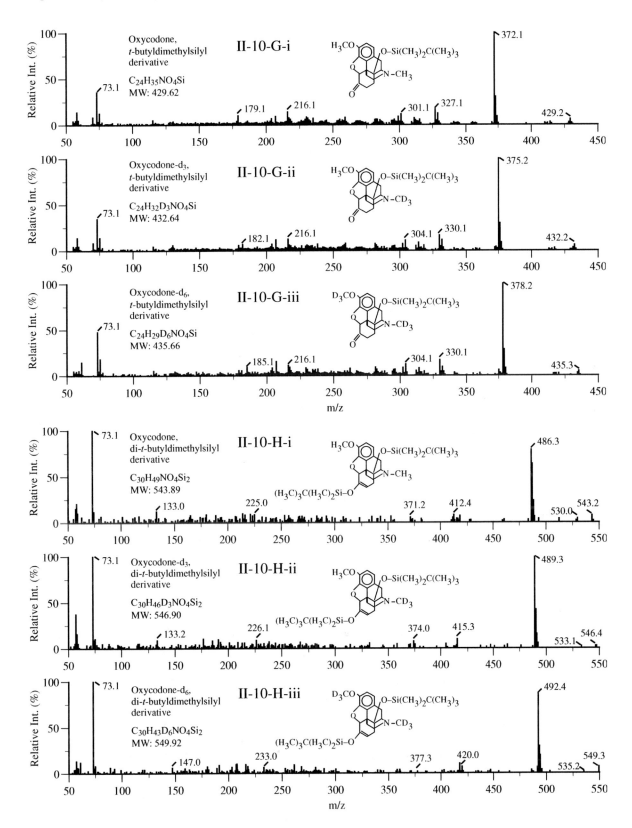

Figure II — Opioids

Figure II-10. (Continued)

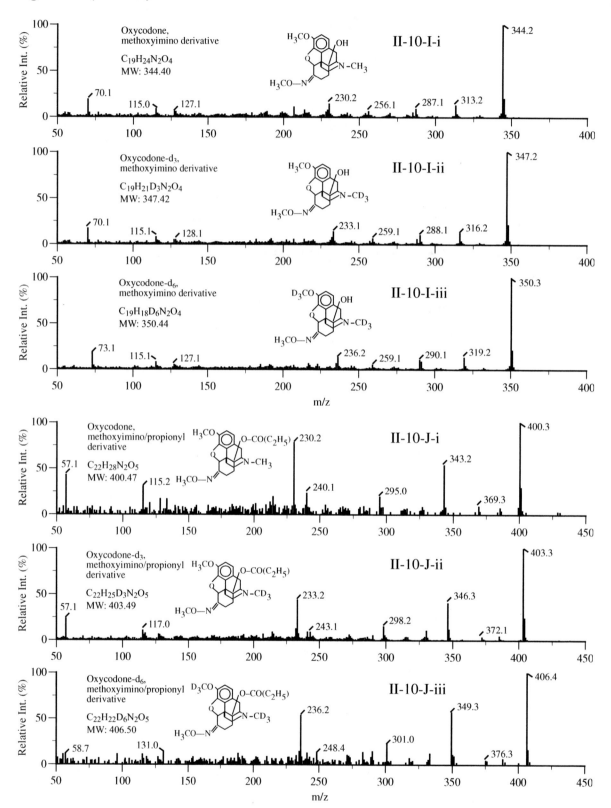

Figure II-10. (Continued)

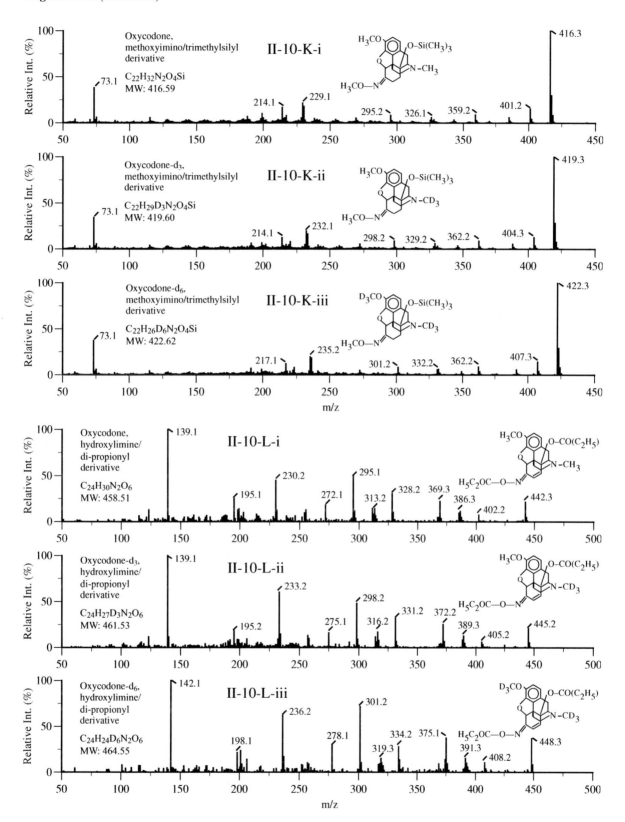

Figure II — Opioids

Figure II-10. (Continued)

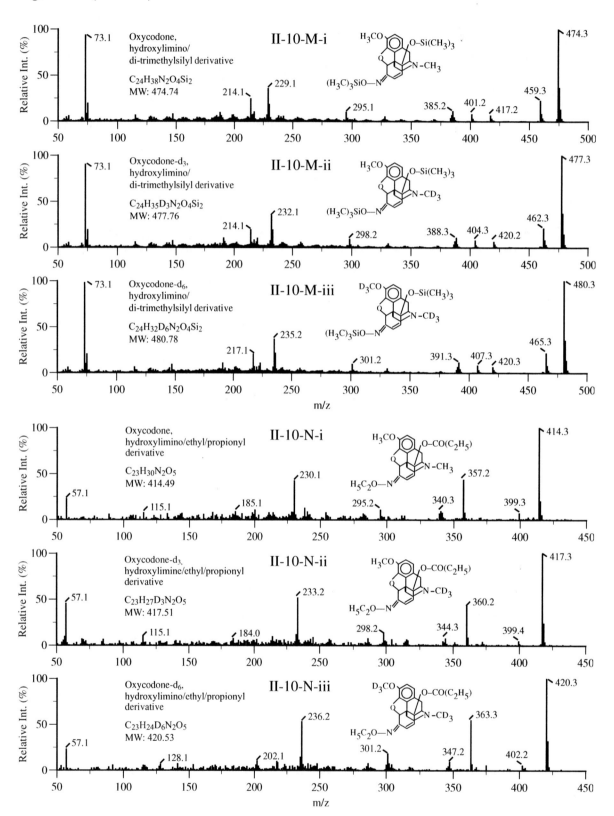

Figure II-11. Mass spectra of noroxycodone and its deuterated analogs (noroxycodone-d₃): (A) underivatized; (B) [acetyl]₂-derivatized; (C) [TFA]₃-derivatized; (D) propionyl-derivatized; (E) [PFP]₂-derivatized; (F) [HFB]₂-derivatized; (G) [TMS]₂-derivatized; (H) [TMS]₃-derivatized; (I) MA/ethyl-derivatized; (J) MA/acetyl-derivatized; (K) MA/[TFA]₂-derivatized; (L) MA/propionyl-derivatized; (M) MA/PFP-derivatized; (N) MA/[HFB]₂-derivatized; (O) MA/[TMS]₂-derivatized; (P) MA/*t*-BDMS-derivatized; (Q) MA/ethyl/propionyl-derivatized; (R) MA/ethyl/TMS-derivatized; (S) MA/ethyl/*t*-BDMS-derivatized; (T) MA/acetyl/TMS-derivatized; (U) MA/propionyl/TMS-derivatized; (V) HA/[ethyl]₂/TMS-derivatized.

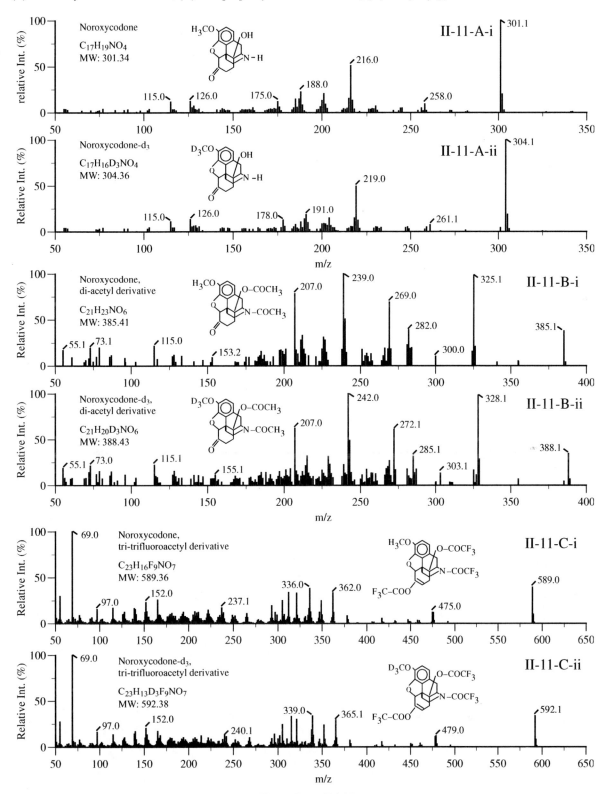

Figure II — Opioids

Figure II-11. (Continued)

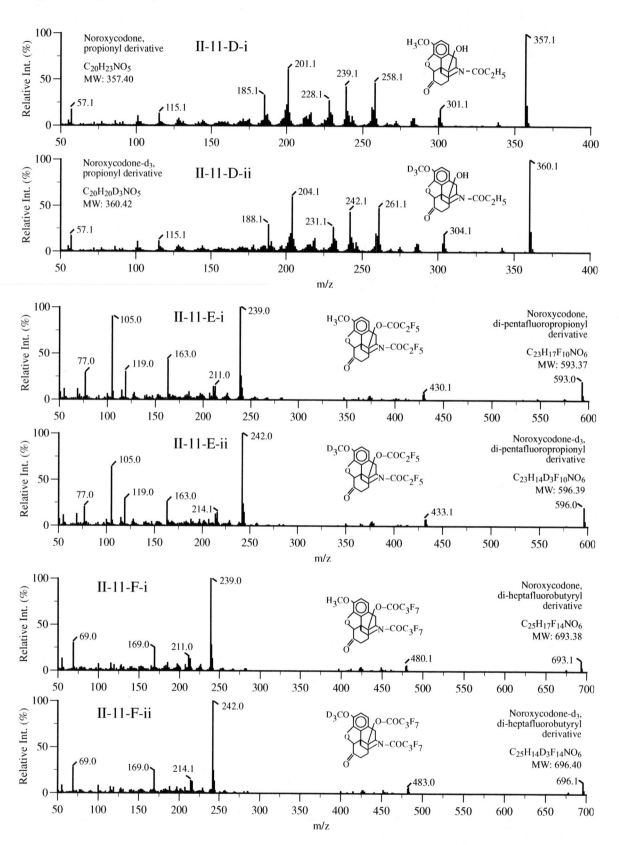

Figure II-11. (Continued)

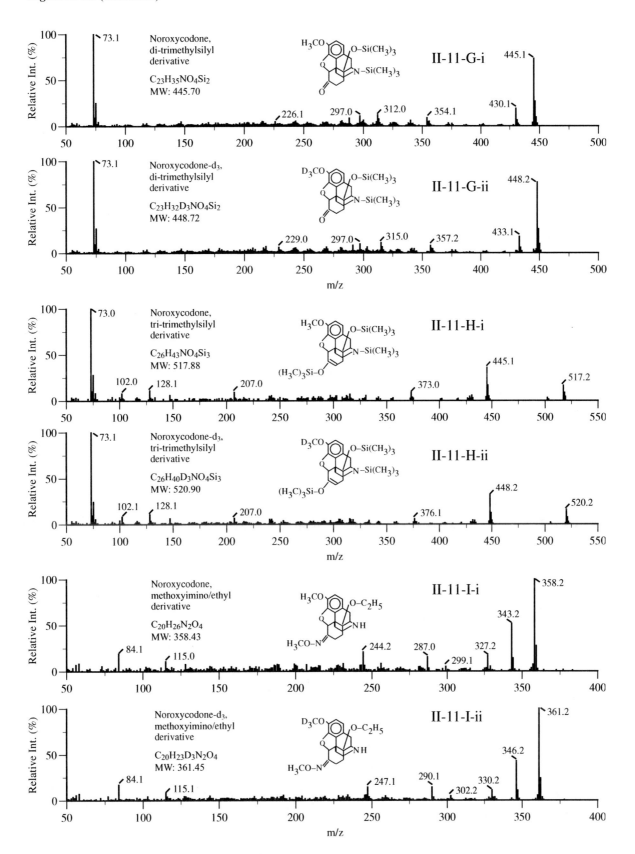

Figure II — Opioids

Figure II-11. (Continued)

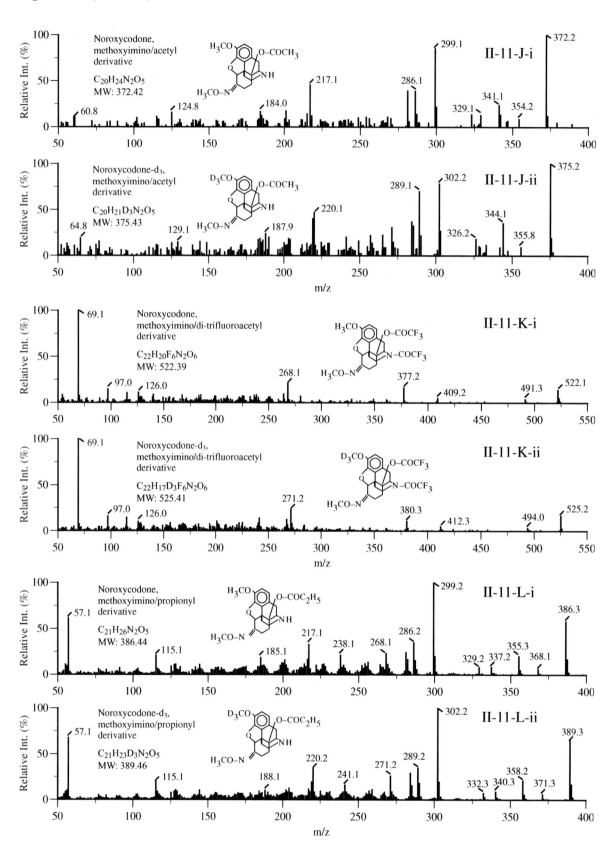

Figure II-11. (Continued)

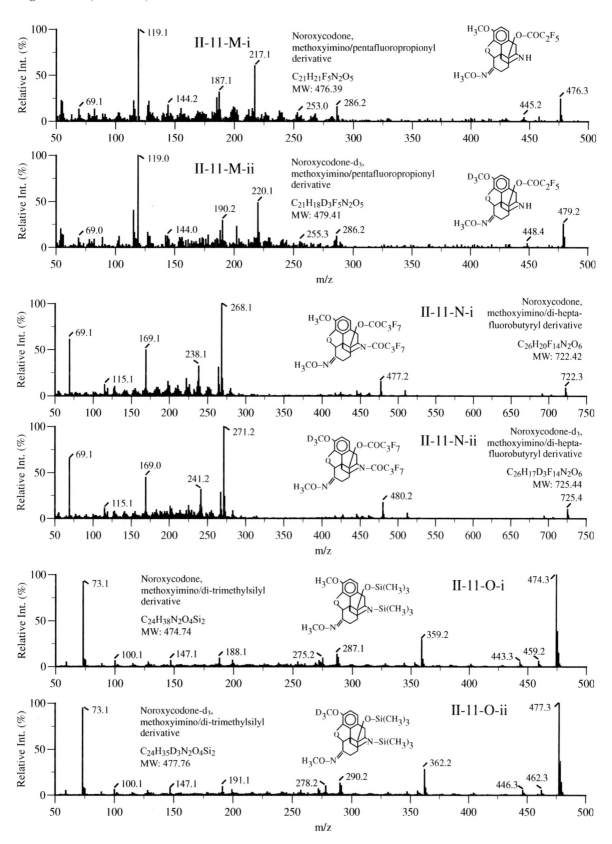

II-11-M-i — Noroxycodone, methoxyimino/pentafluoropropionyl derivative — $C_{21}H_{21}F_5N_2O_5$ — MW: 476.39

II-11-M-ii — Noroxycodone-d$_3$, methoxyimino/pentafluoropropionyl derivative — $C_{21}H_{18}D_3F_5N_2O_5$ — MW: 479.41

II-11-N-i — Noroxycodone, methoxyimino/di-heptafluorobutyryl derivative — $C_{26}H_{20}F_{14}N_2O_6$ — MW: 722.42

II-11-N-ii — Noroxycodone-d$_3$, methoxyimino/di-heptafluorobutyryl derivative — $C_{26}H_{17}D_3F_{14}N_2O_6$ — MW: 725.44

II-11-O-i — Noroxycodone, methoxyimino/di-trimethylsilyl derivative — $C_{24}H_{38}N_2O_4Si_2$ — MW: 474.74

II-11-O-ii — Noroxycodone-d$_3$, methoxyimino/di-trimethylsilyl derivative — $C_{24}H_{35}D_3N_2O_4Si_2$ — MW: 477.76

Figure II — Opioids

Figure II-11. (Continued)

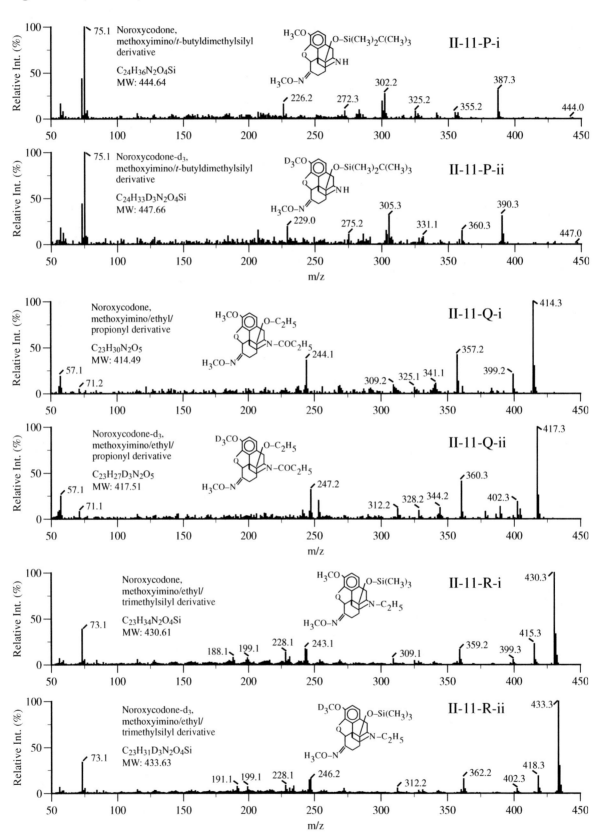

Figure II-11. (Continued)

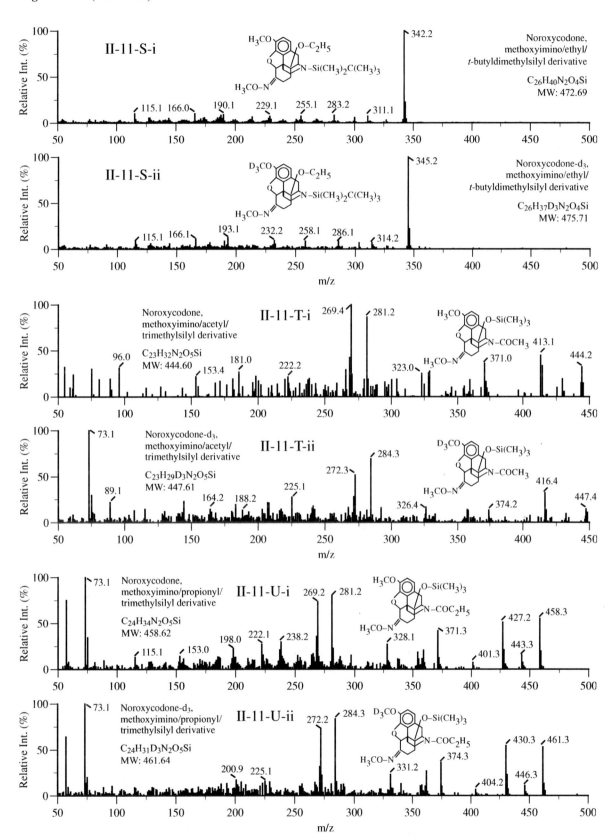

Figure II — Opioids

Figure II-11. (Continued)

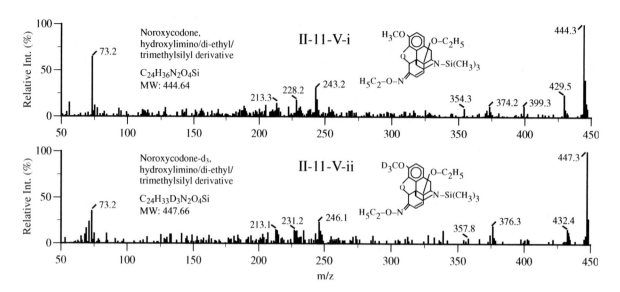

Figure II-12. Mass spectra of buprenorphine and its deuterated analogs (buprenorphine-d₄): (A) methyl-derivatized; (B) ethyl-derivatized; (C) acetyl-derivatized; (D) MBTFA-derivatized; (E)PFP-derivatized; (F) HFB-derivatized; (G) TMS-derivatized; (H) [TMS]₂-derivatized; (I) t-BDMS-derivatized.

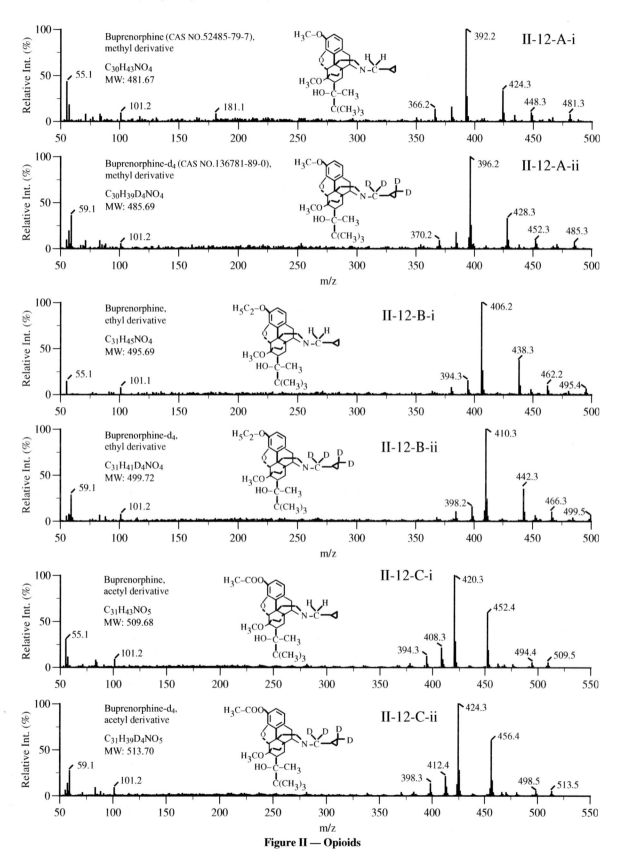

Figure II — Opioids

Figure II-12. (Continued)

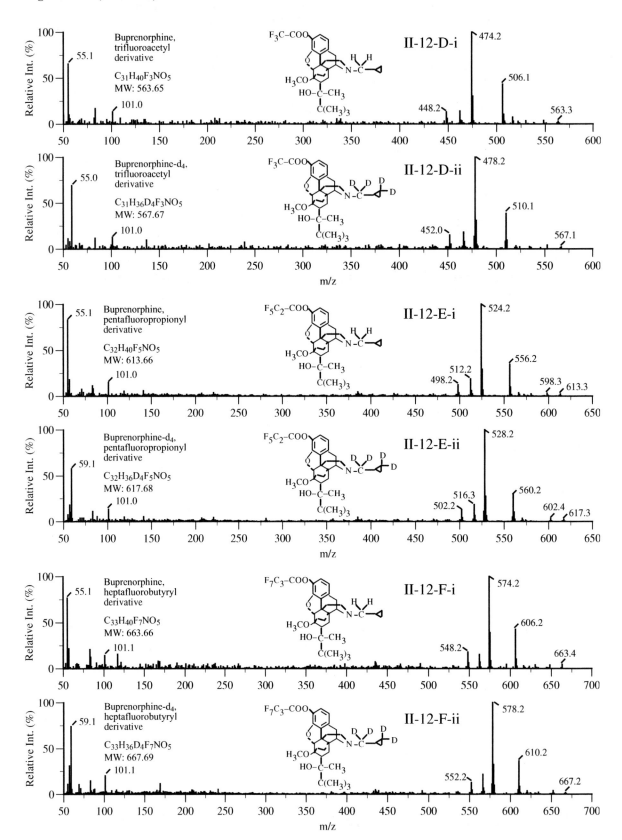

Figure II-12. (Continued)

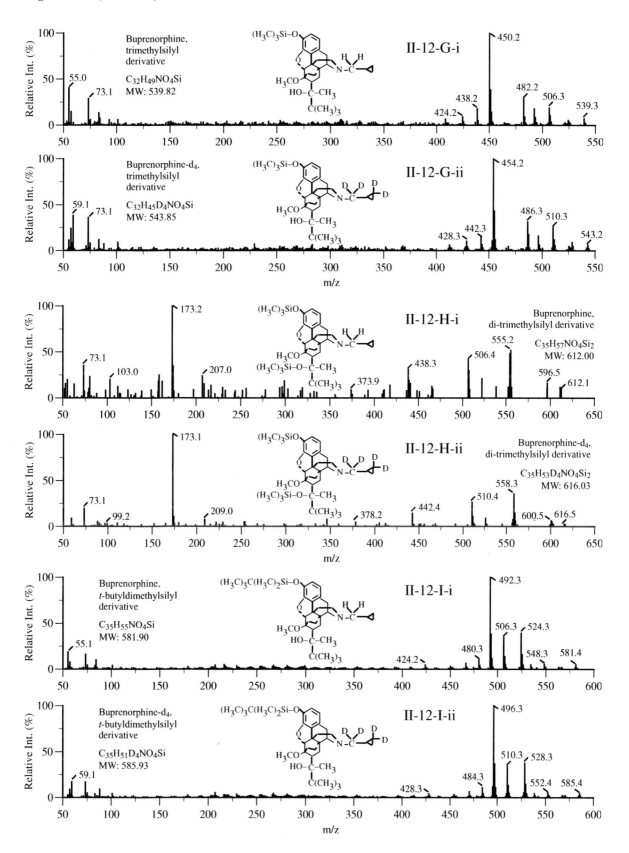

Figure II — Opioids

Figure II-13. Mass spectra of norbuprenorphine and its deuterated analogs (norbuprenorphine-d₃): (A) [methyl]₂-derivatized; (B) [ethyl]₂-derivatized; (C) [acetyl]₂-derivatized; (D) [MBTFA]₂-derivatized; (E) [PFP]₂-derivatized; (F) [HFB]₂-derivatized; (G) [TMS]₂-derivatized; (H) [TMS]₃-derivatized; (I) *t*-BDMS-derivatized.

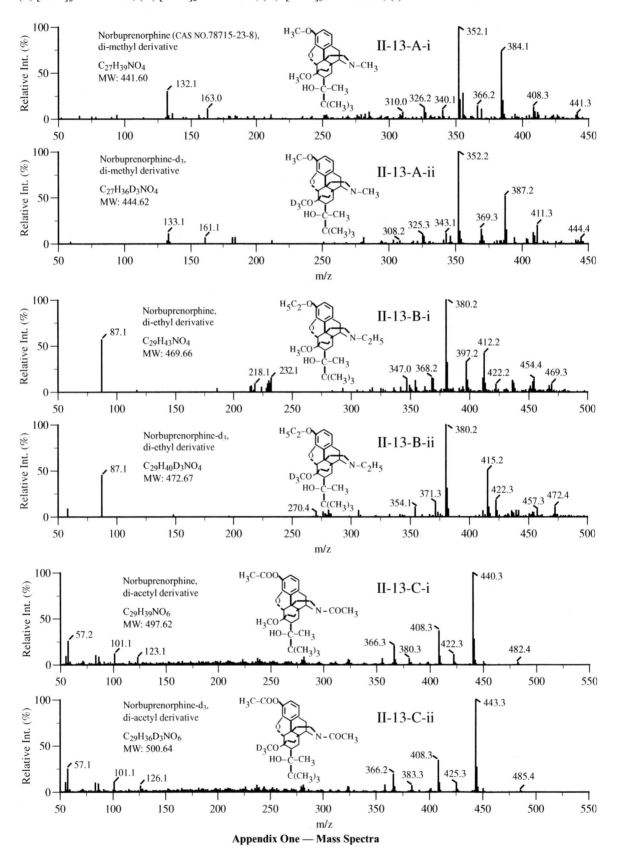

Figure II-13. (Continued)

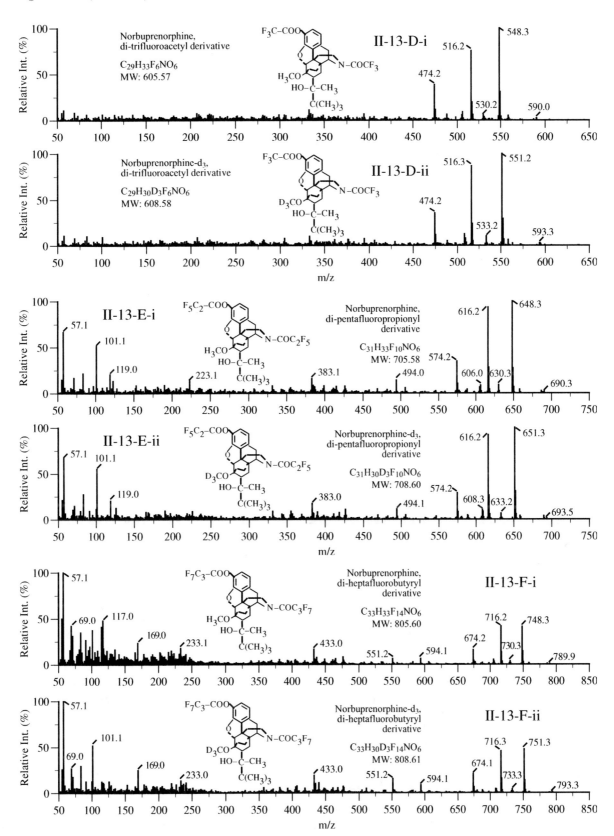

Figure II — Opioids

202

Figure II-13. (Continued)

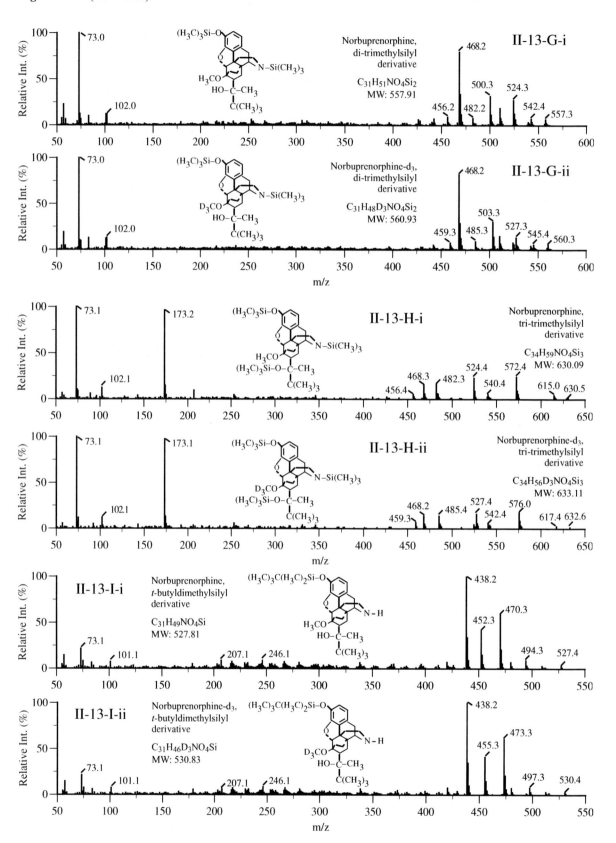

Appendix One — Mass Spectra

Figure II-14. Mass spectra of fentanyl and its deuterated analogs (fentanyl-d₅).

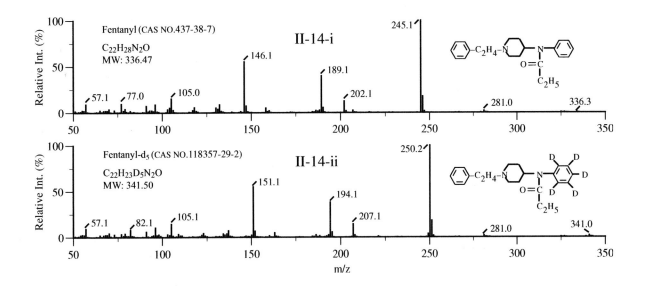

Figure II — Opioids

Figure II-15. Mass spectra of norfentanyl and its deuterated analogs (norfentanyl-d_5): (A) underivatized; (B) acetyl-derivatized; (C) TCA-derivatized; (D) TFA-derivatized; (E) PFP-derivatized; (F) HFB-derivatized; (G) 4-CB-derivatized; (H) TMS-derivatized; (I) *t*-BDMS-derivatized.

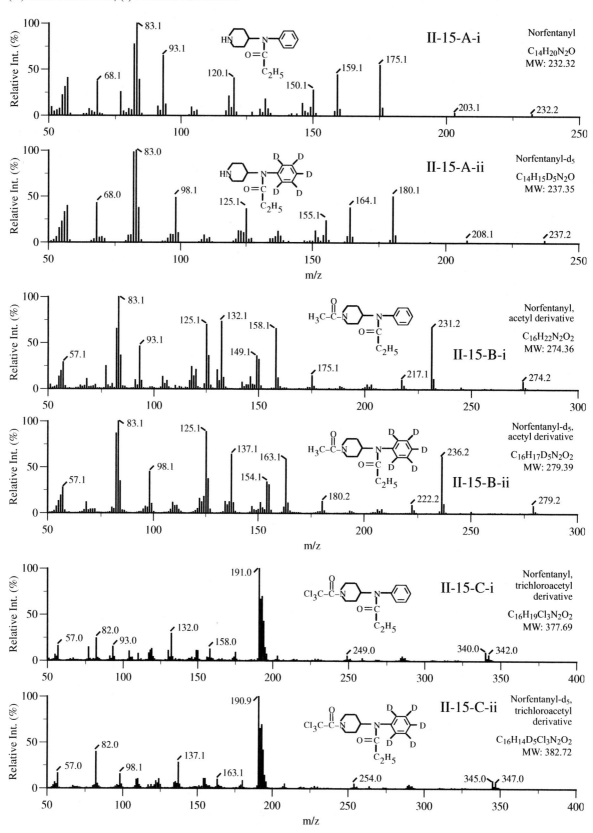

205

Figure II-15. (Continued)

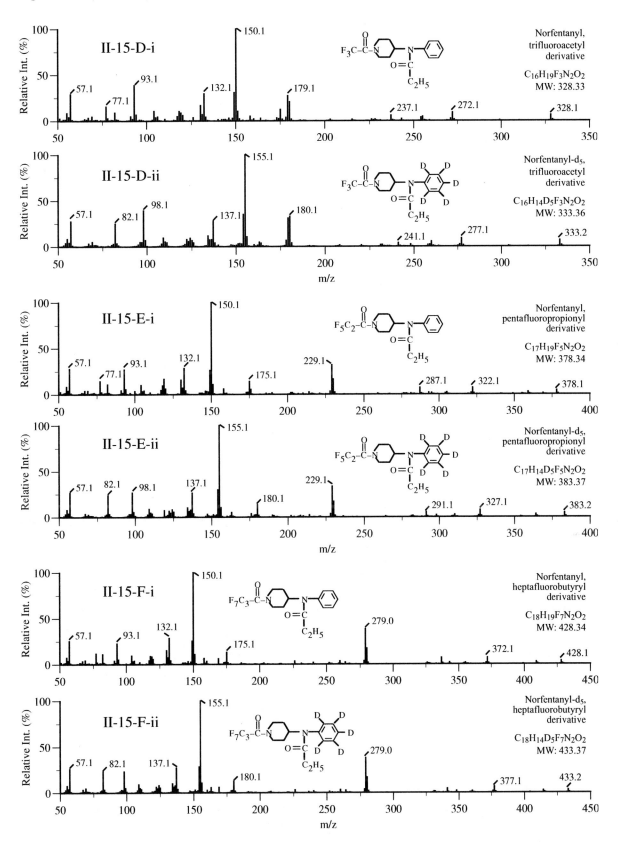

Figure II — Opioids

206

Figure II-15. (Continued)

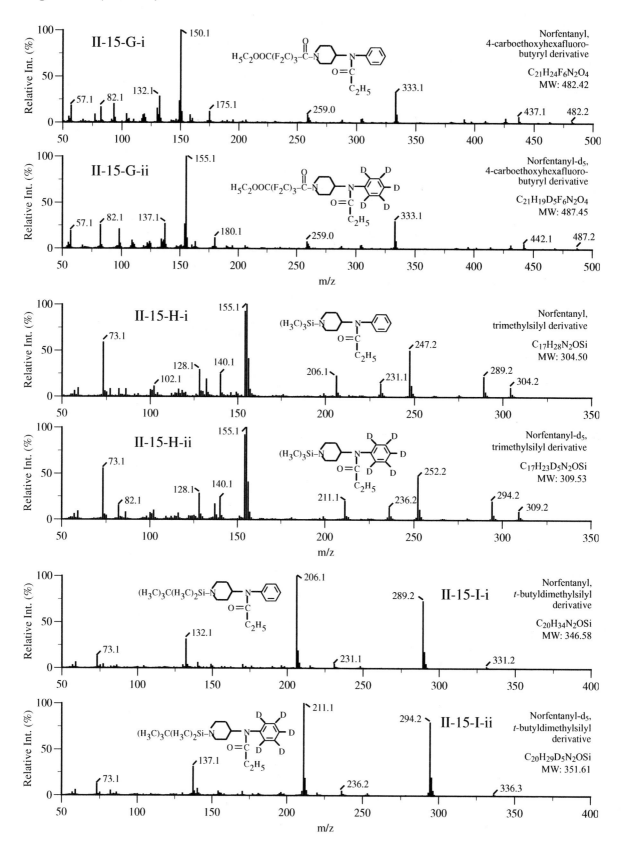

Figure II-16. Mass spectra of methadone and its deuterated analogs (methadone-d₃, -d₉).

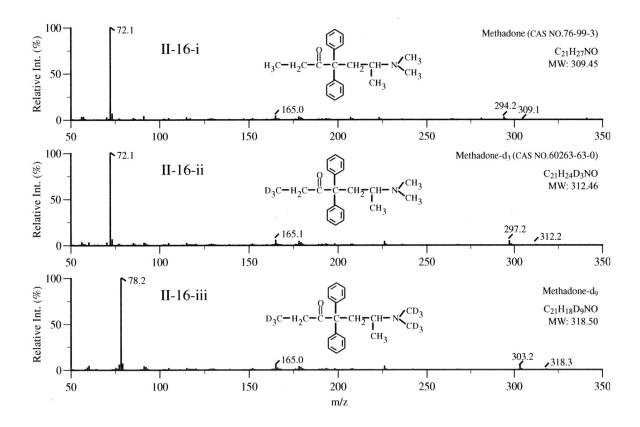

Figure II — Opioids

Figure II-17. Mass spectra of EDDP and its deuterated analogs (EDDP-d₃).

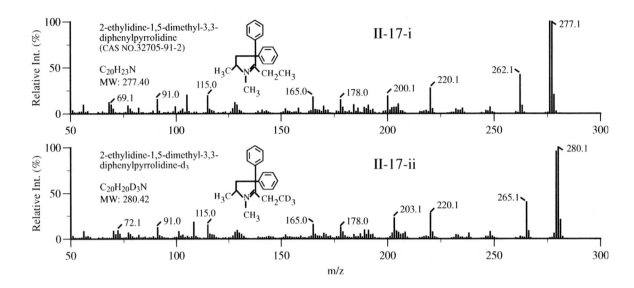

Figure II-18. Mass spectra of propoxyphene and its deuterated analogs (propoxyphene-d$_5$, -d$_7$, -d$_{11}$).

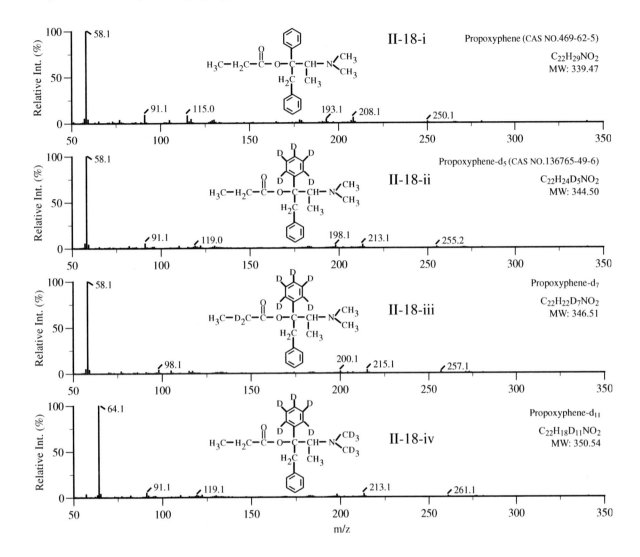

Figure II — Opioids

Figure II-19. Mass spectra of norpropoxyphene and its deuterated analogs (norpropoxyphene-d$_5$).

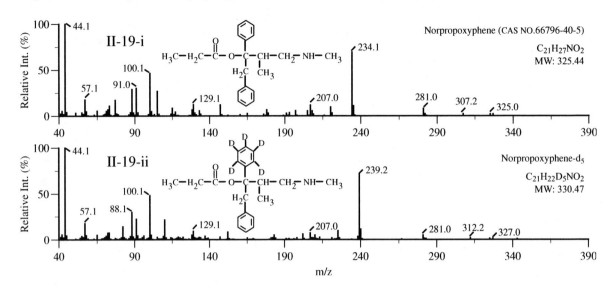

Figure II-20. Mass spectra of meperidine and its deuterated analogs (meperidine-d₄).

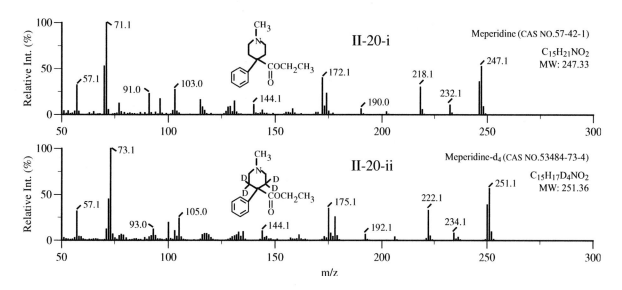

Figure II — Opioids

Figure II-21. Mass spectra of normeperidine and its deuterated analogs (normeperidine-d₄): (A) underivatized-derivatized; (B) ethyl-derivatized; (C) propyl-derivatized; (D) butyl-derivatizedacetyl; (E) acetyl-derivatized; (F) TCA-derivatized; (G) TFA-derivatized; (H) PFP-derivatized; (I) HFB-derivatized; (J) 4-CB-derivatized; (K) TMS-derivatized; (L) *t*-BDMS-derivatized.

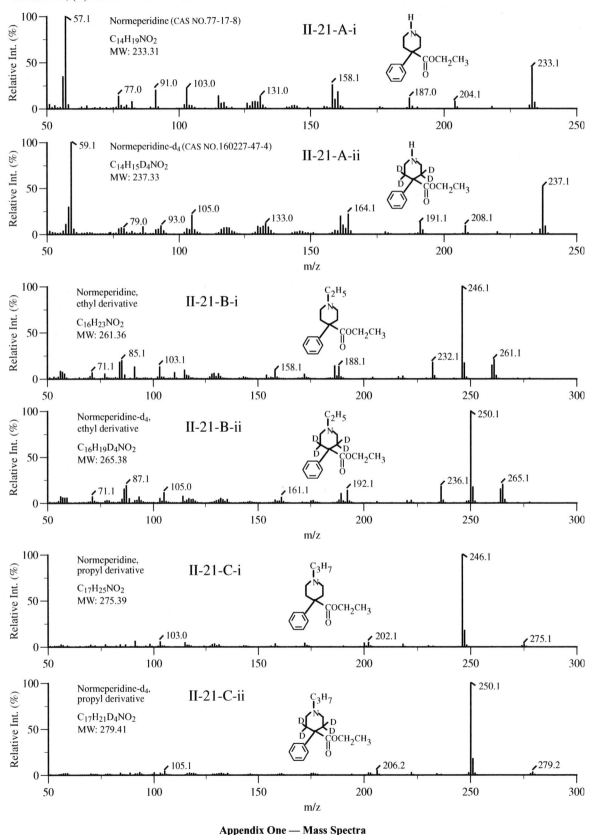

Figure II-21. (Continued)

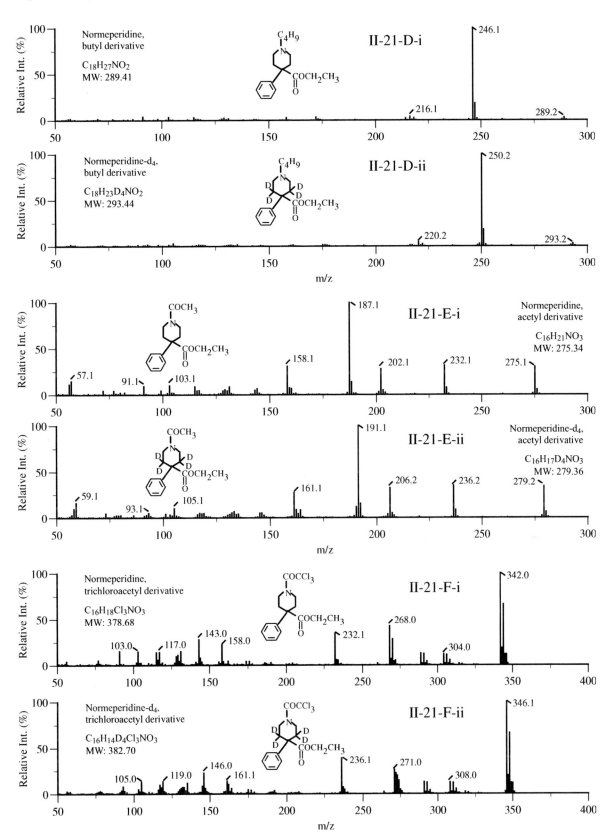

II-21-D-i

Normeperidine, butyl derivative

$C_{18}H_{27}NO_2$
MW: 289.41

II-21-D-ii

Normeperidine-d$_4$, butyl derivative

$C_{18}H_{23}D_4NO_2$
MW: 293.44

II-21-E-i

Normeperidine, acetyl derivative

$C_{16}H_{21}NO_3$
MW: 275.34

II-21-E-ii

Normeperidine-d$_4$, acetyl derivative

$C_{16}H_{17}D_4NO_3$
MW: 279.36

Normeperidine, trichloroacetyl derivative

$C_{16}H_{18}Cl_3NO_3$
MW: 378.68

II-21-F-i

Normeperidine-d$_4$, trichloroacetyl derivative

$C_{16}H_{14}D_4Cl_3NO_3$
MW: 382.70

II-21-F-ii

Figure II — Opioids

Figure II-21. (Continued)

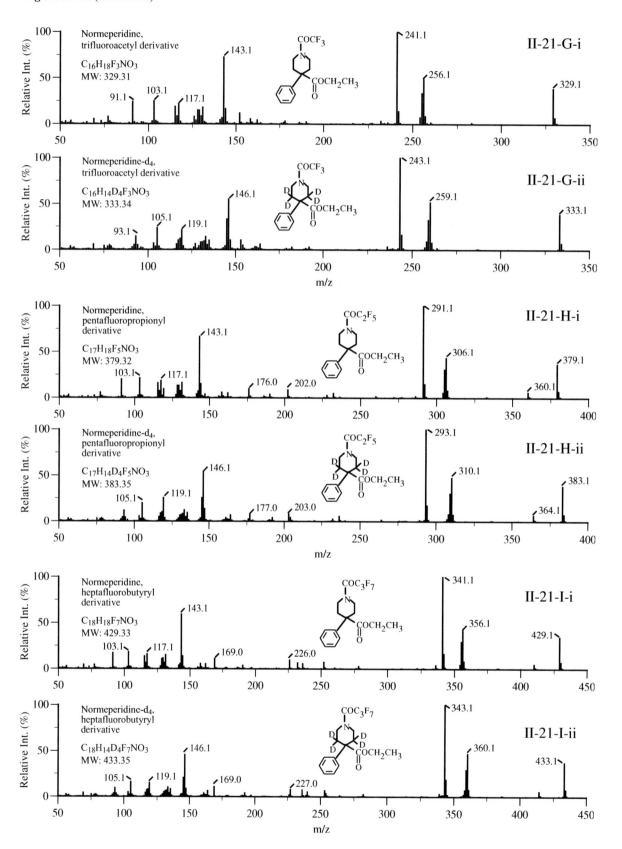

Figure II-21. (Continued)

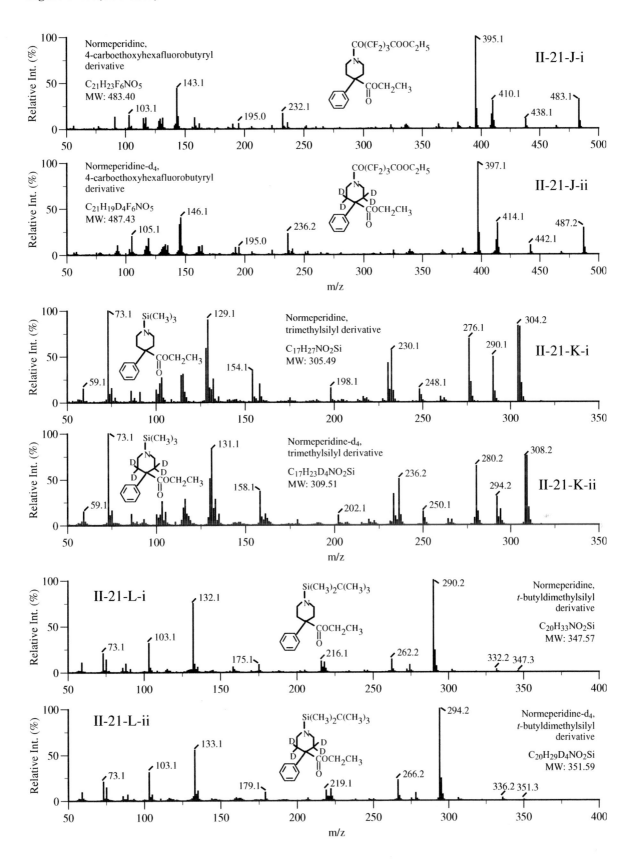

Figure II — Opioids

Summary of Drugs, Isotopic Analogs, and Chemical Derivatization Groups Included in Figure III (Hallucinogens)

Compound	Isotopic analog	Chemical derivatization group (no. of spectra)	Figure #
Cannabinol	d_3	Methyl, ethyl, propyl, butyl, propionyl (10)	III-1
Tetrahydrocannabinol	d_3	Methyl, ethyl, propyl, butyl, TFA, propionyl, PFP, HFB, TMS, t-BDMS (20)	III-2
THC-OH	d_3	[Methyl]$_2$, [ethyl]$_2$, [propyl]$_2$, [butyl]$_2$, [TFA]$_2$, propionyl, [PFP]$_2$, [HFB]$_2$, [TMS]$_2$, [t-BDMS]$_2$ (20)	III-3
THC-COOH	d_3, d_9	[Methyl]$_2$, [ethyl]$_2$, [propyl]$_2$, [butyl]$_2$, propionyl, [TMS]$_2$, [t-BDMS]$_2$, methyl/TFA, PFPoxy/PFP, HFPoxy/HFB (30)	III-4
Ketamine	d_4	None, acetyl, TFA, HFB, PFB, TMS (12)	III-5
Norketamine	d_4	None, acetyl, TCA, TFA, PFP, HFB, 4-CB, PFB, TMS, TFA/t-BDMS, PFP/t-BDMS, HFB/t-BDMS (24)	III-6
Phencyclidine	d_5	None (2)	III-7
LSD	d_3	None, TMS (4)	III-8
Mescaline	d_9	Acetyl, TCA, TFA, PFP, HFB, 4-CB, [TMS]$_2$, t-BDMS, TFA/TMS, TFA/t-BDMS, PFP/TMS, PFP/t-BDMS, HFB/TMS, HFB/t-BDMS (28)	III-9
Psilocin	d_{10}	None, acetyl, [acetyl]$_2$, [TMS]$_2$, t-BDMS, [t-BDMS]$_2$ (12)	III-10

Total no. of mass spectra: 162

Figure III — Hallucinogens

Appendix One — Figure III
Mass Spectra of Commonly Abused Drugs and Their Isotopically Labeled Analogs
in Various Derivatization Forms — Hallucinogens

Figure III — Hallucinogens

Figure III-1. Mass spectra of cannabinol and its deuterated analogs (cannabinol-d₃): (A) methyl-derivatized; (B) ethyl-derivatized; (C) propyl-derivatized; (D) butyl-derivatized; (E) propionyl-derivatized.

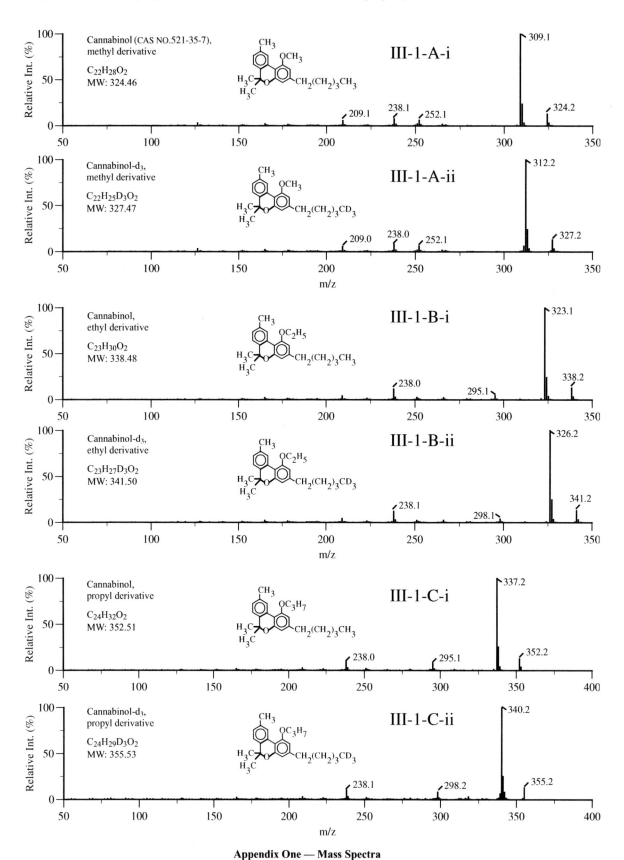

Figure III-1. (Continued)

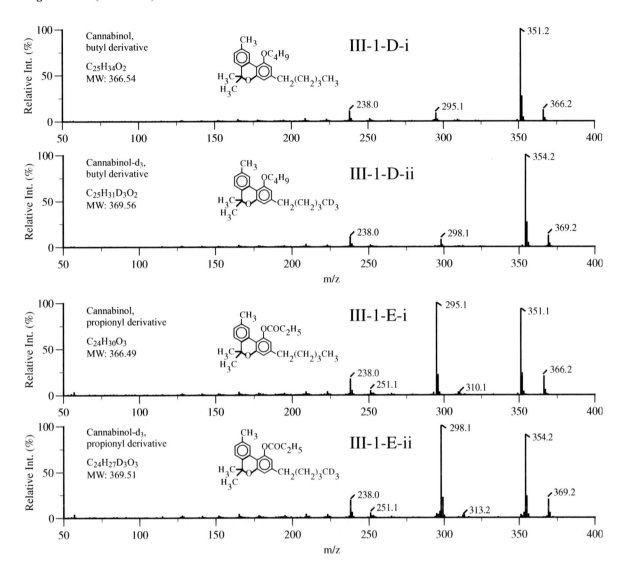

Figure III — Hallucinogens

Figure III-2. Mass spectra of tetrahydrocannabinol and its deuterated analogs (tetrahydrocannabinol-d₃): (A) methyl-derivatized; (B) ethyl-derivatized; (C) propyl-derivatized; (D) butyl-derivatized; (E) TFA-derivatized; (F) propionyl-derivatized; (G) PFP-derivatized; (H) HFB-derivatized; (I) TMS-derivatized; (J) *t*-BDMS-derivatized.

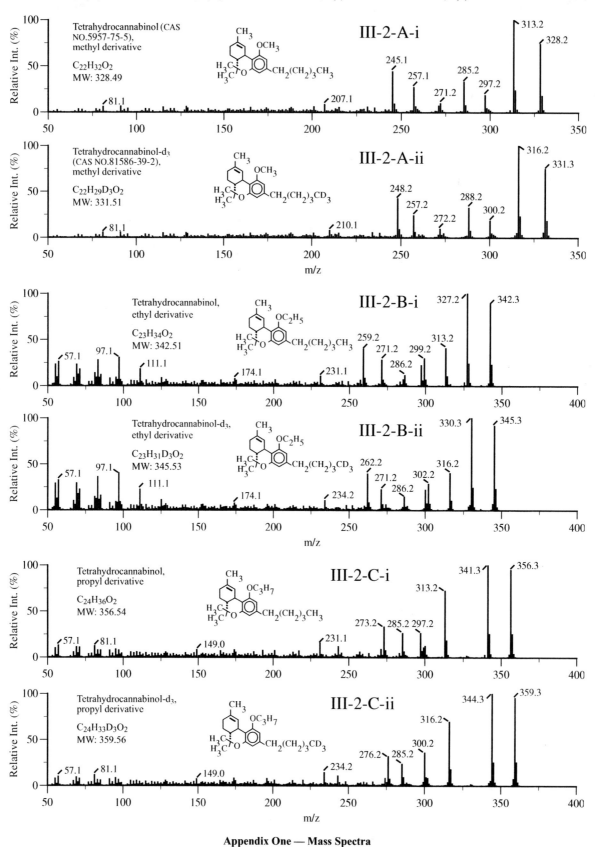

Figure III-2. (Continued)

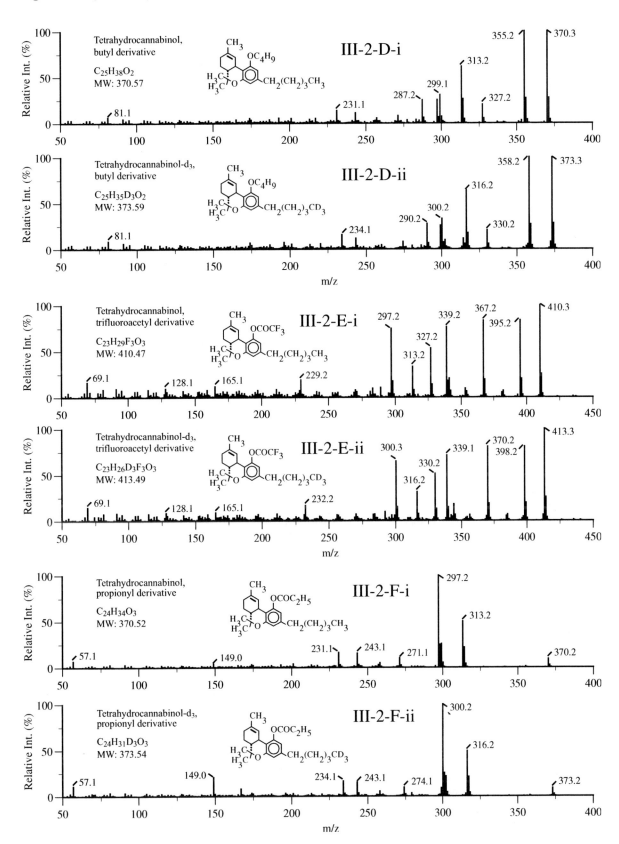

Figure III — Hallucinogens

Figure III-2. (Continued)

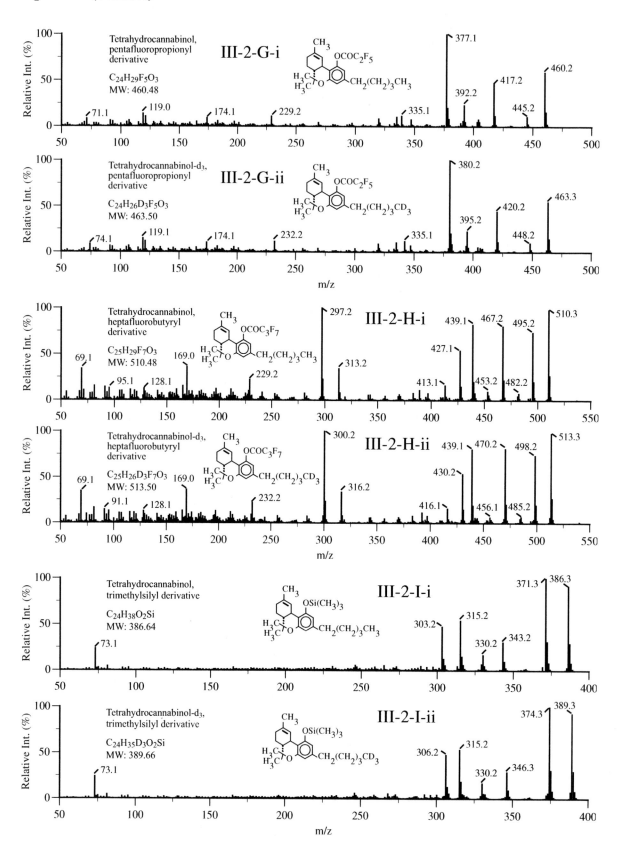

Figure III-2. (Continued)

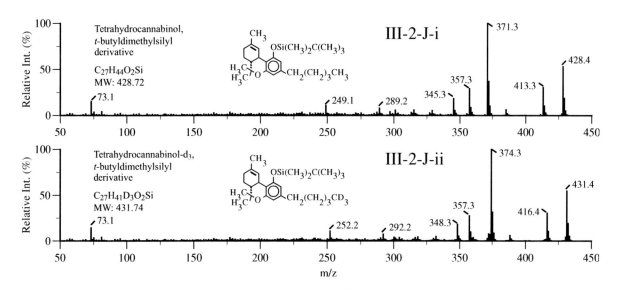

Figure III — Hallucinogens

Figure III-3. Mass spectra of THC-OH and its deuterated analogs (THC-OH-d₃): (A) [methyl]₂-derivatized; (B) [ethyl]₂-derivatized; (C) [propyl]₂-derivatized; (D) [butyl]₂-derivatized; (E) [TFA]₂-derivatized; (F) propionyl-derivatized; (G) [PFP]₂-derivatized; (H) [HFB]₂-derivatized; (I) [TMS]₂-derivatized; (J) [t-BDMS]₂-derivatized.

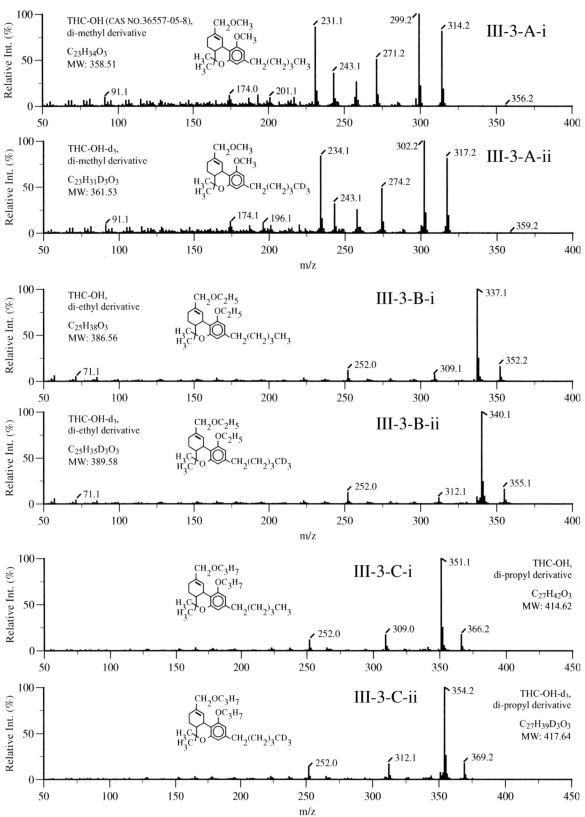

Figure III-3. (Continued)

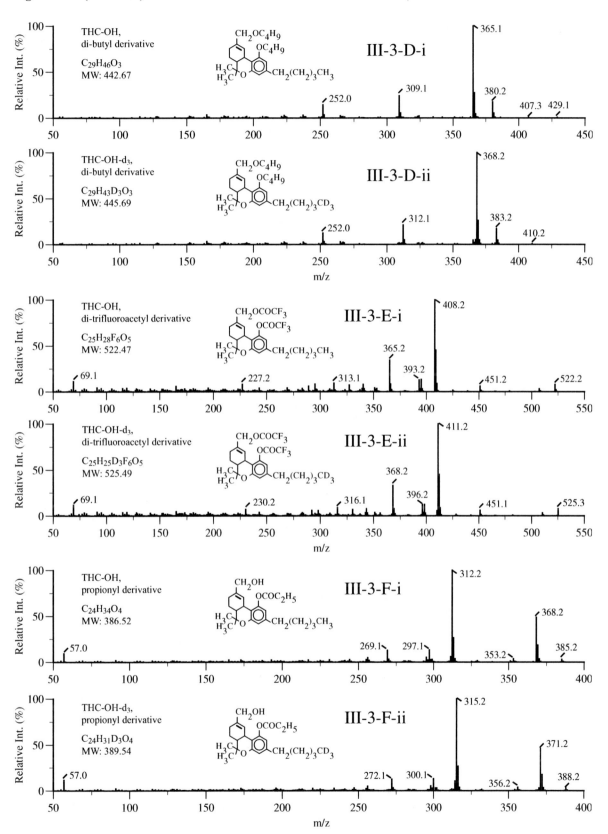

Figure III — Hallucinogens

Figure III-3. (Continued)

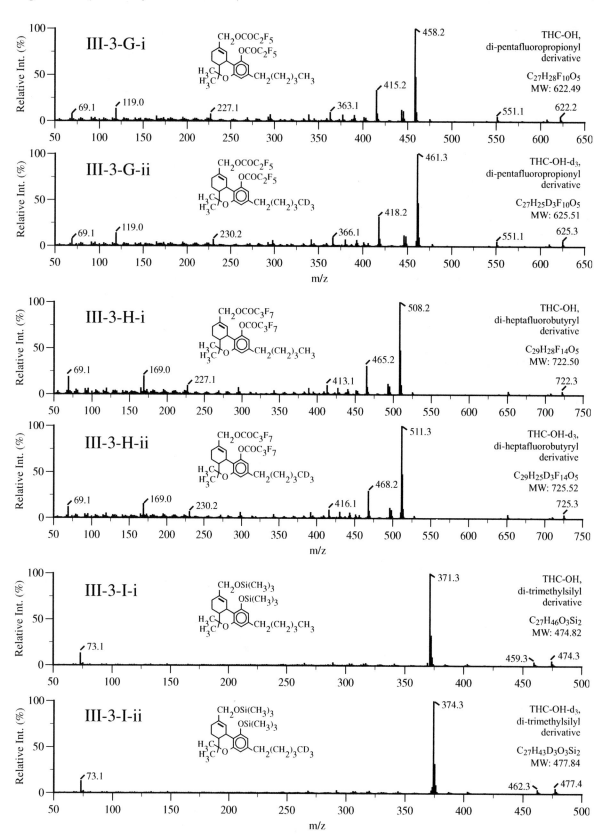

Figure III-3. (Continued)

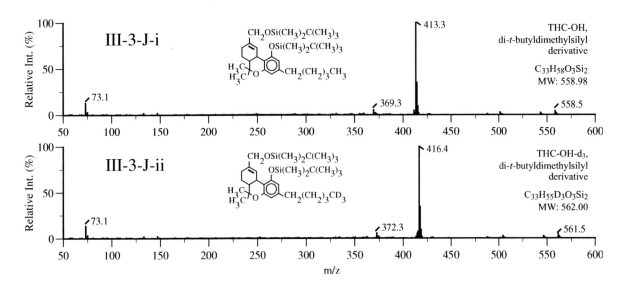

III-3-J-i

CH$_2$OSi(CH$_3$)$_2$C(CH$_3$)$_3$
OSi(CH$_3$)$_2$C(CH$_3$)$_3$
H$_3$C
H$_3$C O
CH$_2$(CH$_2$)$_3$CH$_3$

73.1
369.3
413.3
558.5

THC-OH,
di-*t*-butyldimethylsilyl
derivative

C$_{33}$H$_{58}$O$_3$Si$_2$
MW: 558.98

III-3-J-ii

CH$_2$OSi(CH$_3$)$_2$C(CH$_3$)$_3$
OSi(CH$_3$)$_2$C(CH$_3$)$_3$
H$_3$C
H$_3$C O
CH$_2$(CH$_2$)$_3$CD$_3$

73.1
372.3
416.4
561.5

THC-OH-d$_3$,
di-*t*-butyldimethylsilyl
derivative

C$_{33}$H$_{55}$D$_3$O$_3$Si$_2$
MW: 562.00

m/z

Figure III — Hallucinogens

Figure III-4. Mass spectra of THC-COOH and its deuterated analogs (THC-COOH-d₃, -d₉): (A) [methyl]₂-derivatized; (B) [ethyl]₂-derivatized; (C) [propyl]₂-derivatized; (D) [butyl]₂-derivatized; (E) propionyl-derivatized; (F) [TMS]₂-derivatized; (G) [*t*-BDMS]₂-derivatized; (H) methyl/TFA-derivatized; (I) PFPoxy/PFP-derivatized; (J) HFPoxy/HFB-derivatized.

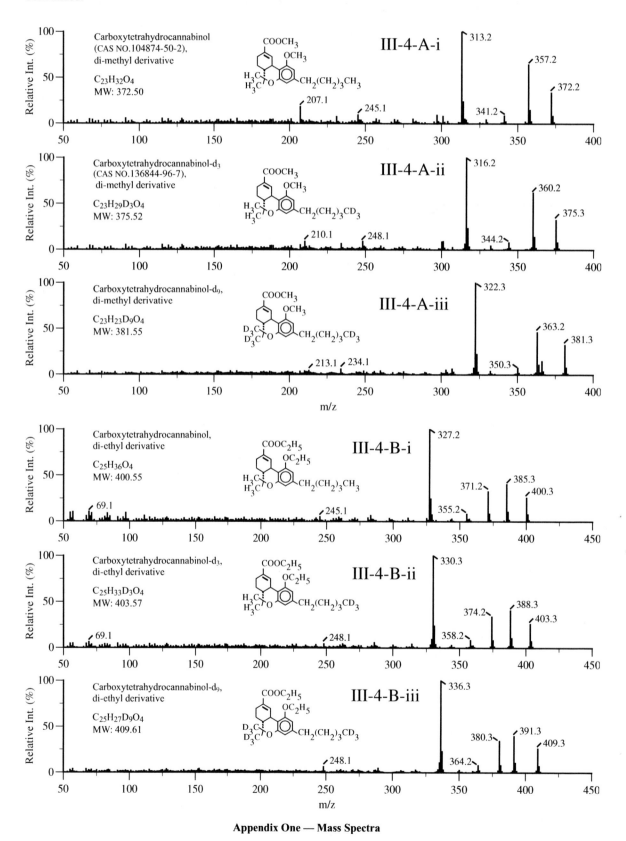

Figure III-4. (Continued)

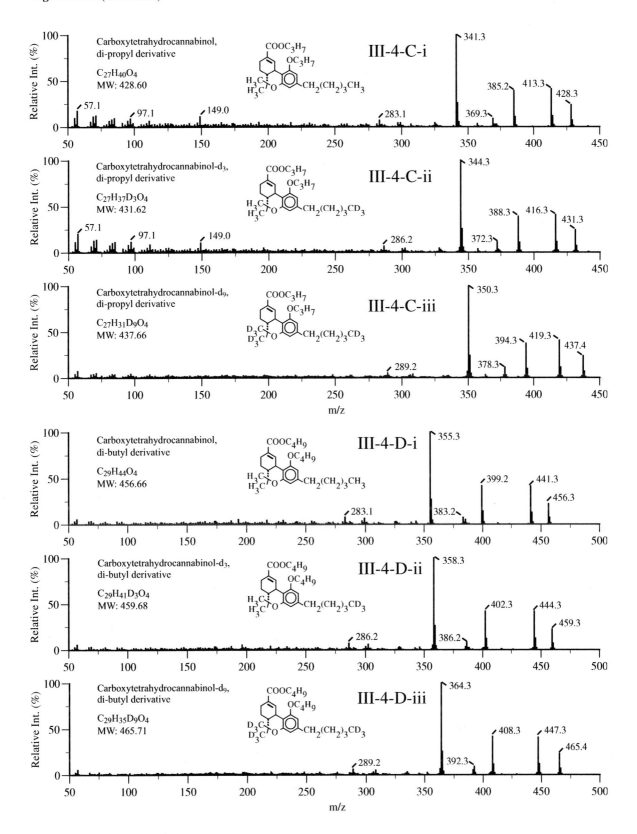

Figure III — Hallucinogens

Figure III-4. (Continued)

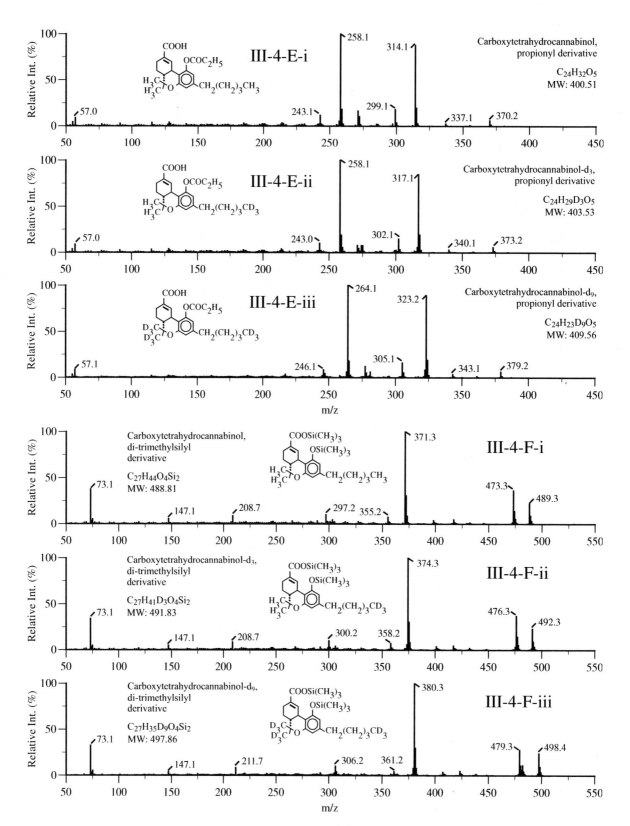

Figure III-4. (Continued)

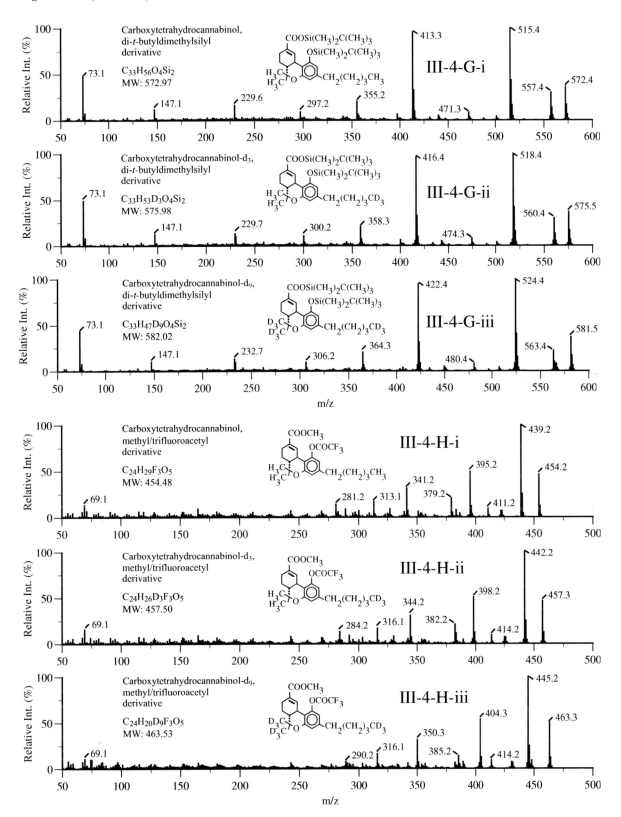

Figure III — Hallucinogens

Figure III-4. (Continued)

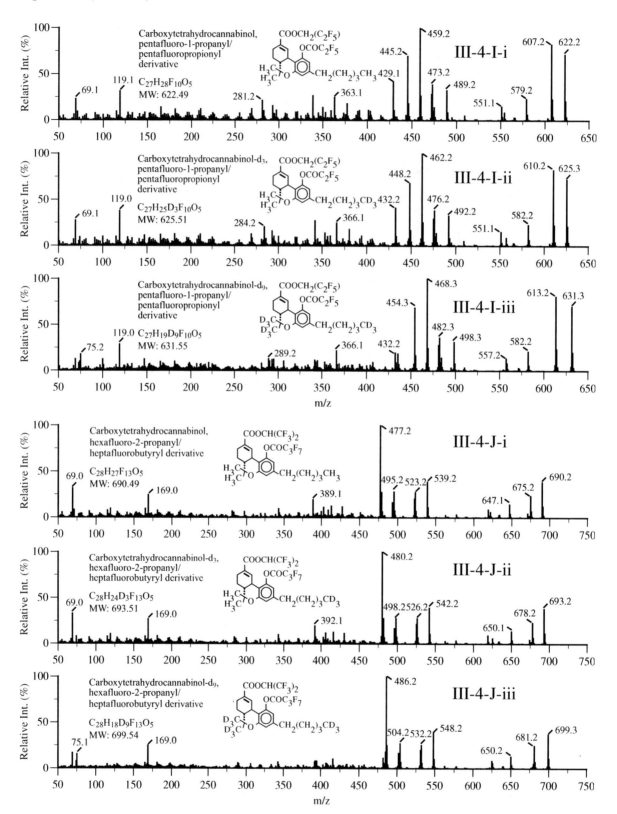

Figure III-5. Mass spectra of ketamine and its deuterated analogs (ketamine-d₄): (A) underivatized; (B) acetyl-derivatized; (C) TFA-derivatized; (D) HFB-derivatized; (E) PFB-derivatized; (F) TMS-derivatized.

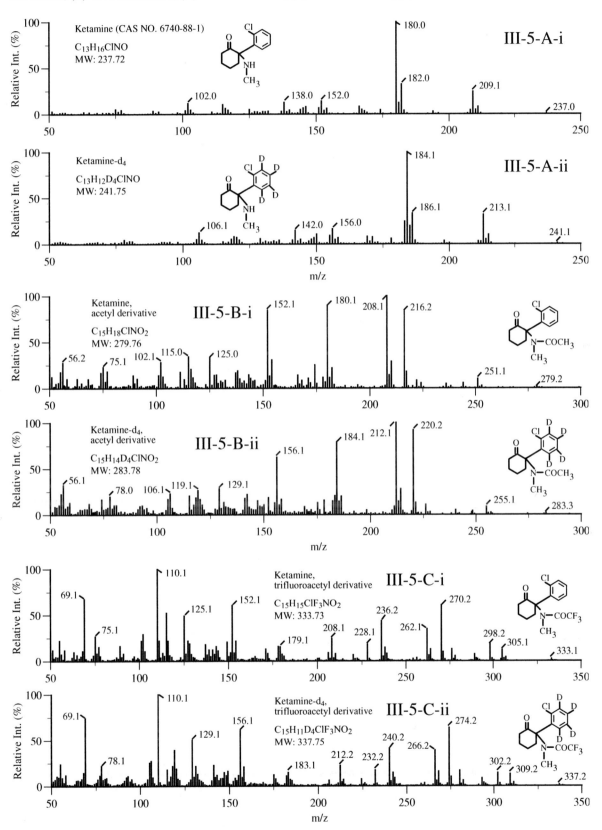

Figure III — Hallucinogens

Figure III-5. (Continued)

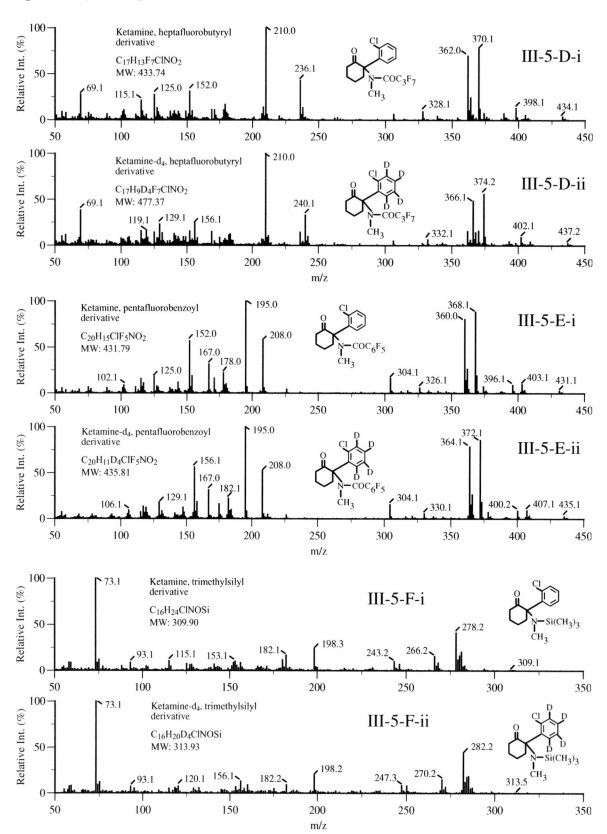

Ketamine, heptafluorobutyryl derivative
$C_{17}H_{13}F_7ClNO_2$
MW: 433.74
III-5-D-i

Ketamine-d_4, heptafluorobutyryl derivative
$C_{17}H_9D_4F_7ClNO_2$
MW: 477.37
III-5-D-ii

Ketamine, pentafluorobenzoyl derivative
$C_{20}H_{15}ClF_5NO_2$
MW: 431.79
III-5-E-i

Ketamine-d_4, pentafluorobenzoyl derivative
$C_{20}H_{11}D_4ClF_5NO_2$
MW: 435.81
III-5-E-ii

Ketamine, trimethylsilyl derivative
$C_{16}H_{24}ClNOSi$
MW: 309.90
III-5-F-i

Ketamine-d_4, trimethylsilyl derivative
$C_{16}H_{20}D_4ClNOSi$
MW: 313.93
III-5-F-ii

Appendix One — Mass Spectra

Figure III-6. Mass spectra of norketamine and its deuterated analogs (norketamine-d$_4$): (A) underivatized; (B) acetyl-derivatized; (C) TCA-derivatized; (D) TFA-derivatized; (E) PFP-derivatized; (F) HFB-derivatized; (G) 4-CB-derivatized; (H) PFB-derivatized; (I) TMS-derivatized; (J) TFA/*t*-BDMS-derivatized; (K) PFP/*t*-BDMS-derivatized ; (L) HFB/*t*-BDMS-derivatized.

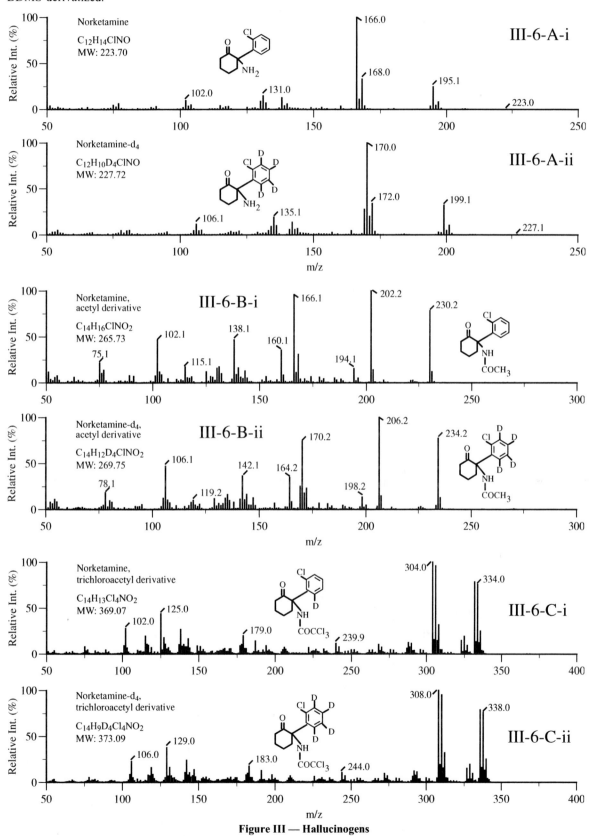

Figure III — Hallucinogens

Figure III-6. (Continued)

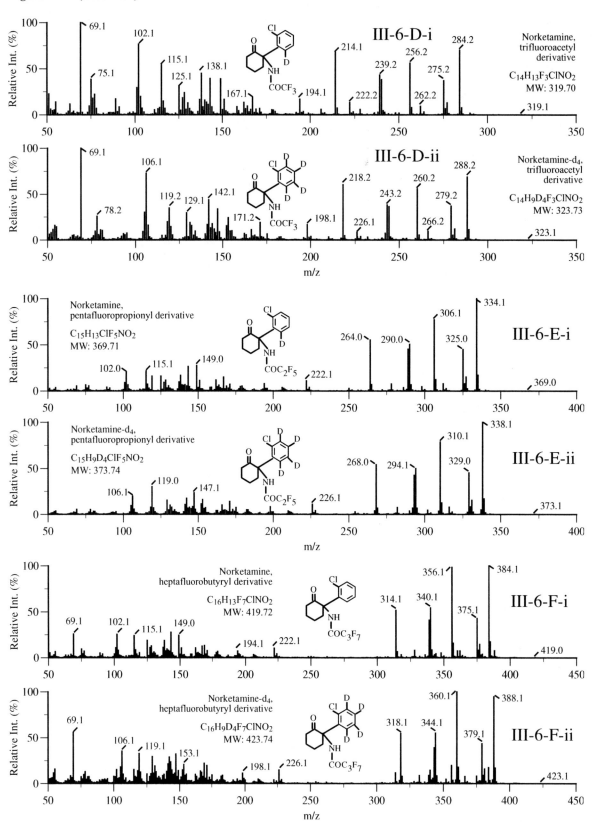

Figure III-6. (Continued)

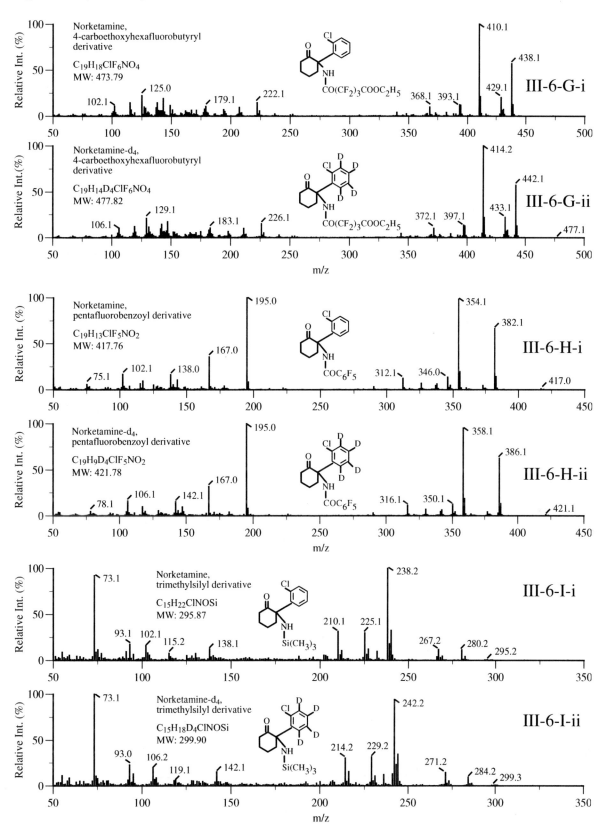

Figure III — Hallucinogens

Figure III-6. (Continued)

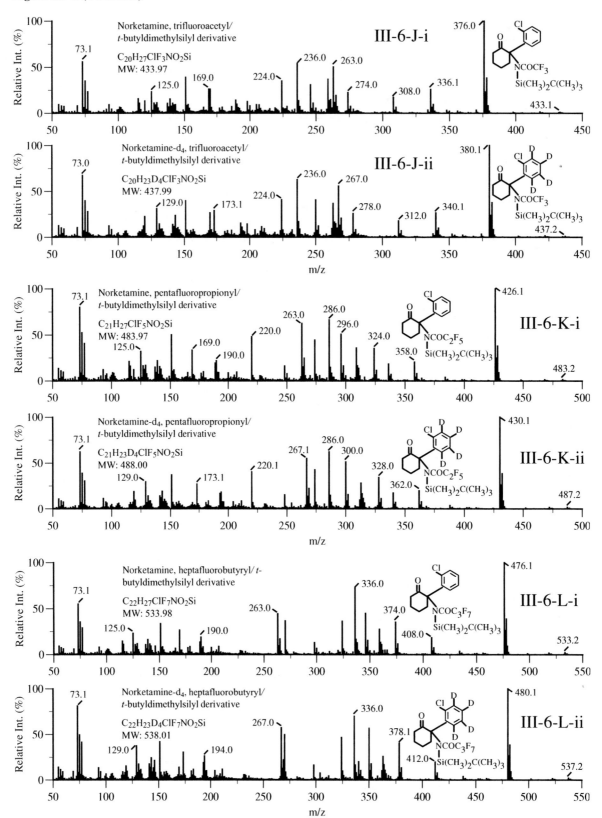

Norketamine, trifluoroacetyl/ *t*-butyldimethylsilyl derivative

$C_{20}H_{27}ClF_3NO_2Si$
MW: 433.97

III-6-J-i

Norketamine-d$_4$, trifluoroacetyl/ *t*-butyldimethylsilyl derivative

$C_{20}H_{23}D_4ClF_3NO_2Si$
MW: 437.99

III-6-J-ii

Norketamine, pentafluoropropionyl/ *t*-butyldimethylsilyl derivative

$C_{21}H_{27}ClF_5NO_2Si$
MW: 483.97

III-6-K-i

Norketamine-d$_4$, pentafluoropropionyl/ *t*-butyldimethylsilyl derivative

$C_{21}H_{23}D_4ClF_5NO_2Si$
MW: 488.00

III-6-K-ii

Norketamine, heptafluorobutyryl/ *t*-butyldimethylsilyl derivative

$C_{22}H_{27}ClF_7NO_2Si$
MW: 533.98

III-6-L-i

Norketamine-d$_4$, heptafluorobutyryl/ *t*-butyldimethylsilyl derivative

$C_{22}H_{23}D_4ClF_7NO_2Si$
MW: 538.01

III-6-L-ii

Appendix One — Mass Spectra

Figure III-7. Mass spectra of phencyclidine and its deuterated analogs (phencyclidine-d$_5$): (A) underivatized.

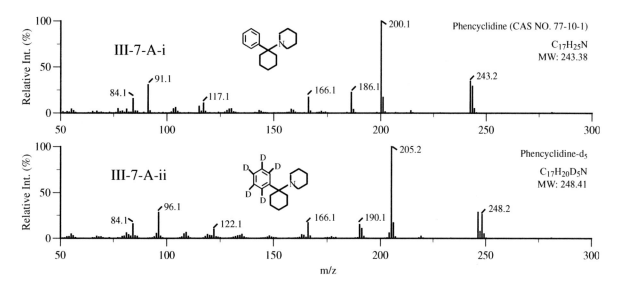

Figure III — Hallucinogens

Figure III-8. Mass spectra of LSD and its deuterated analogs (LSD-d₃): (A) underivatized-derivatized; (B) TMS-derivatized.

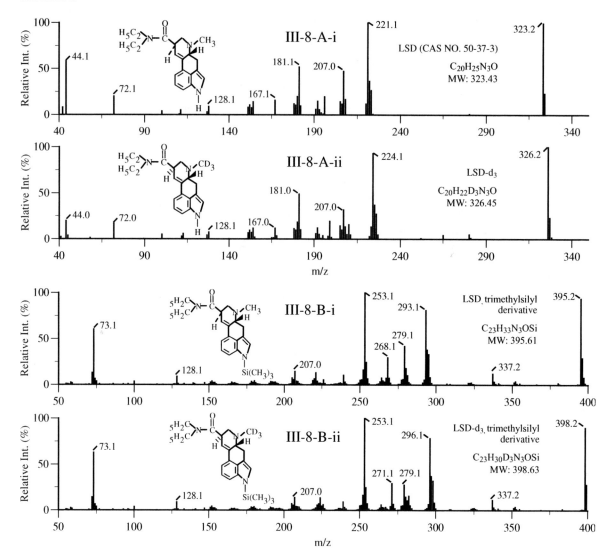

Figure III-9. Mass spectra of mescaline and its deuterated analogs (mescaline-d$_9$): (A) acetyl-derivatized; (B) TCA-derivatized; (C) TFA-derivatized; (D) PFP-derivatized; (E) HFB-derivatized; (F) 4-CB-derivatized; (G) [TMS]$_2$-derivatized; (H) t-BDMS-derivatized; (I) TFA/TMS-derivatized; (J) TFA/t-BDMS-derivatized; (K) PFP/TMS-derivatized; (L) PFP/t-BDMS-derivatized; (M) HFB/TMS-derivatized; (N) HFB/t-BDMS-derivatized.

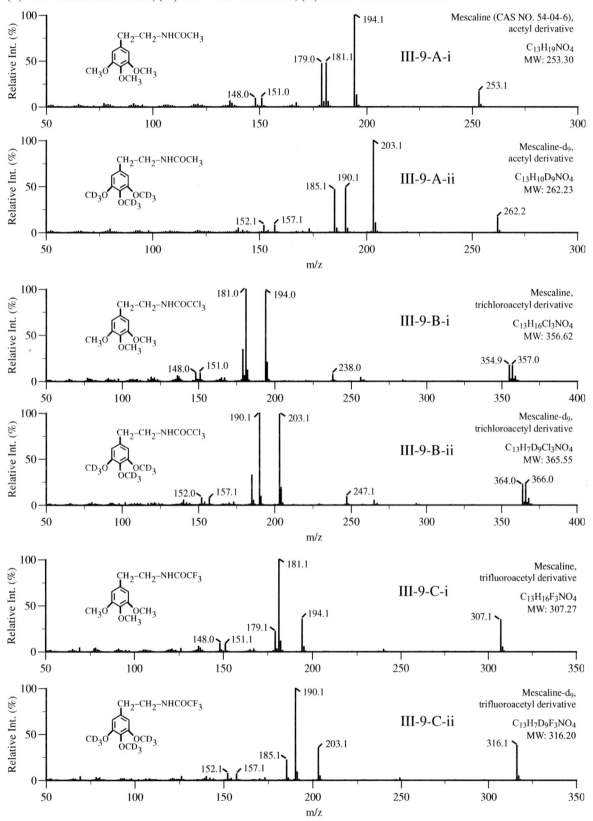

Figure III — Hallucinogens

244

Figure III-9. (Continued)

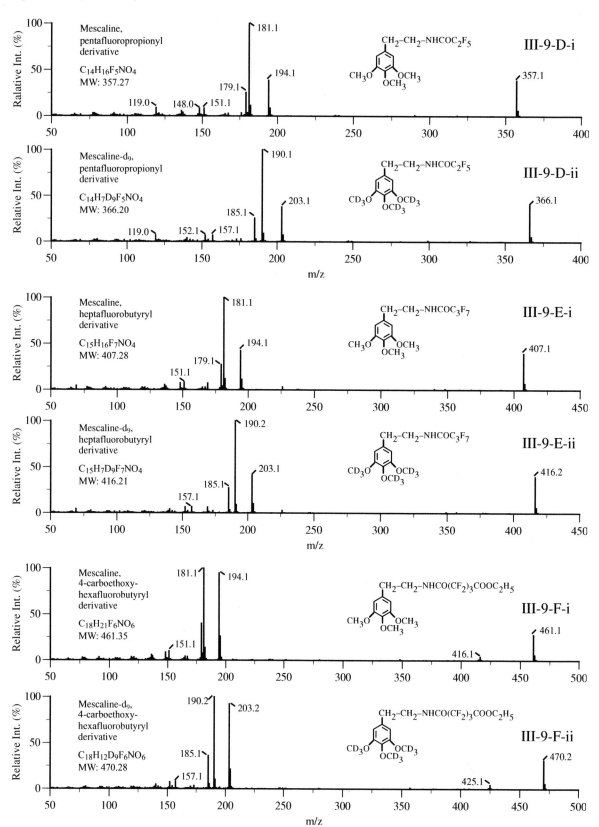

Appendix One — Mass Spectra

Figure III-9. (Continued)

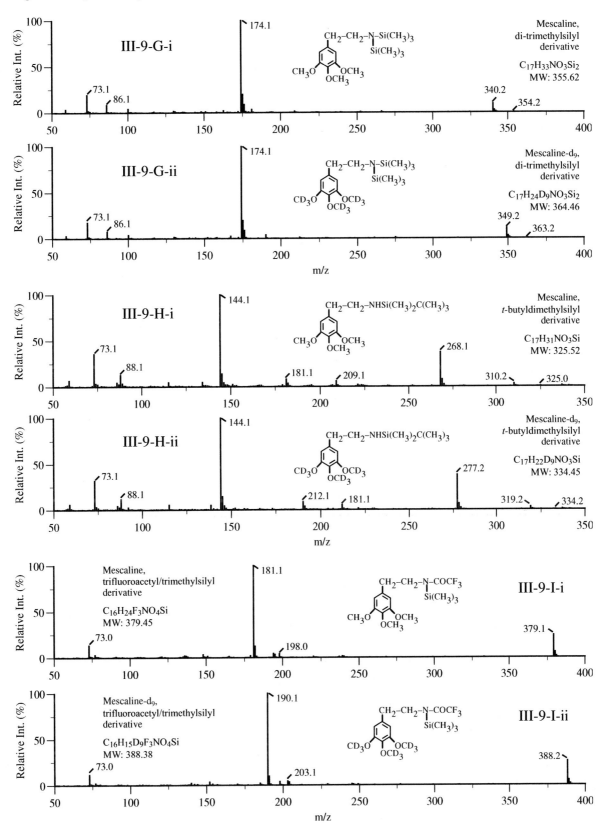

III-9-G-i

Mescaline, di-trimethylsilyl derivative

$C_{17}H_{33}NO_3Si_2$
MW: 355.62

III-9-G-ii

Mescaline-d₉, di-trimethylsilyl derivative

$C_{17}H_{24}D_9NO_3Si_2$
MW: 364.46

III-9-H-i

Mescaline, t-butyldimethylsilyl derivative

$C_{17}H_{31}NO_3Si$
MW: 325.52

III-9-H-ii

Mescaline-d₉, t-butyldimethylsilyl derivative

$C_{17}H_{22}D_9NO_3Si$
MW: 334.45

III-9-I-i

Mescaline, trifluoroacetyl/trimethylsilyl derivative

$C_{16}H_{24}F_3NO_4Si$
MW: 379.45

III-9-I-ii

Mescaline-d₉, trifluoroacetyl/trimethylsilyl derivative

$C_{16}H_{15}D_9F_3NO_4Si$
MW: 388.38

Figure III — Hallucinogens

Figure III-9. (Continued)

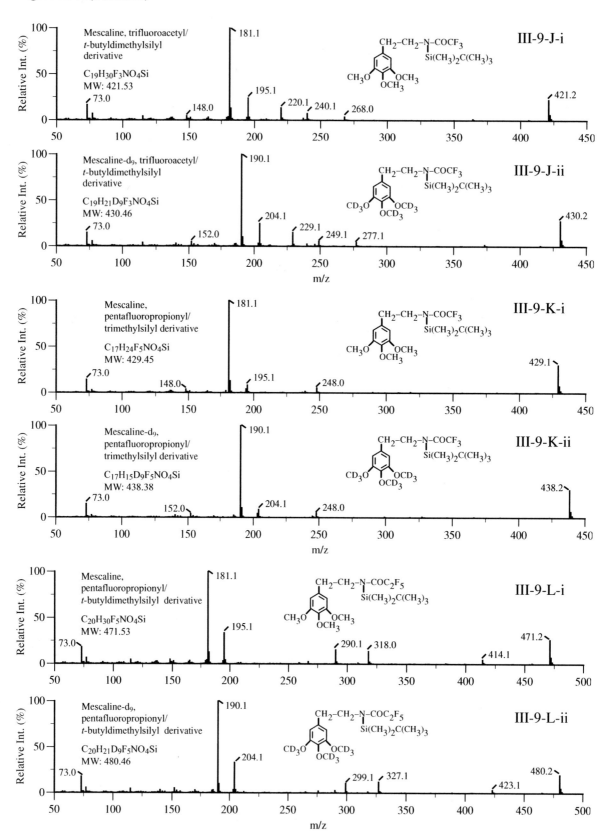

Figure III-9. (Continued)

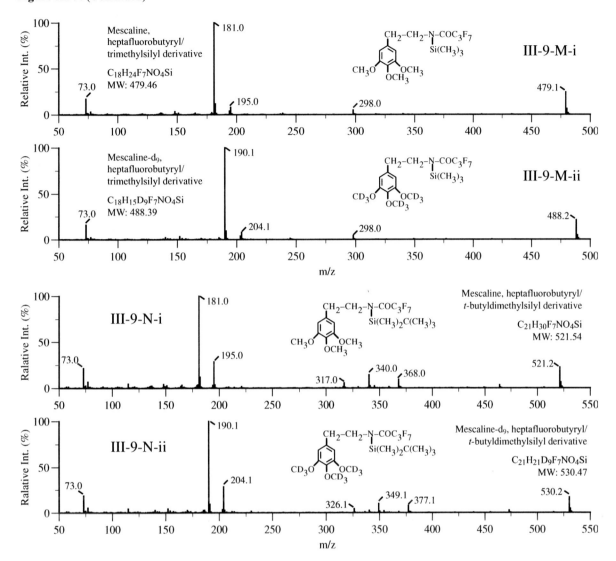

Mescaline, heptafluorobutyryl/ trimethylsilyl derivative

$C_{18}H_{24}F_7NO_4Si$
MW: 479.46

III-9-M-i

Mescaline-d$_9$, heptafluorobutyryl/ trimethylsilyl derivative

$C_{18}H_{15}D_9F_7NO_4Si$
MW: 488.39

III-9-M-ii

III-9-N-i

Mescaline, heptafluorobutyryl/ t-butyldimethylsilyl derivative

$C_{21}H_{30}F_7NO_4Si$
MW: 521.54

III-9-N-ii

Mescaline-d$_9$, heptafluorobutyryl/ t-butyldimethylsilyl derivative

$C_{21}H_{21}D_9F_7NO_4Si$
MW: 530.47

Figure III — Hallucinogens

Figure III-10. Mass spectra of psilocin and its deuterated analogs (psilocin-d$_{10}$): (A) underivatized; (B) acetyl-derivatized; (C) [acetyl]$_2$-derivatized; (D) [TMS]$_2$-derivatized; (E) *t*-BDMS-derivatized; (F) [*t*-BDMS]$_2$-derivatized.

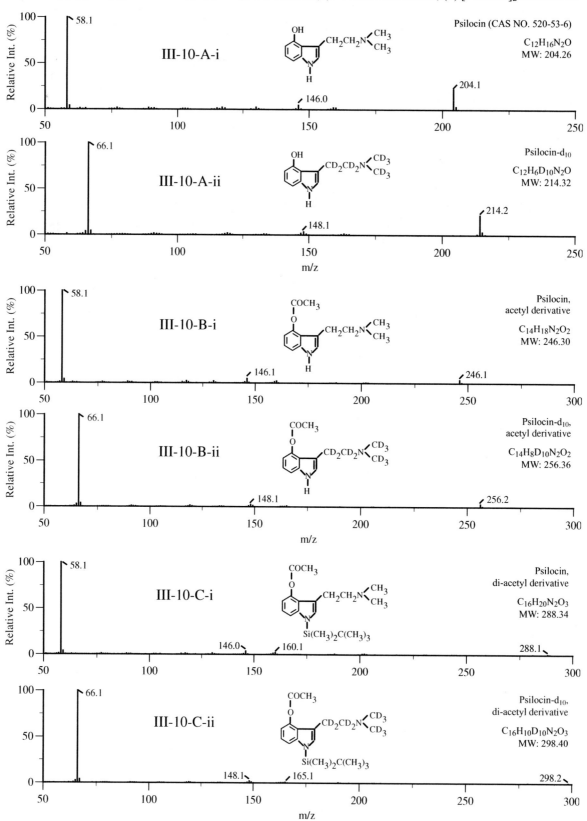

III-10-A-i — Psilocin (CAS NO. 520-53-6), C$_{12}$H$_{16}$N$_2$O, MW: 204.26. 58.1, 146.0, 204.1

III-10-A-ii — Psilocin-d$_{10}$, C$_{12}$H$_6$D$_{10}$N$_2$O, MW: 214.32. 66.1, 148.1, 214.2

III-10-B-i — Psilocin, acetyl derivative, C$_{14}$H$_{18}$N$_2$O$_2$, MW: 246.30. 58.1, 146.1, 246.1

III-10-B-ii — Psilocin-d$_{10}$, acetyl derivative, C$_{14}$H$_8$D$_{10}$N$_2$O$_2$, MW: 256.36. 66.1, 148.1, 256.2

III-10-C-i — Psilocin, di-acetyl derivative, C$_{16}$H$_{20}$N$_2$O$_3$, MW: 288.34. 58.1, 146.0, 160.1, 288.1

III-10-C-ii — Psilocin-d$_{10}$, di-acetyl derivative, C$_{16}$H$_{10}$D$_{10}$N$_2$O$_3$, MW: 298.40. 66.1, 148.1, 165.1, 298.2

Figure III-10. (Continued)

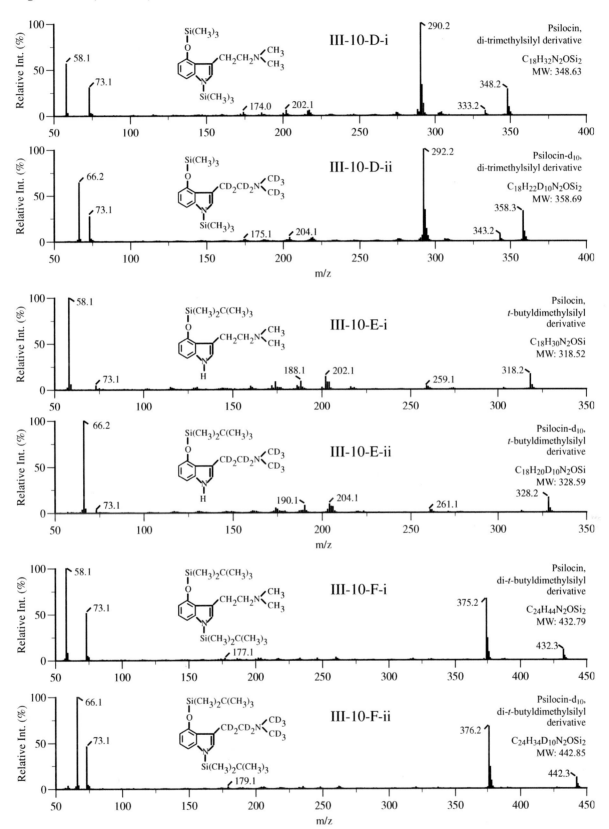

Figure III — Hallucinogens

Summary of Drugs, Isotopic Analogs, and Chemical Derivatization Groups Included in Figure IV (Depressants/Hypnotics)

Compound	Isotopic analog	Chemical derivatization group (no. of spectra)	Figure #
Pentobarbital	d_5	None, [methyl]$_2$, [ethyl]$_2$, [propyl]$_2$, [butyl]$_2$, [TMS]$_2$, [t-BDMS]$_2$ (14)	IV-1
Phenobarbital	d_5, d_5 (ring)	[Methyl]$_2$, [ethyl]$_2$, [propyl]$_2$, [butyl]$_2$, [TMS]$_2$, [t-BDMS]$_2$ (18)	IV-2
Butabital	d_5, $^{13}C_4$	None, [methyl]$_2$, [ethyl]$_2$, [propyl]$_2$, [butyl]$_2$, [TMS]$_2$, [t-BDMS]$_2$ (21)	IV-3
Sceobarbital	d_5, $^{13}C_4$	None, [methyl]$_2$, [ethyl]$_2$, [propyl]$_2$, [butyl]$_2$, [TMS]$_2$, [t-BDMS]$_2$ (21)	IV-4
Methohexital	d_5	None, methyl, ethyl, propyl, butyl, TMS, t-BDMS (14)	IV-5
γ-Hydroxybutyric acid	d_6	[TMS]$_2$, [t-BDMS]$_2$ (4)	IV-6
γ-Butyrolactone	d_6	None (2)	IV-7
		Total no. of mass spectra: 94	

Figure IV — Depressants/Hypnotics

Appendix One — Figure IV
Mass Spectra of Commonly Abused Drugs and Their Isotopically Labeled Analogs
in Various Derivatization Forms — Depressants/Hypnotics

Figure IV — Depressants/Hypnotics

Figure IV-1. Mass spectra of pentobarbital and its deuterated analogs (pentobarbital-d$_5$): (A) underivatized; (B) [methyl]$_2$-derivatized; (C) [ethyl]$_2$-derivatized; (D) [propyl]$_2$-derivatized; (E) [butyl]$_2$-derivatized; (F) [TMS]$_2$-derivatized; (G) [*t*-BDMS]$_2$-derivatized.

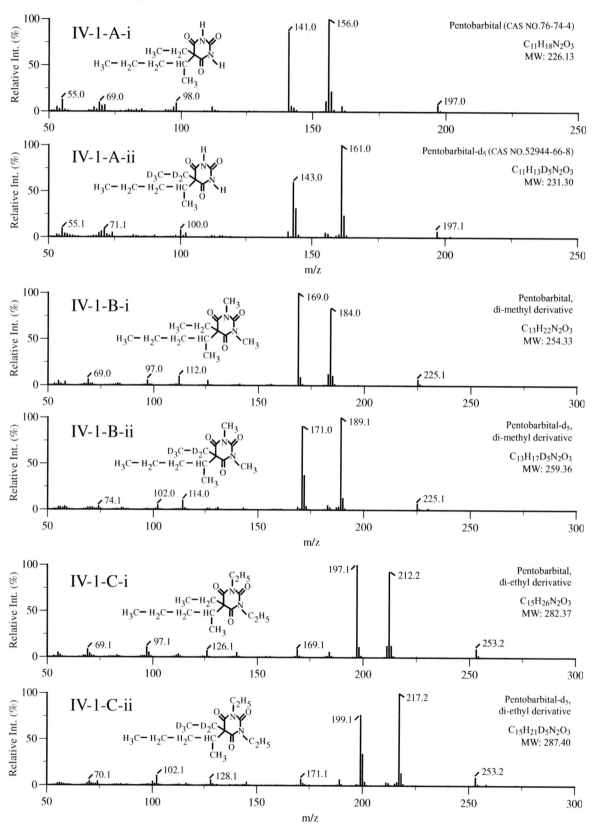

Figure IV-1. (Continued)

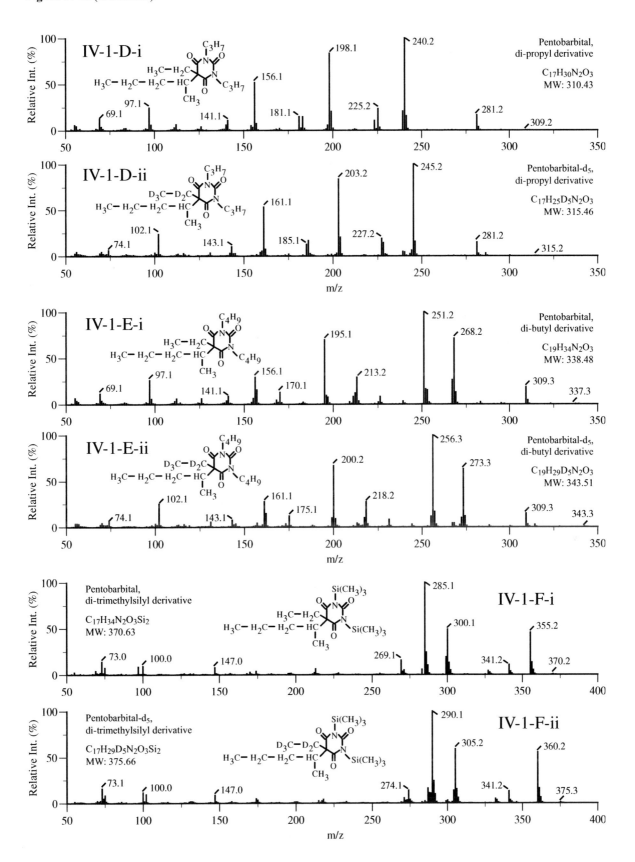

IV-1-D-i
Pentobarbital, di-propyl derivative
$C_{17}H_{30}N_2O_3$
MW: 310.43

IV-1-D-ii
Pentobarbital-d$_5$, di-propyl derivative
$C_{17}H_{25}D_5N_2O_3$
MW: 315.46

IV-1-E-i
Pentobarbital, di-butyl derivative
$C_{19}H_{34}N_2O_3$
MW: 338.48

IV-1-E-ii
Pentobarbital-d$_5$, di-butyl derivative
$C_{19}H_{29}D_5N_2O_3$
MW: 343.51

Pentobarbital, di-trimethylsilyl derivative
$C_{17}H_{34}N_2O_3Si_2$
MW: 370.63
IV-1-F-i

Pentobarbital-d$_5$, di-trimethylsilyl derivative
$C_{17}H_{29}D_5N_2O_3Si_2$
MW: 375.66
IV-1-F-ii

Figure IV — Depressants/Hypnotics

Figure IV-1. (Continued)

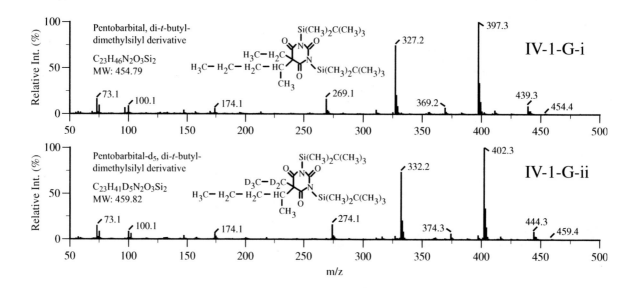

Figure IV-2. Mass spectra of phenobarbital and its deuterated analogs (phenobarbital-d$_5$, -d$_5$ ring): (A) [methyl]$_2$-derivatized; (B) [ethyl]$_2$-derivatized; (C) [propyl]$_2$-derivatized; (D) [butyl]$_2$-derivatized; (E) [TMS]$_2$-derivatized; (F) [t-BDMS]$_2$-derivatized.

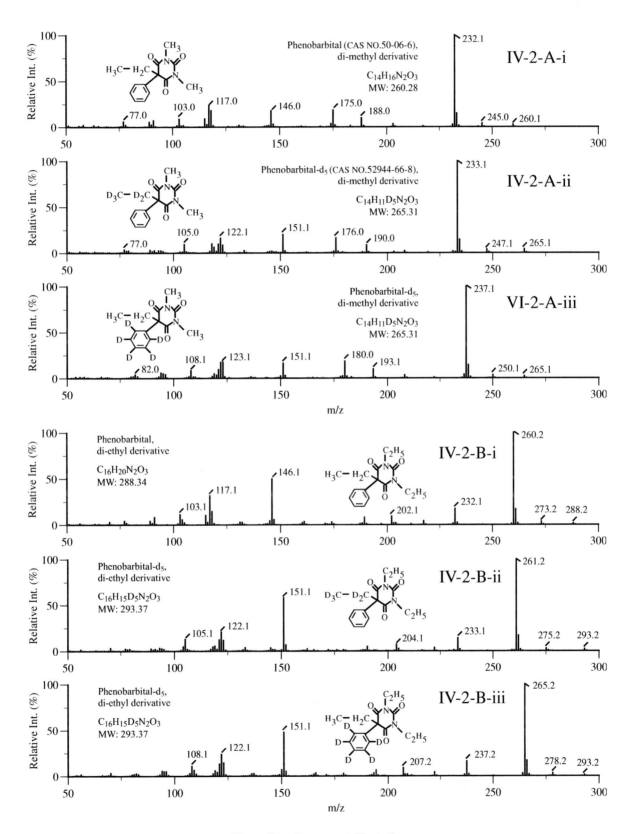

Figure IV — Depressants/Hypnotics

Figure IV-2. (Continued)

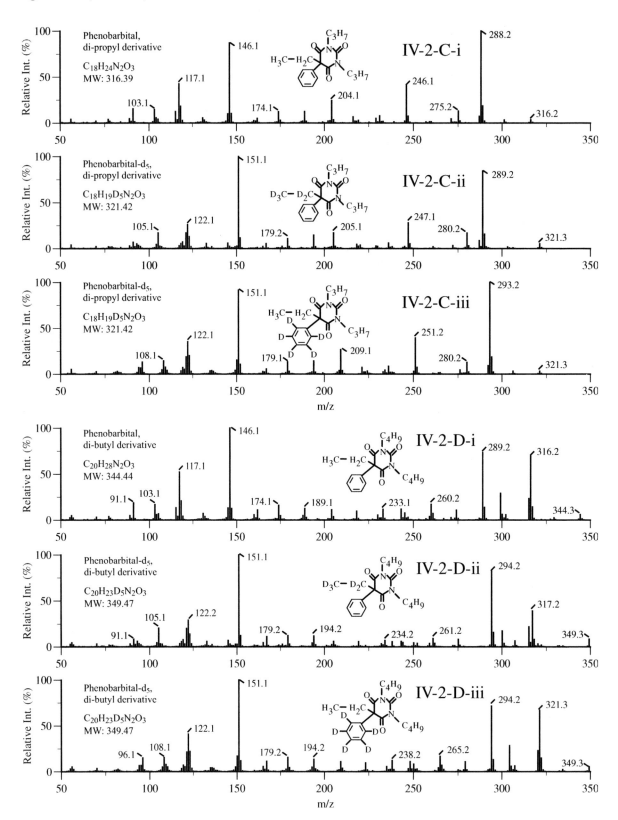

Figure IV-2. (Continued)

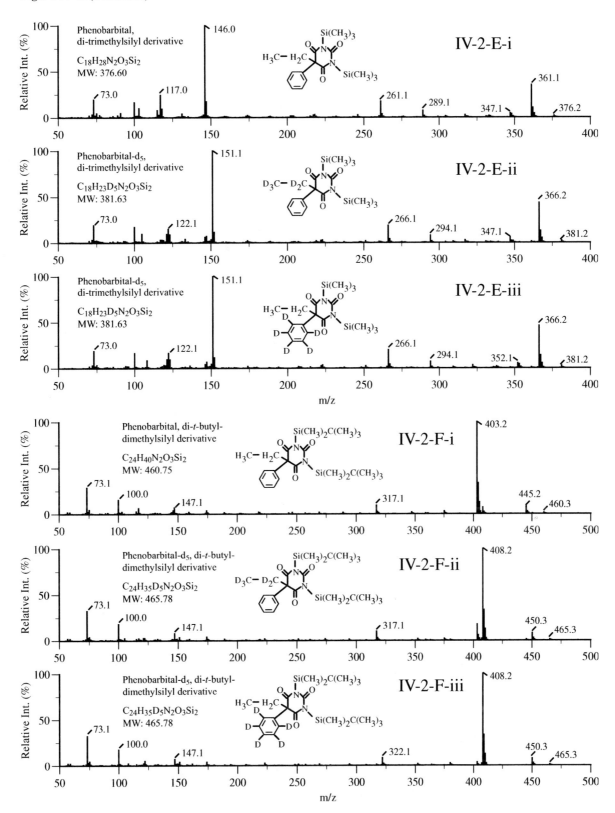

Figure IV — Depressants/Hypnotics

Figure IV-3. Mass spectra of butabital and its deuterated analogs (butabital-d$_5$, -^{13}C$_4$): (A) underivatized; (B) [methyl]$_2$-derivatized; (C) [ethyl]$_2$-derivatized; (D) [propyl]$_2$-derivatized; (E) [butyl]$_2$-derivatized; (F) [TMS]$_2$-derivatized; (G) [*t*-BDMS]$_2$-derivatized.

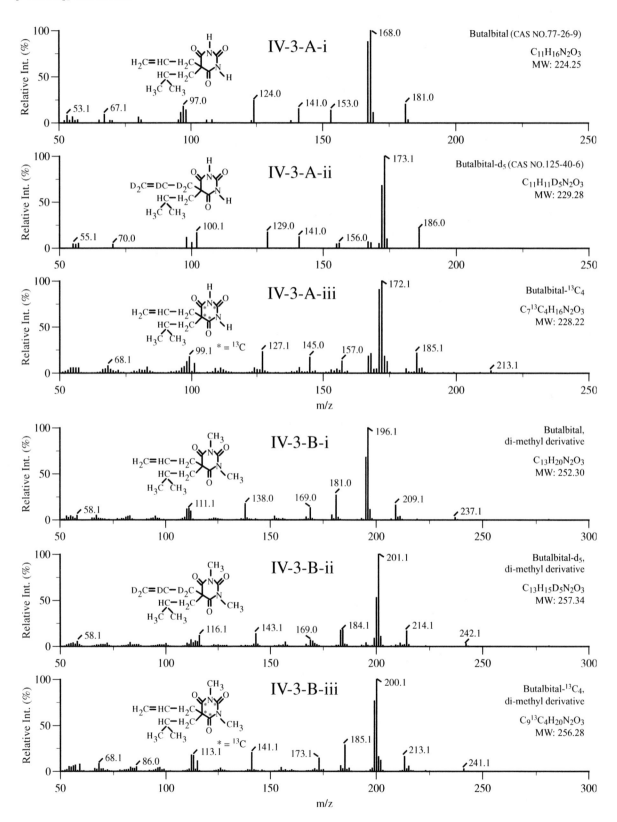

Figure IV-3. (Continued)

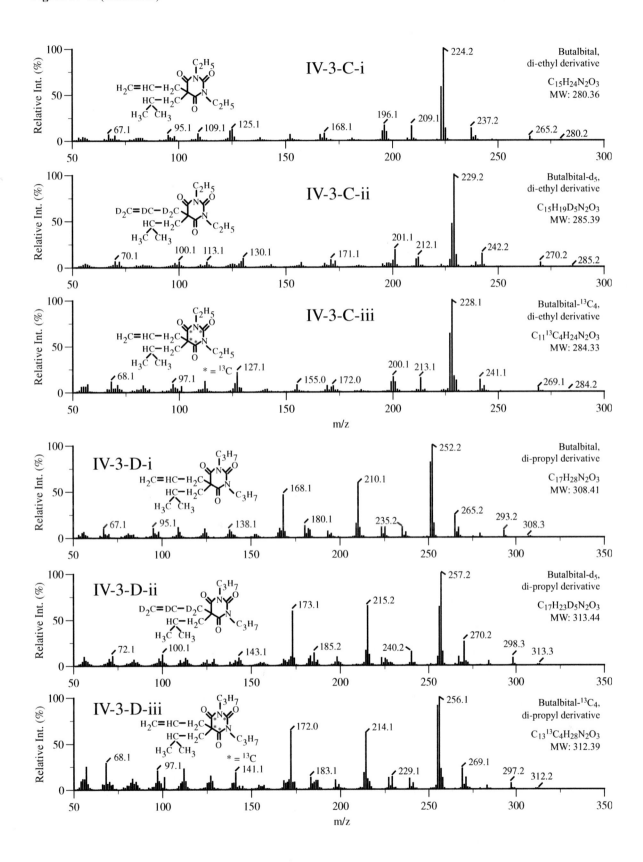

Figure IV — Depressants/Hypnotics

262

Figure IV-3. (Continued)

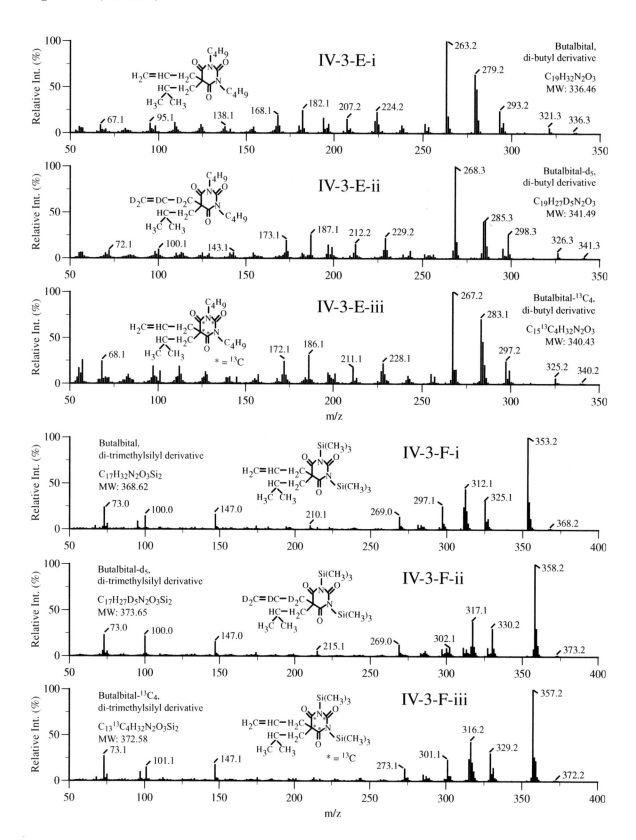

Appendix One — Mass Spectra

Figure IV-3. (Continued)

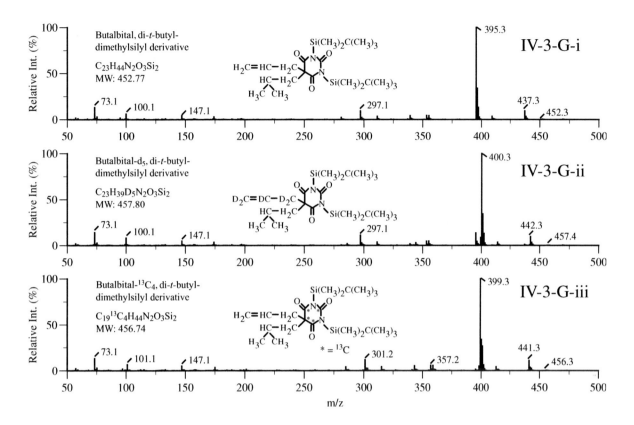

IV-3-G-i

Butalbital, di-*t*-butyl-
dimethylsilyl derivative

C$_{23}$H$_{44}$N$_2$O$_3$Si$_2$
MW: 452.77

IV-3-G-ii

Butalbital-d$_5$, di-*t*-butyl-
dimethylsilyl derivative

C$_{23}$H$_{39}$D$_5$N$_2$O$_3$Si$_2$
MW: 457.80

IV-3-G-iii

Butalbital-^{13}C$_4$, di-*t*-butyl-
dimethylsilyl derivative

C$_{19}$^{13}C$_4$H$_{44}$N$_2$O$_3$Si$_2$
MW: 456.74

* = ^{13}C

m/z

Figure IV — Depressants/Hypnotics

Figure IV-4. Mass spectra of secobarbital and its deuterated analogs (secobarbital-d₅, -¹³C₄): (A) underivatized; (B) [methyl]₂-derivatized; (C) [ethyl]₂-derivatized; (D) [propyl]₂-derivatized; (E) [butyl]₂-derivatized; (F) [TMS]₂-derivatized; (G) [t-BDMS]₂-derivatized.

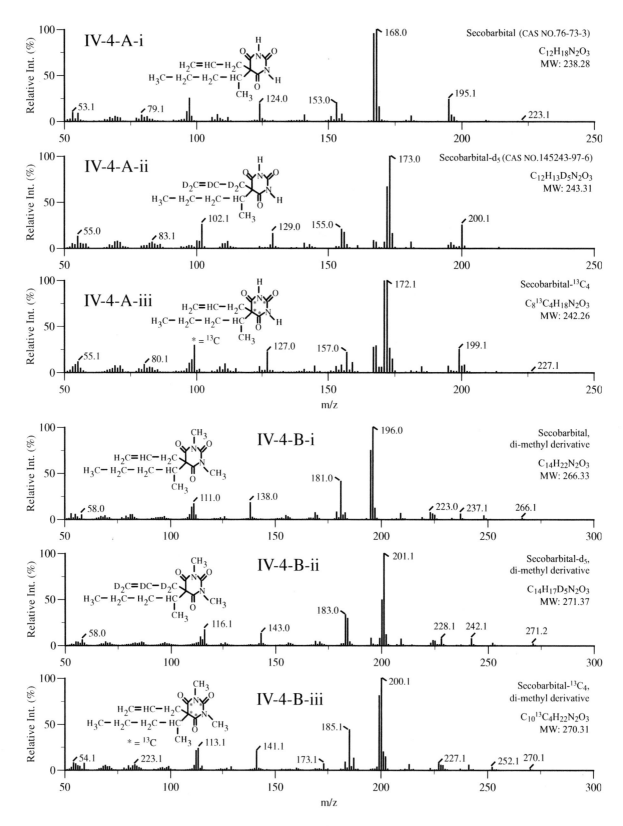

Figure IV-4. (Continued)

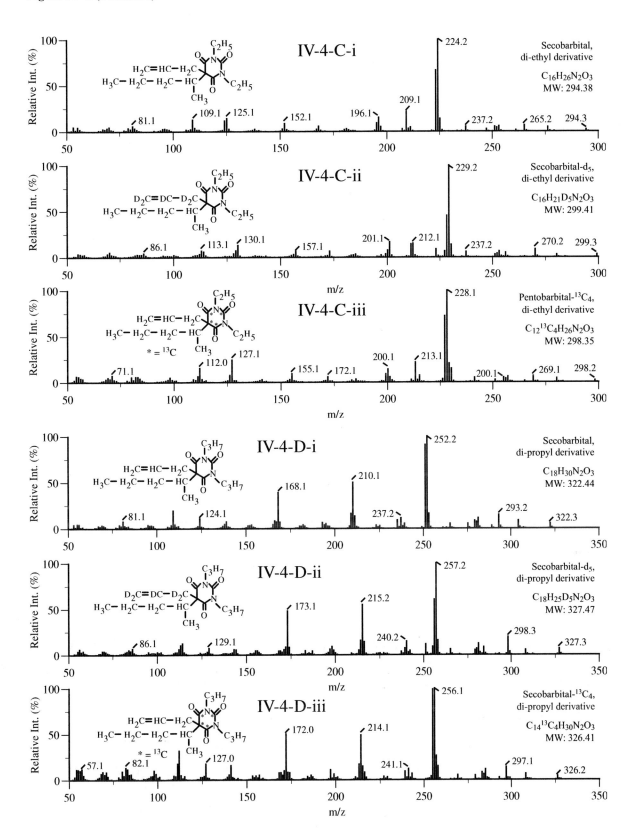

IV-4-C-i — Secobarbital, di-ethyl derivative, $C_{16}H_{26}N_2O_3$, MW: 294.38

IV-4-C-ii — Secobarbital-d_5, di-ethyl derivative, $C_{16}H_{21}D_5N_2O_3$, MW: 299.41

IV-4-C-iii — Pentobarbital-$^{13}C_4$, di-ethyl derivative, $C_{12}{}^{13}C_4H_{26}N_2O_3$, MW: 298.35

IV-4-D-i — Secobarbital, di-propyl derivative, $C_{18}H_{30}N_2O_3$, MW: 322.44

IV-4-D-ii — Secobarbital-d_5, di-propyl derivative, $C_{18}H_{25}D_5N_2O_3$, MW: 327.47

IV-4-D-iii — Secobarbital-$^{13}C_4$, di-propyl derivative, $C_{14}{}^{13}C_4H_{30}N_2O_3$, MW: 326.41

Figure IV — Depressants/Hypnotics

Figure IV-4. (Continued)

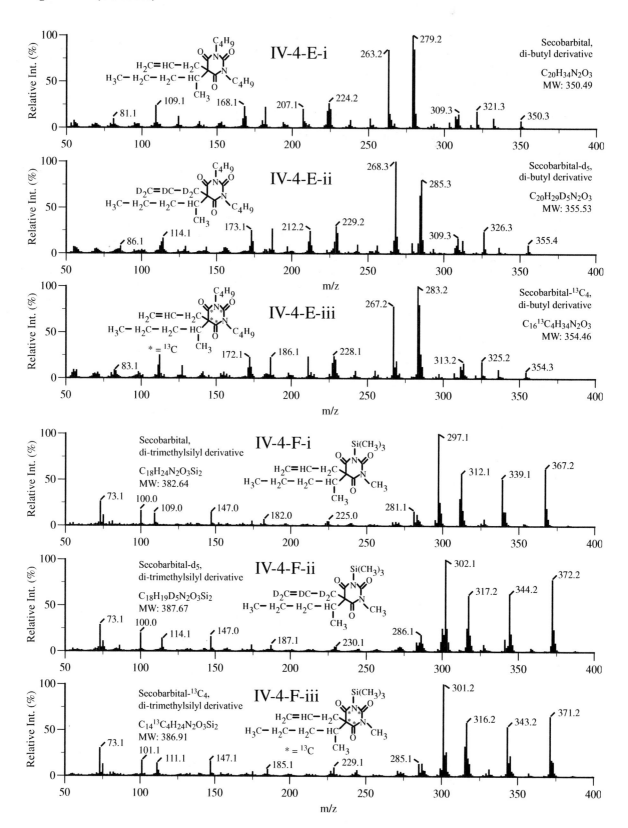

Figure IV-4. (Continued)

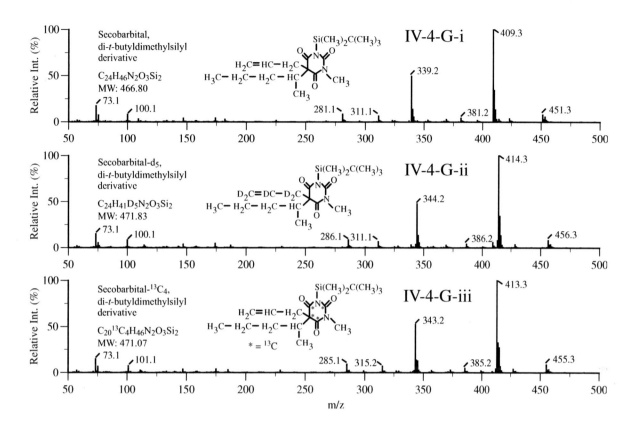

Figure IV — Depressants/Hypnotics

268

Figure IV-5. Mass spectra of methohexital and its deuterated analogs (methohexital-d₅): (A) underivatized; (B) methyl-derivatized; (C) ethyl-derivatized; (D) propyl-derivatized; (E) butyl-derivatized; (F) TMS-derivatized; (G) *t*-BDMS-derivatized.

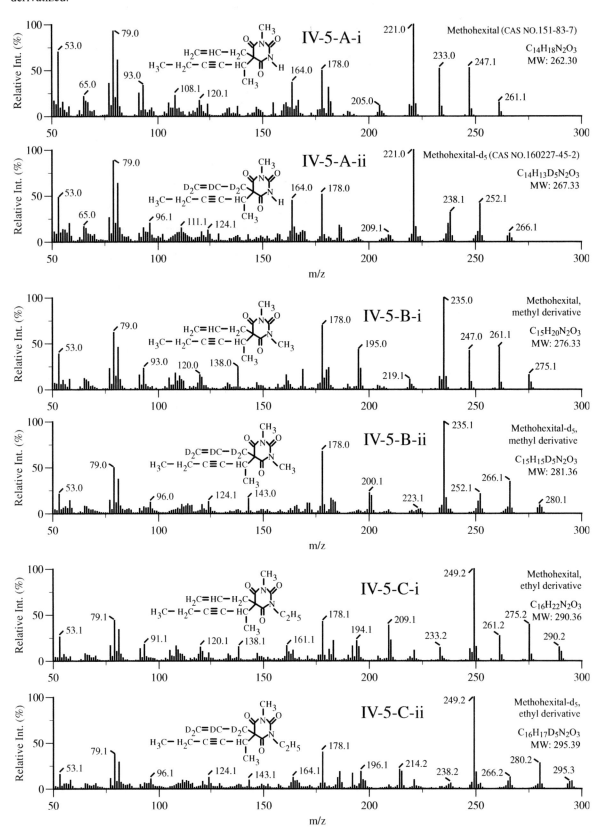

Figure IV-5. (Continued)

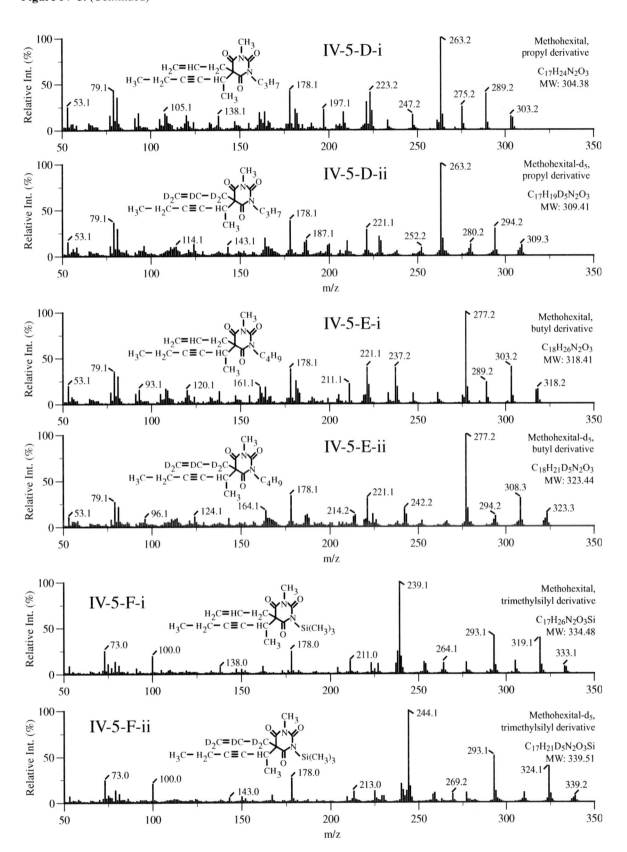

Figure IV — Depressants/Hypnotics

Figure IV-5. (Continued)

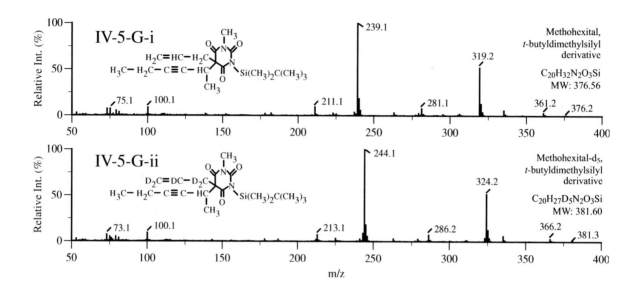

Figure IV-6. Mass spectra of γ-hydroxybutyric acid (GHB) and its deuterated analogs (GHB-d$_6$): (A) [TMS]$_2$-derivatized; (B) [*t*-BDMS]$_2$-derivatized.

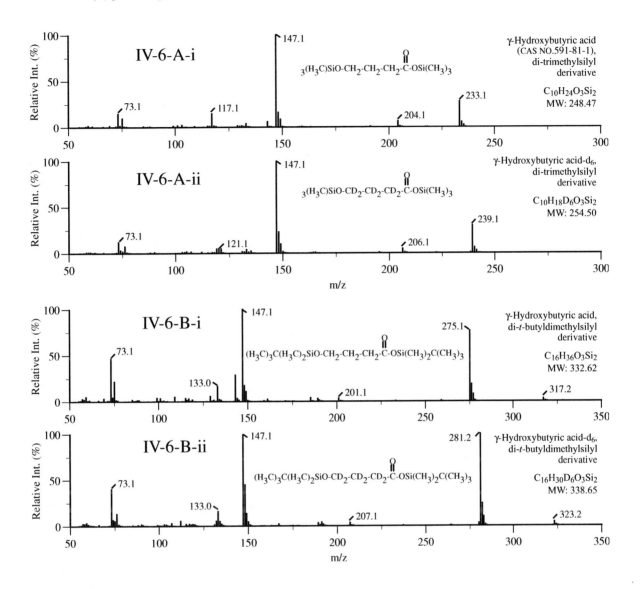

Figure IV — Depressants/Hypnotics

Figure IV-7. Mass spectra of γ-butyrolactone (GBL) and its deuterated analogs (GBL-d$_6$).

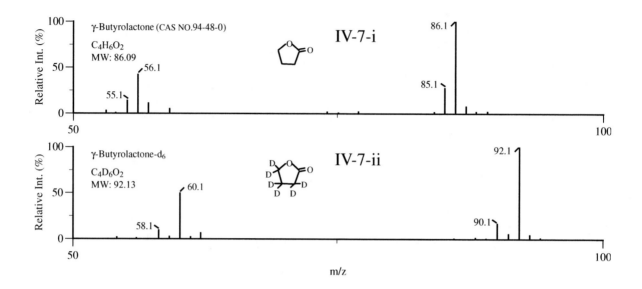

Summary of Drugs, Isotopic Analogs, and Chemical Derivatization Groups Included in Figure V (Antianxiety Agents)

Compound	Isotopic analog	Chemical derivatization group (no. of spectra)	Figure #
Oxazepam	d_5	None, [methyl]$_2$, [ethyl]$_2$, [propyl]$_2$, [butyl]$_2$, [TMS]$_2$, t-BDMS (14)	V-1
Diazepam	d_3, d_5	None (3)	V-2
Nordiazepam	d_5	None, methyl, ethyl, propyl, butyl, TMS, t-BDMS (14)	V-3
Nitrazepam	d_5	Methyl, ethyl, propyl, butyl, TMS, t-BDMS (12)	V-4
Temazepam	d_5	None, methyl, ethyl, propyl, butyl, acetyl, TMS, t-BDMS (16)	V-5
Clonazepam	d_4	Methyl, ethyl, propyl, butyl, TMS, t-BDMS (12)	V-6
7-Aminoclonazepam	d_4	[Methyl]$_3$, [ethyl]$_2$, [ethyl]$_3$, propyl, [propyl]$_2$, butyl, [butyl]$_2$, PFP, HFB, [TMS]$_2$, t-BDMS, [t-BDMS]$_2$, TFA/[TMS]$_2$, [TFA]$_2$/t-BDMS, TFA/[t-BDMS]$_2$, PFP/TMS, PFP/[TMS]$_2$, PFP/[t-BDMS]$_2$, HFB/[t-BDMS]$_2$ (38)	V-7
Prazepam	d_5	None (2)	V-8
Lorazepam	d_4	[Methyl]$_2$, [ethyl]$_2$, [propyl]$_2$, [butyl]$_2$, HFB, [TMS]$_2$, [t-BDMS]$_2$ (14)	V-9
Flunitrazepam	d_3, d_7	None (3)	V-10
7-Aminoflunitrazepam	d_3, d_7	None, [methyl]$_2$, ethyl, [ethyl]$_2$, propyl, butyl, acetyl, TFA, PFP, HFB, TMS, TFA/TMS, TFA/t-BDMS, PFP/TMS, PFP/t-BDMS, HFB/TMS, HFB/t-BDMS (51)	V-11
N-Desalkylflurazepam	d_4	None, methyl, [methyl]$_2$, ethyl, propyl, butyl, acetyl, TMS, t-BDMS (18)	V-12
N-Desmethylflunitrazepam	d_4	[Methyl]$_2$, ethyl, propyl, butyl, acetyl, TMS, t-BDMS (14)	V-13
2-Hydroxyethylflurazepam	d_4	None, butyl, TMS, t-BDMS (8)	V-14
Estazolam	d_5	None (2)	V-15
Alprazolam	d_5	None (2)	V-16
α-Hydroxyalprazolam	d_5	TMS, t-BDMS (4)	V-17
α-Hydroxytriazolam	d_4	TMS, t-BDMS (4)	V-18
Mianserin	d_3	None (2)	V-19
Methaqualone	d_7	None (2)	V-20
Haloperidol	d_4	TMS (2)	V-21

Total no. of mass spectra: 237

Figure V — Antianxiety Agent

Appendix One — Figure V
Mass Spectra of Commonly Abused Drugs and Their Isotopically Labeled Analogs in Various Derivatization Forms — Antianxiety Agents

Figure V — Antianxiety Agent

Figure V-1. Mass spectra of oxazepam and its deuterated analogs (oxazepam-d₅): (A) underivatized; (B) [methyl]₂-derivatized; (C) [ethyl]₂-derivatized; (D) [propyl]₂-derivatized; (E) [butyl]₂-derivatized; (F) [TMS]₂-derivatized; (G) [*t*-BDMS]₂-derivatized.

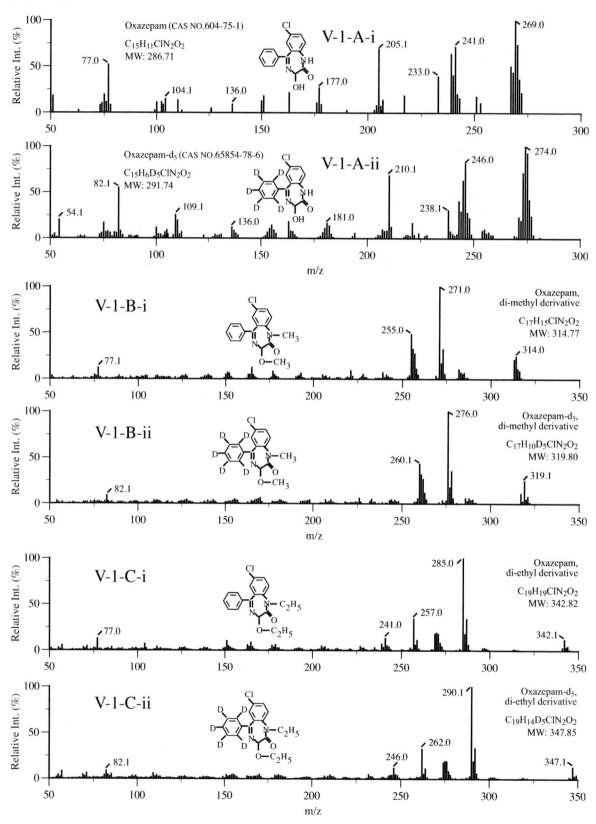

Figure V-1. (Continued)

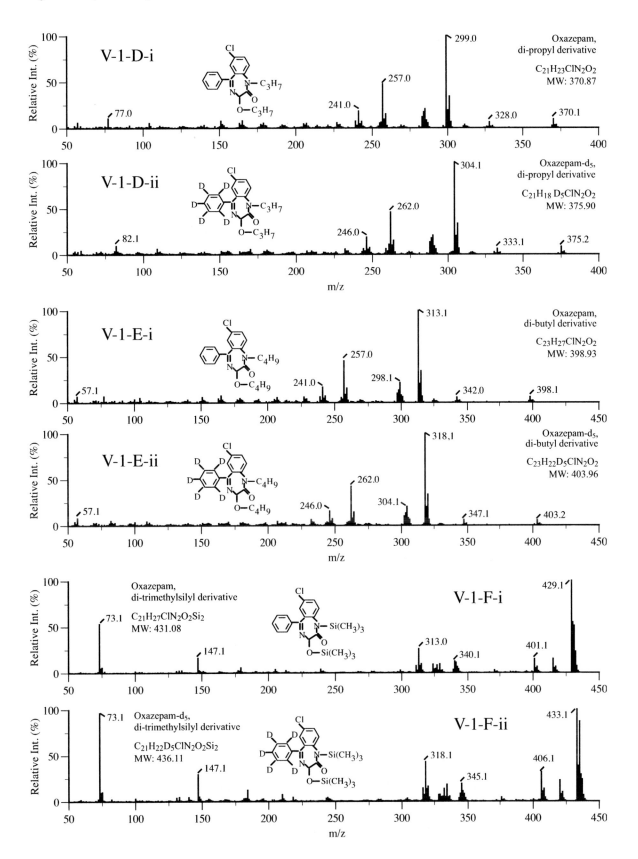

Figure V — Antianxiety Agent

Figure V-1. (Continued)

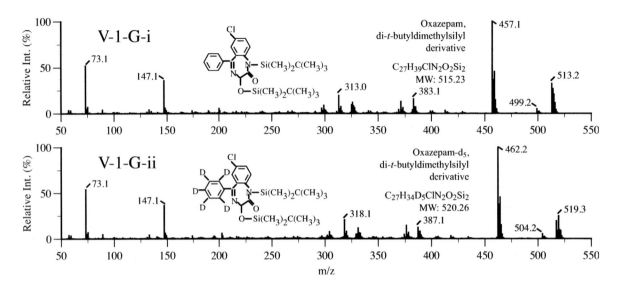

Figure V-2. Mass spectra of diazepam and its deuterated analogs (diazepam-d_3, -d_5).

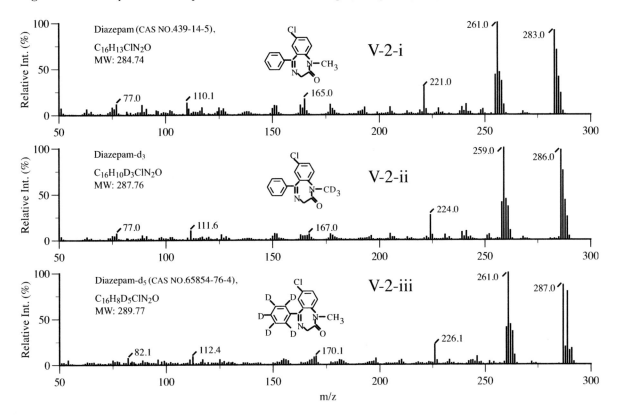

m/z

Figure V — Antianxiety Agent

280

Figure V-3. Mass spectra of nordiazepam and its deuterated analogs (nordiazepam-d$_5$): (A) underivatized; (B) methyl-derivatized; (C) ethyl-derivatized; (D) propyl-derivatized; (E) butyl-derivatized; (F) TMS-derivatized; (G) *t*-BDMS-derivatized.

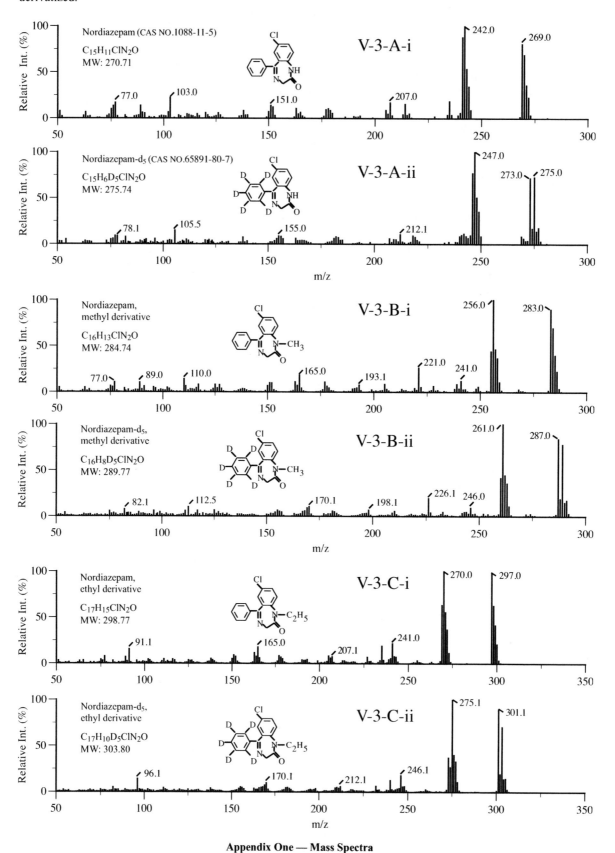

Figure V-3. (Continued)

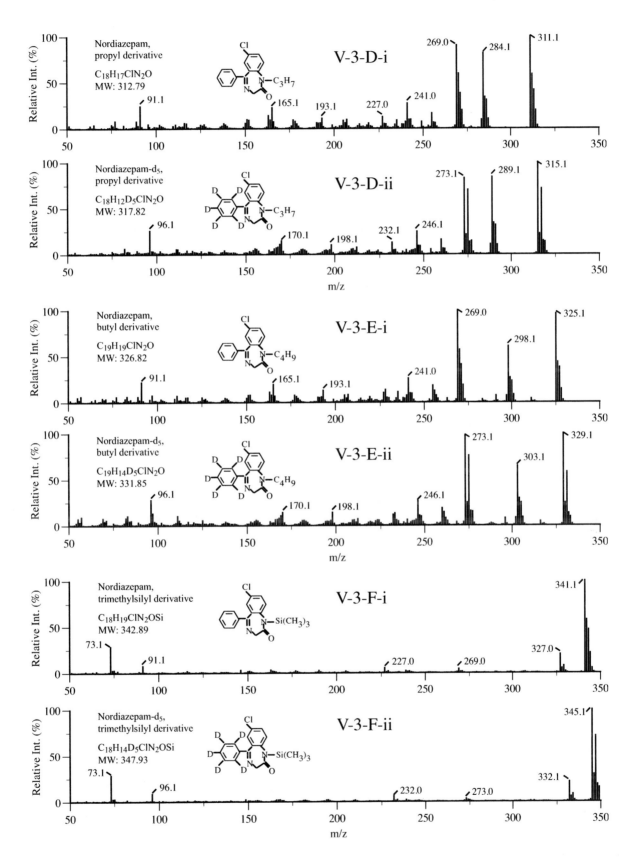

Figure V — Antianxiety Agent

Figure V-3. (Continued)

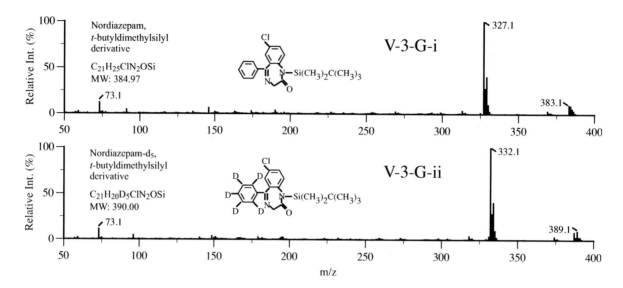

Figure V-4. Mass spectra of nitrazepam and its deuterated analogs (nitrazepam-d_5): (A) methyl-derivatized; (B) ethyl-derivatized; (C) propyl-derivatized; (D) butyl-derivatized; (E) TMS-derivatized; (F) *t*-BDMS-derivatized.

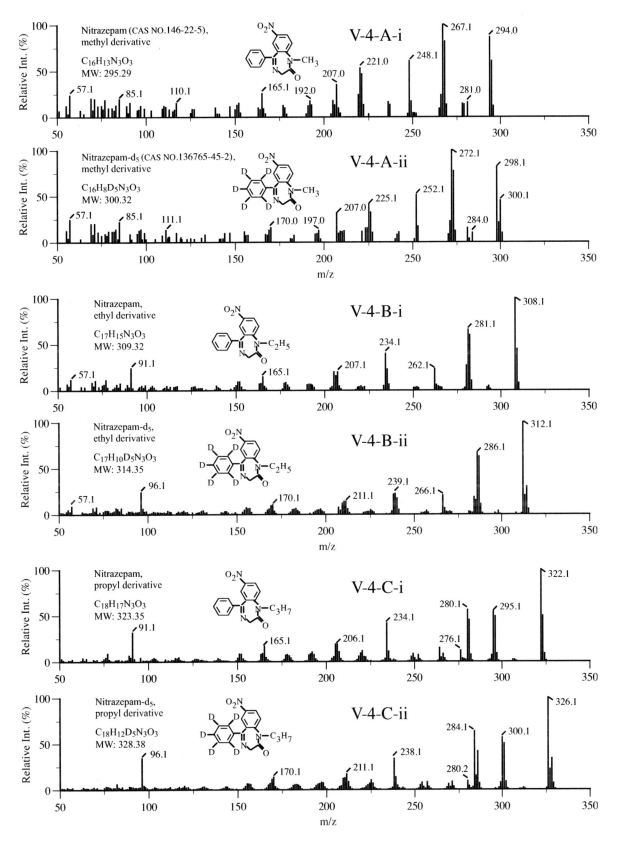

Figure V — Antianxiety Agent

284

Figure V-4. (Continued)

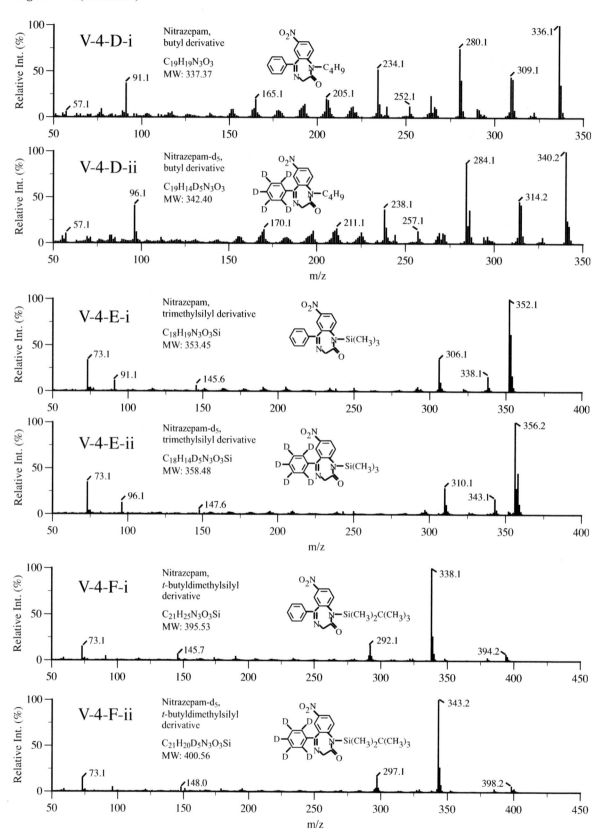

Appendix One — Mass Spectra

Figure V-5. Mass spectra of temazepam and its deuterated analogs (temazepam-d5): (A) underivatized; (B) methyl-derivatized; (C) ethyl-derivatized; (D) propyl-derivatized; (E) butyl-derivatized; (F) acetyl-derivatized; (G) TMS-derivatized; (H) *t*-BDMS-derivatized.

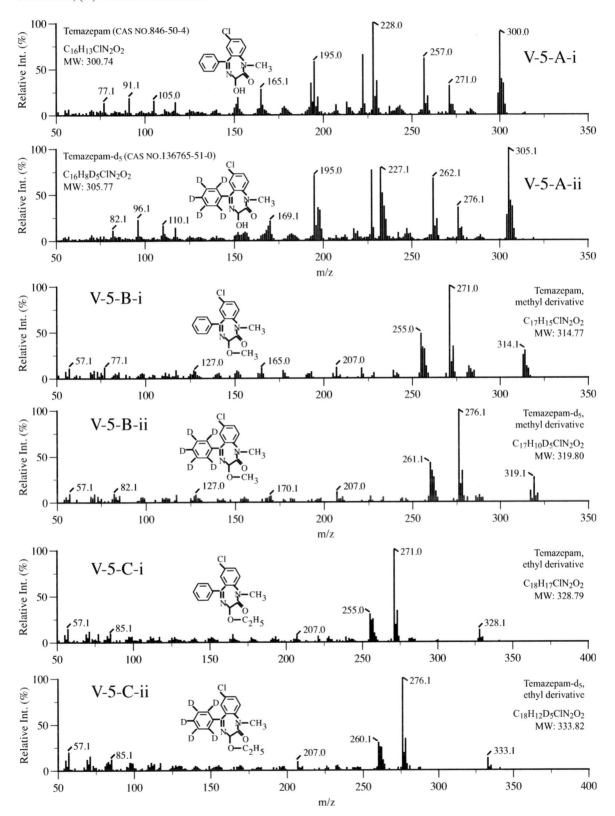

Figure V — Antianxiety Agent

Figure V-5. (Continued)

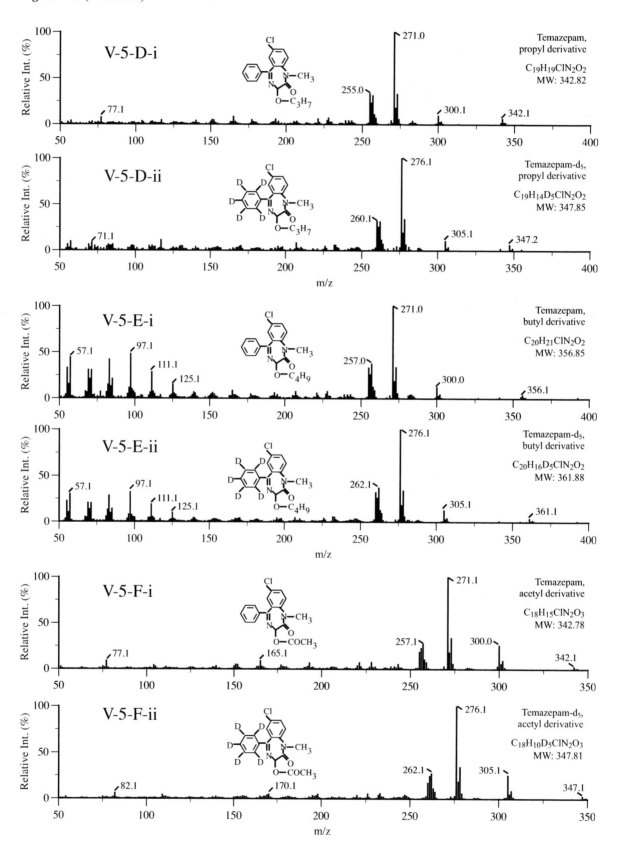

Figure V-5. (Continued)

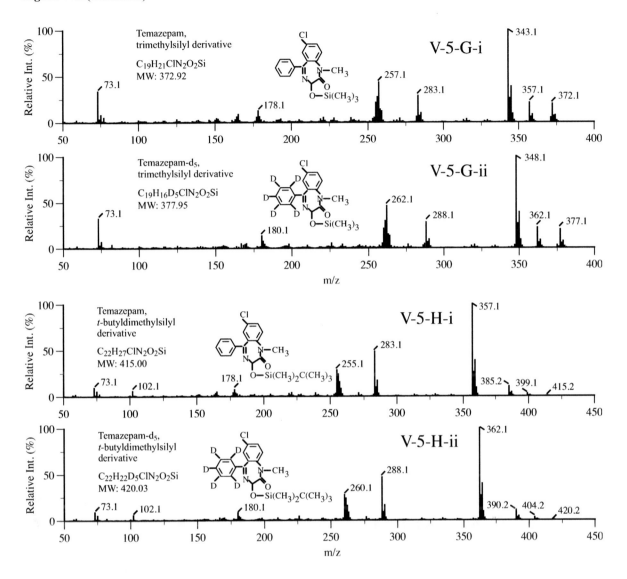

Figure V — Antianxiety Agent

288

Figure V-6. Mass spectra of clonazepam and its deuterated analogs (clonazepam-d_4): (A) methyl-derivatized; (B) ethyl-derivatized; (C) propyl-derivatized; (D) butyl-derivatized; (E) TMS-derivatized; (F) *t*-BDMS-derivatized.

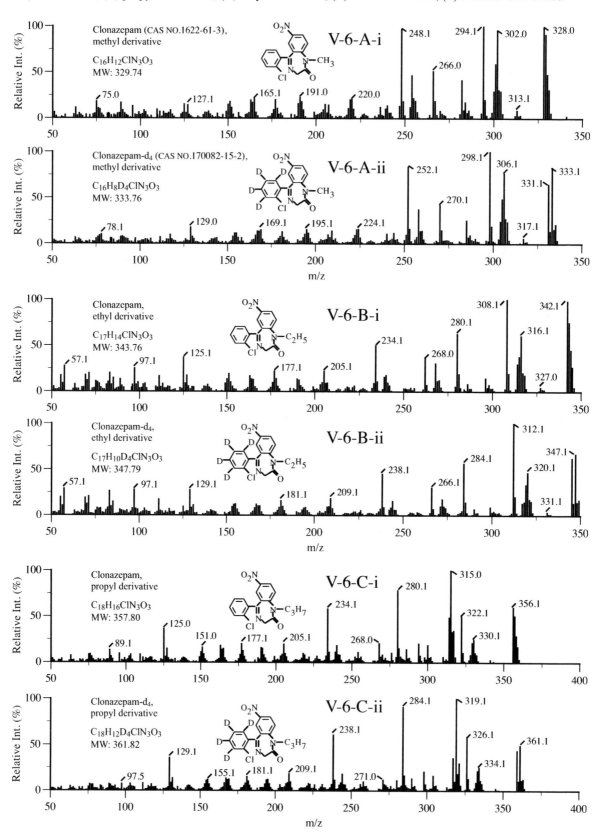

Figure V-6. (Continued)

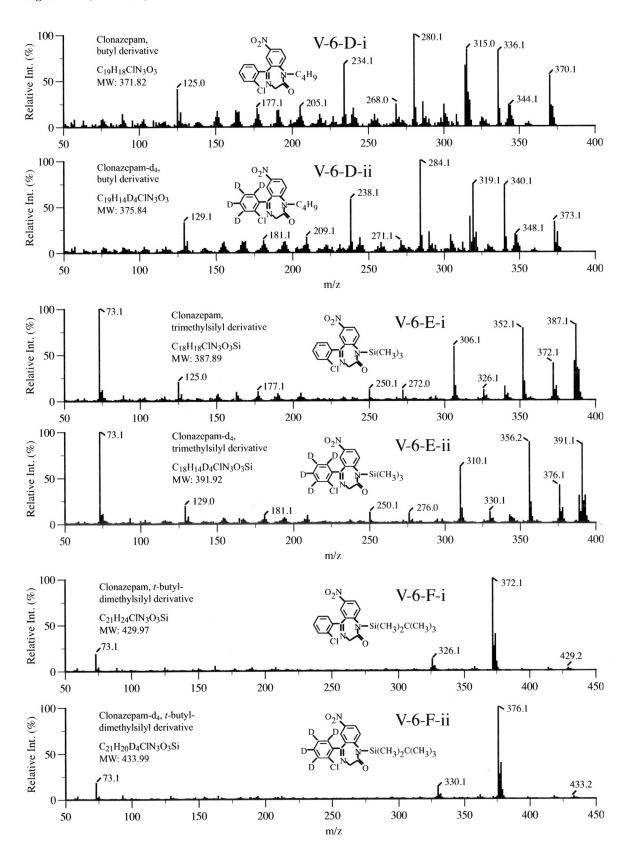

Figure V — Antianxiety Agent

Figure V-7. Mass spectra of 7-aminoclonazepam and its deuterated analogs (7-aminoclonazepam-d₄): (A) [methyl]₃-derivatized; (B) [ethyl]₂-derivatized; (C) [ethyl]₃-derivatized; (D) propyl-derivatized; (E) [propyl]₂-derivatized; (F) butyl-derivatized; (G) [butyl]₂-derivatized; (H) PFP-derivatized; (I) HFB-derivatized; (J) [TMS]₂-derivatized; (K) *t*-BDMS-derivatized; (L) [*t*-BDMS]₂-derivatized; (M) TFA/[TMS]₂-derivatized; (N) [TFA]₂/*t*-BDMS-derivatized; (O) TFA/[*t*-BDMS]₂-derivatized; (P) PFP/TMS-derivatized; (Q) PFP/[TMS]₂-derivatized; (R) PFP/[*t*-BDMS]₂-derivatized; (S) HFB/[*t*-BDMS]₂-derivatized.

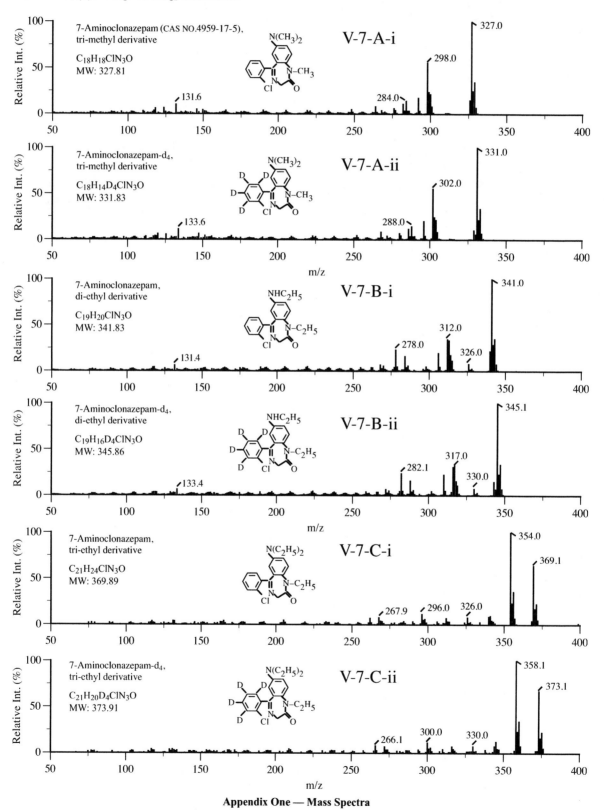

Appendix One — Mass Spectra

Figure V-7. (Continued)

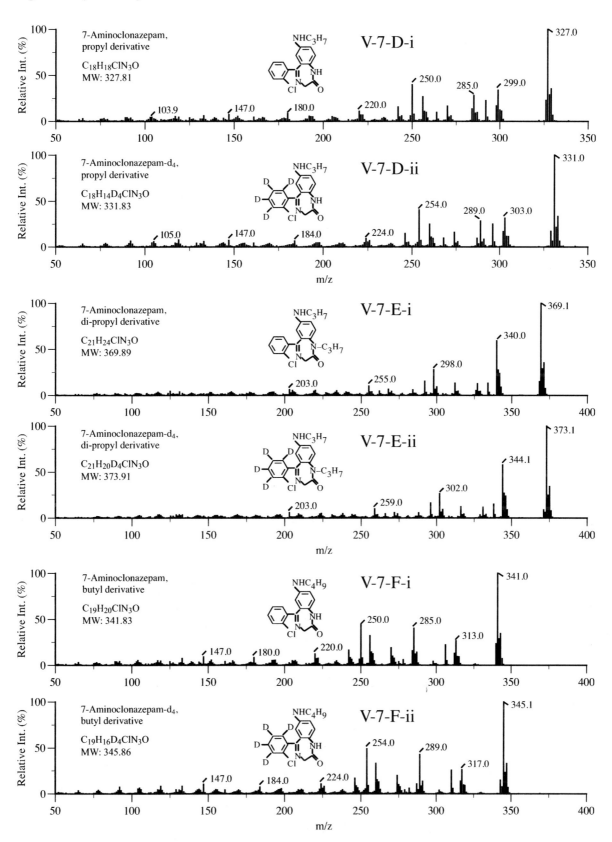

Figure V — Antianxiety Agent

Figure V-7. (Continued)

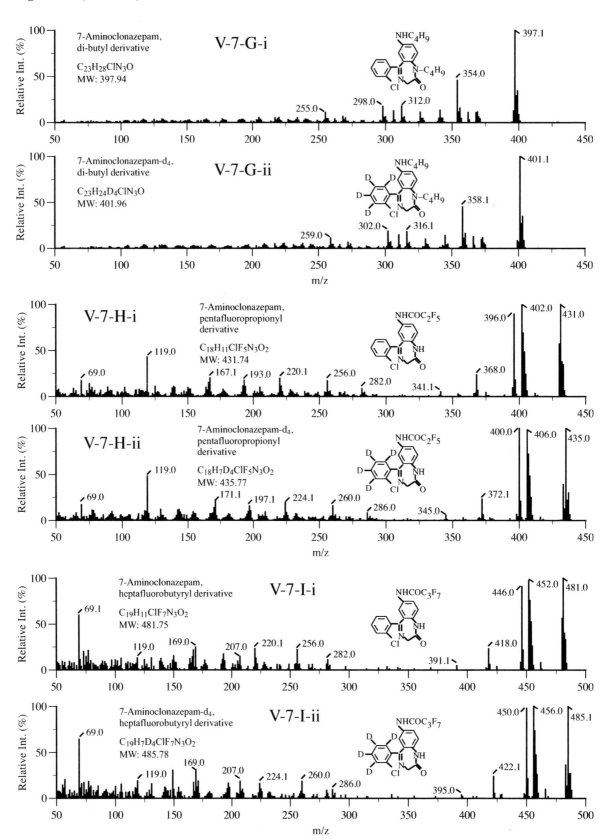

Figure V-7. (Continued)

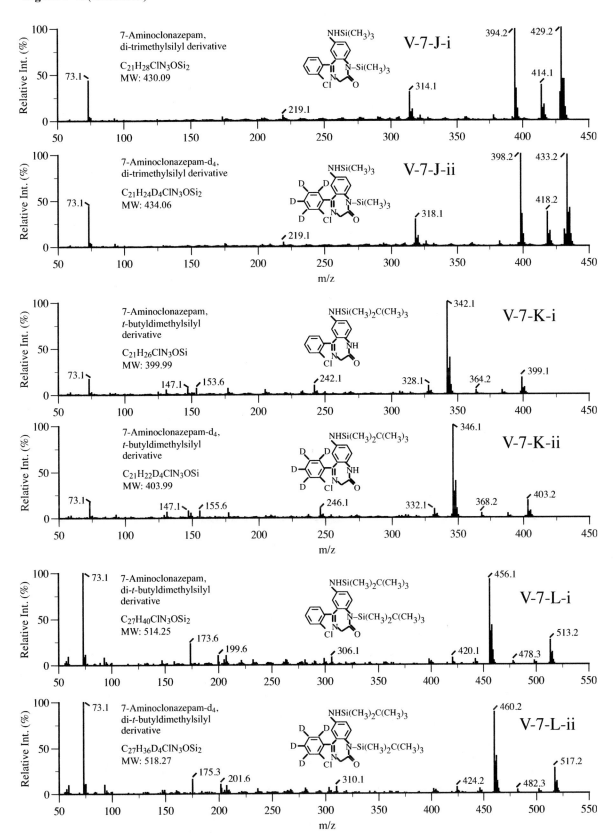

Figure V — Antianxiety Agent

294

Figure V-7. (Continued)

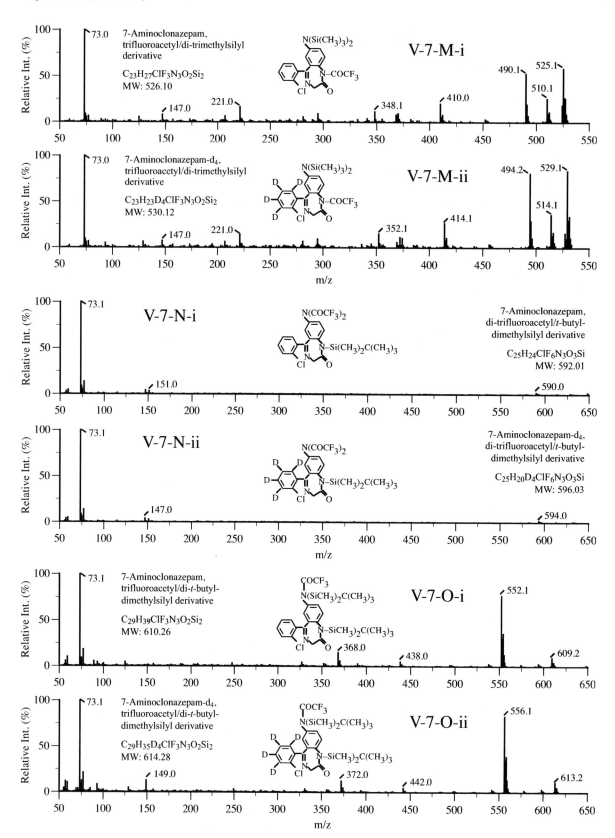

Appendix One — Mass Spectra

Figure V-7. (Continued)

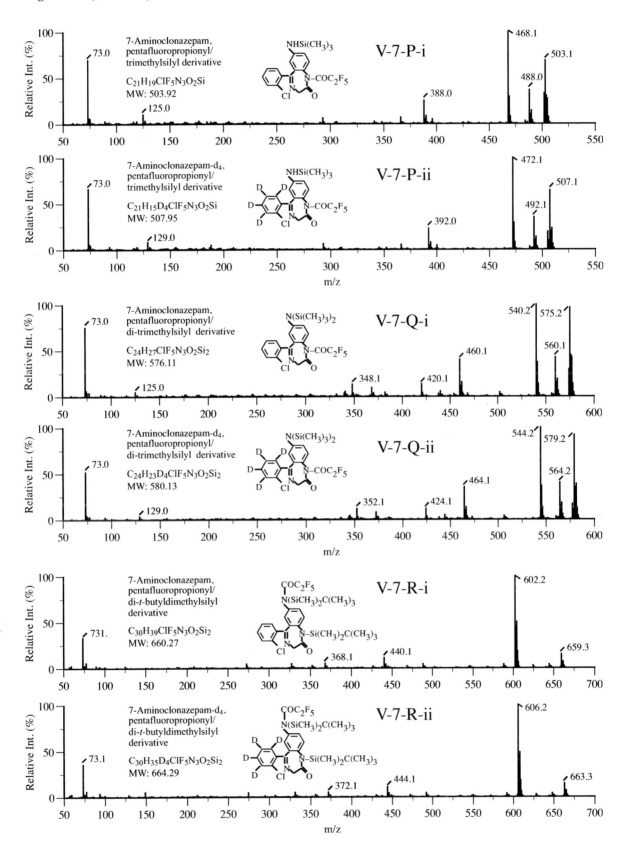

Figure V — Antianxiety Agent

Figure V-7. (Continued)

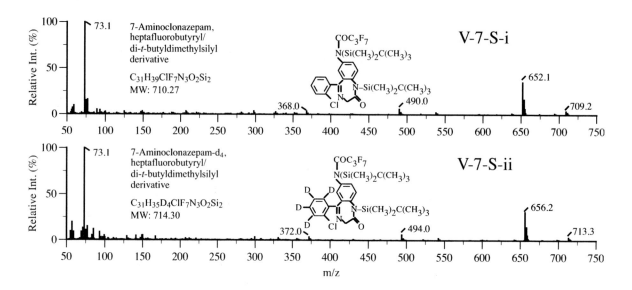

Figure V-8. Mass spectra of prazepam and its deuterated analogs (prazepam-d₅).

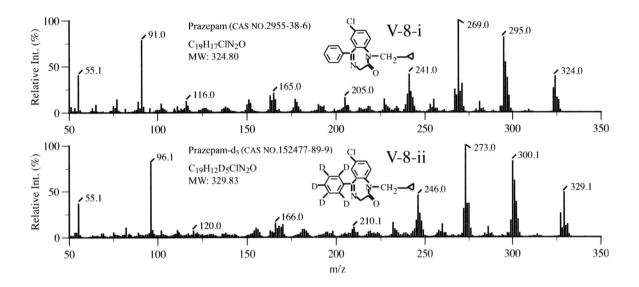

Figure V — Antianxiety Agent

Figure V-9. Mass spectra of lorazepam and its deuterated analogs (lorazepam-d$_4$): (A) [methyl]$_2$-derivatized; (B) [ethyl]$_2$-derivatized; (C) [propyl]$_2$-derivatized; (D) [butyl]$_2$-derivatized; (E) HFB-derivatized; (F) [TMS]$_2$-derivatized; (G) [t-BDMS]$_2$-derivatized.

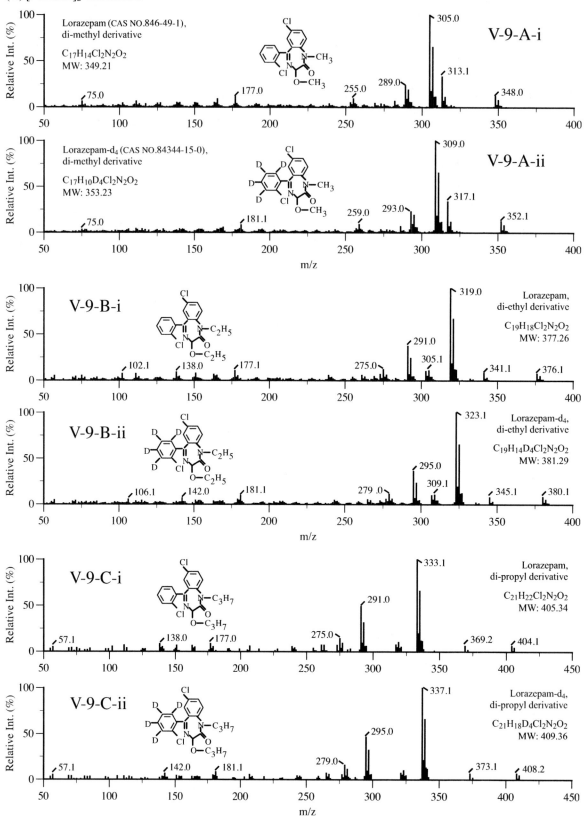

Figure V-9. (Continued)

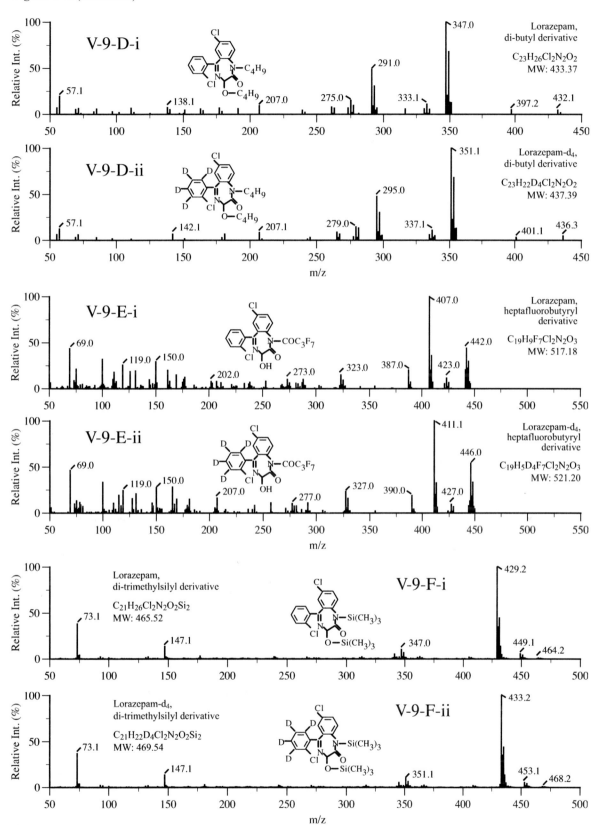

Figure V — Antianxiety Agent

Figure V-9. (Continued)

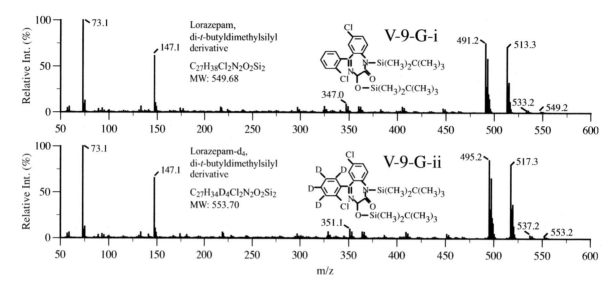

Figure V-10. Mass spectra of flunitrazepam and its deuterated analogs (flunitrazepam-d₃, -d₇).

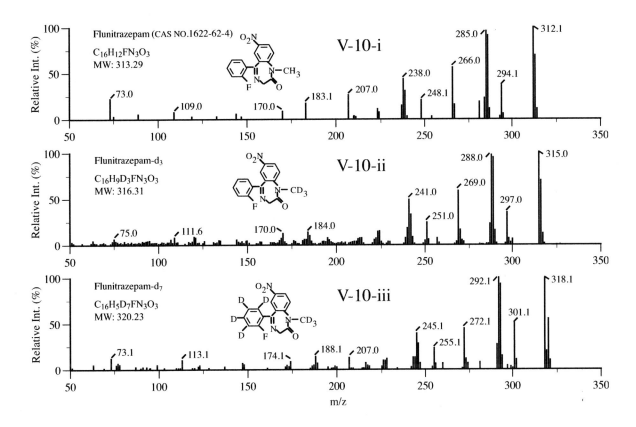

Figure V — Antianxiety Agent

Figure V-11. Mass spectra of 7-aminoflunitrazepam and its deuterated analogs (7-aminoflunitrazepam-d₃, -d₇): (A) underivatized; (B) [methyl]₂-derivatized; (C) ethyl-derivatized; (D) [ethyl]₂-derivatized; (E) propyl-derivatized; (F) butyl-derivatized; (G) acetyl-derivatized; (H) TFA-derivatized; (I) PFP-derivatized; (J) HFB-derivatized; (K) TMS-derivatized; (L) TFA/TMS-derivatized; (M) TFA/*t*-BDMS-derivatized; (N) PFP/TMS-derivatized; (O) PFP/*t*-BDMS-derivatized; (P) HFB/TMS-derivatized; (Q) HFB/*t*-BDMS-derivatized.

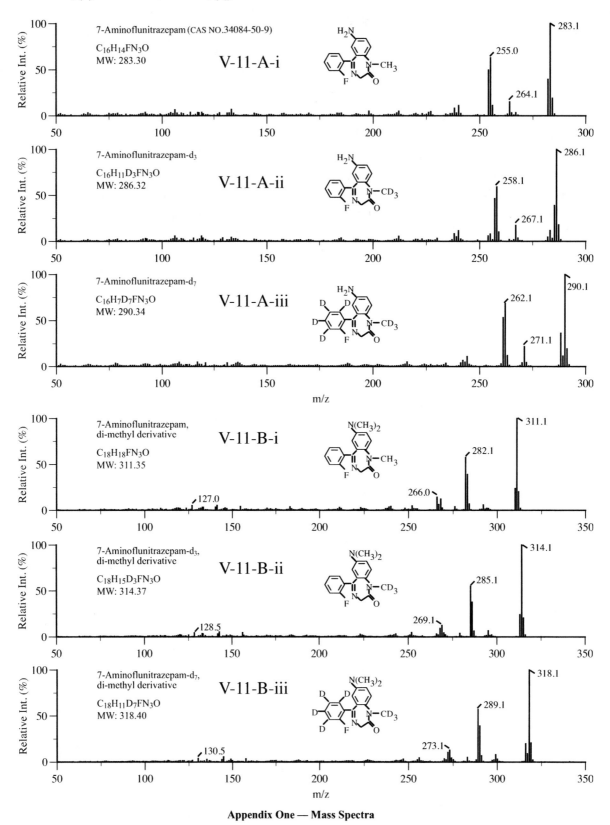

Figure V-11. (Continued)

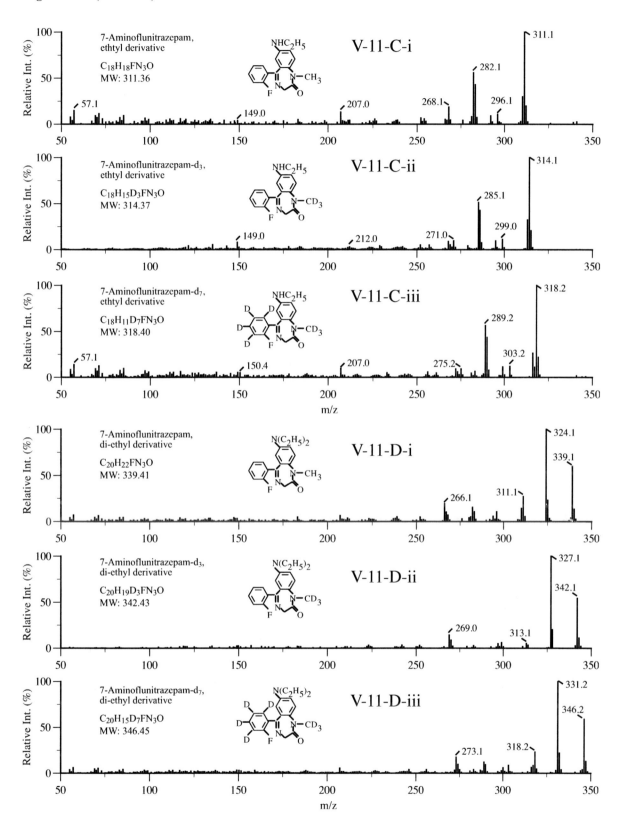

Figure V — Antianxiety Agent

Figure V-11. (Continued)

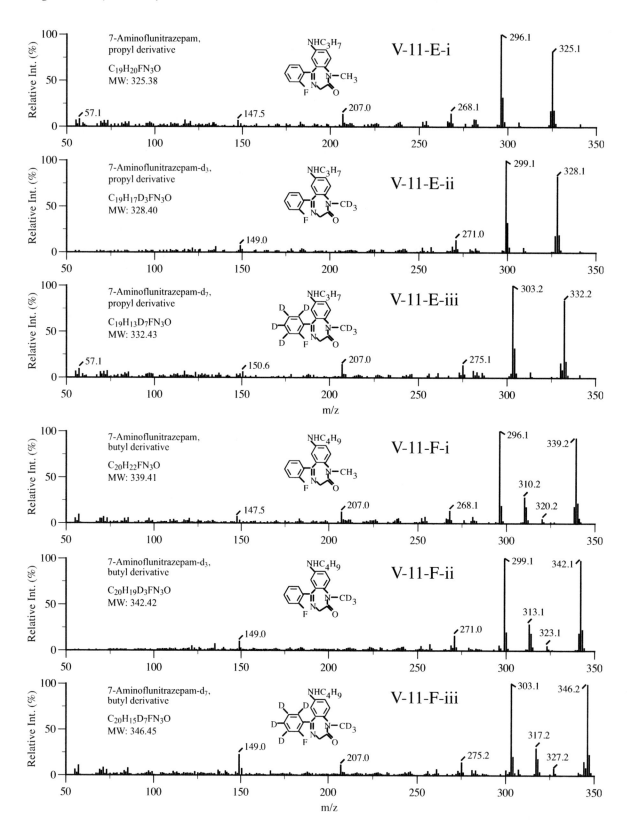

Figure V-11. (Continued)

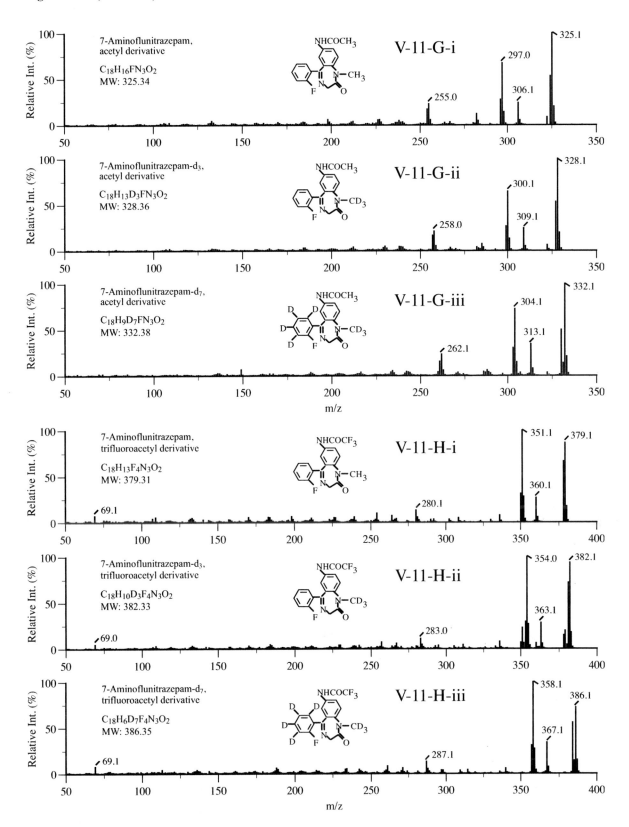

Figure V — Antianxiety Agent

Figure V-11. (Continued)

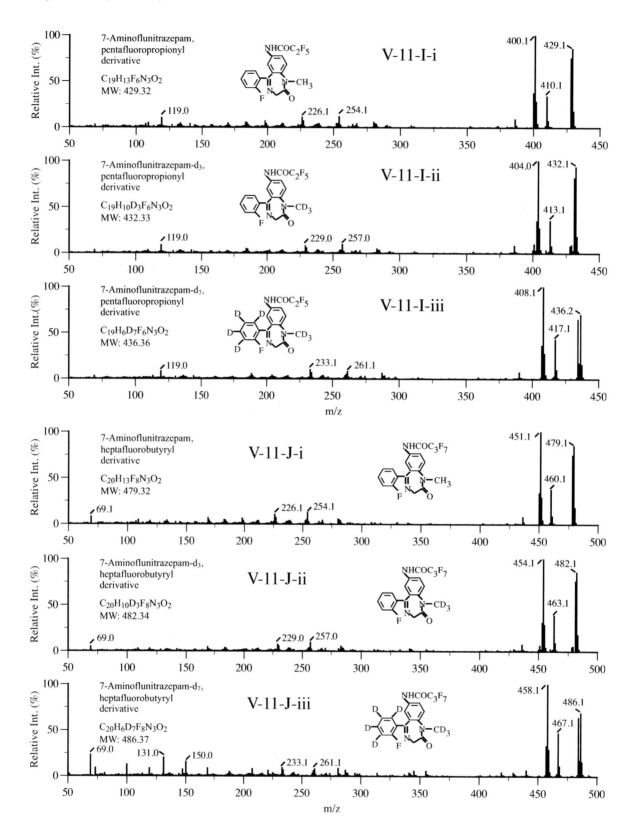

7-Aminoflunitrazepam,
pentafluoropropionyl
derivative

$C_{19}H_{13}F_6N_3O_2$
MW: 429.32

V-11-I-i

7-Aminoflunitrazepam-d$_3$,
pentafluoropropionyl
derivative

$C_{19}H_{10}D_3F_6N_3O_2$
MW: 432.33

V-11-I-ii

7-Aminoflunitrazepam-d$_7$,
pentafluoropropionyl
derivative

$C_{19}H_6D_7F_6N_3O_2$
MW: 436.36

V-11-I-iii

7-Aminoflunitrazepam,
heptafluorobutyryl
derivative

$C_{20}H_{13}F_8N_3O_2$
MW: 479.32

V-11-J-i

7-Aminoflunitrazepam-d$_3$,
heptafluorobutyryl
derivative

$C_{20}H_{10}D_3F_8N_3O_2$
MW: 482.34

V-11-J-ii

7-Aminoflunitrazepam-d$_7$,
heptafluorobutyryl
derivative

$C_{20}H_6D_7F_8N_3O_2$
MW: 486.37

V-11-J-iii

Appendix One — Mass Spectra

307

Figure V-11. (Continued)

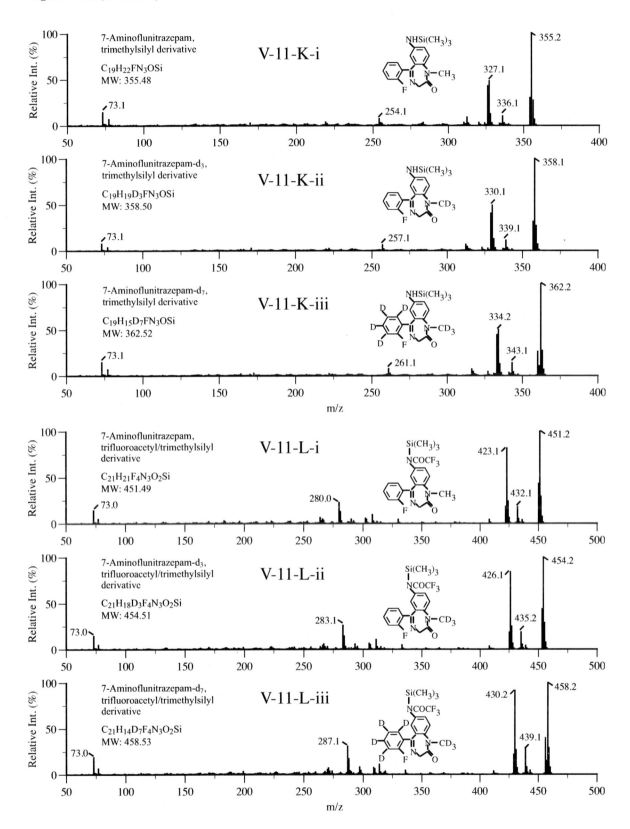

Figure V — Antianxiety Agent

Figure V-11. (Continued)

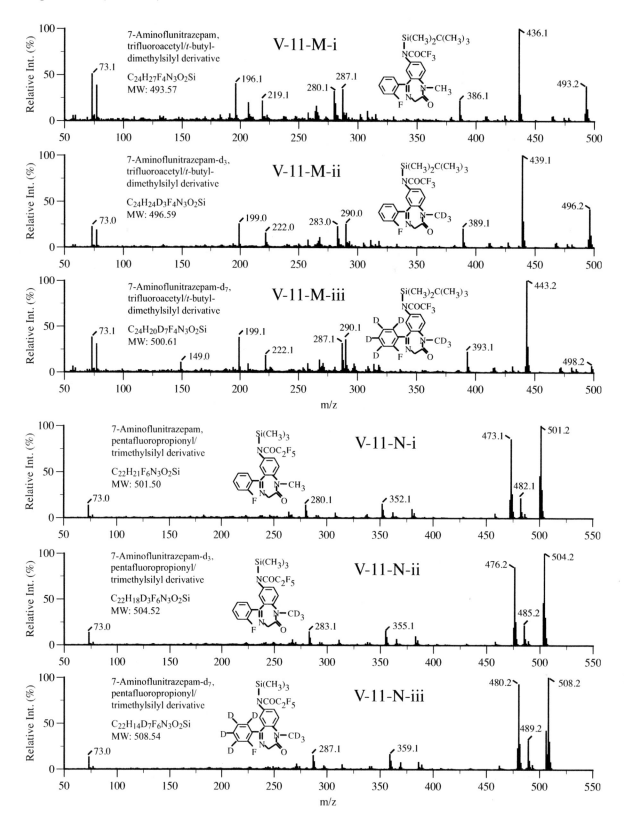

7-Aminoflunitrazepam, trifluoroacetyl/*t*-butyl-dimethylsilyl derivative
C$_{24}$H$_{27}$F$_4$N$_3$O$_2$Si
MW: 493.57

V-11-M-i

7-Aminoflunitrazepam-d$_3$, trifluoroacetyl/*t*-butyl-dimethylsilyl derivative
C$_{24}$H$_{24}$D$_3$F$_4$N$_3$O$_2$Si
MW: 496.59

V-11-M-ii

7-Aminoflunitrazepam-d$_7$, trifluoroacetyl/*t*-butyl-dimethylsilyl derivative
C$_{24}$H$_{20}$D$_7$F$_4$N$_3$O$_2$Si
MW: 500.61

V-11-M-iii

7-Aminoflunitrazepam, pentafluoropropionyl/trimethylsilyl derivative
C$_{22}$H$_{21}$F$_6$N$_3$O$_2$Si
MW: 501.50

V-11-N-i

7-Aminoflunitrazepam-d$_3$, pentafluoropropionyl/trimethylsilyl derivative
C$_{22}$H$_{18}$D$_3$F$_6$N$_3$O$_2$Si
MW: 504.52

V-11-N-ii

7-Aminoflunitrazepam-d$_7$, pentafluoropropionyl/trimethylsilyl derivative
C$_{22}$H$_{14}$D$_7$F$_6$N$_3$O$_2$Si
MW: 508.54

V-11-N-iii

Appendix One — Mass Spectra

Figure V-11. (Continued)

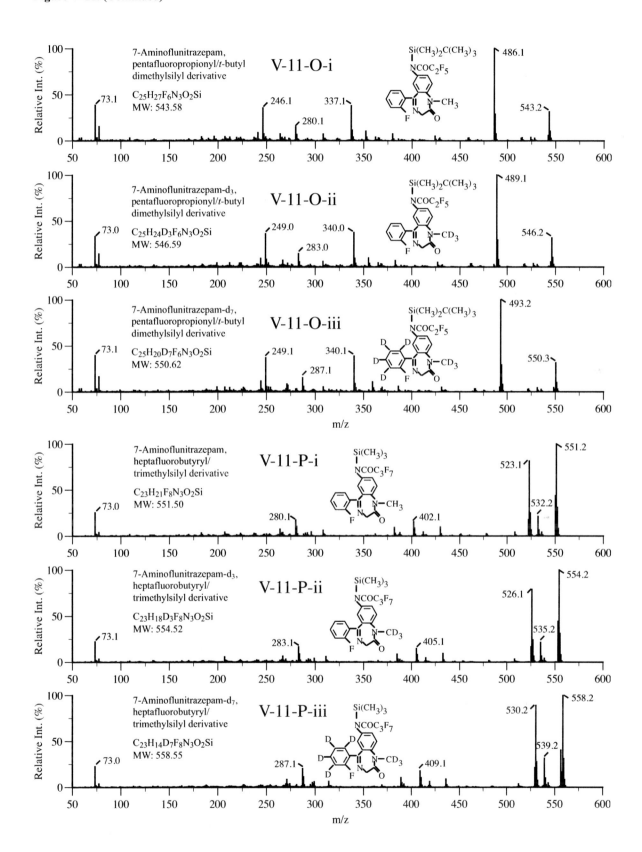

Figure V — Antianxiety Agent

Figure V-11. (Continued)

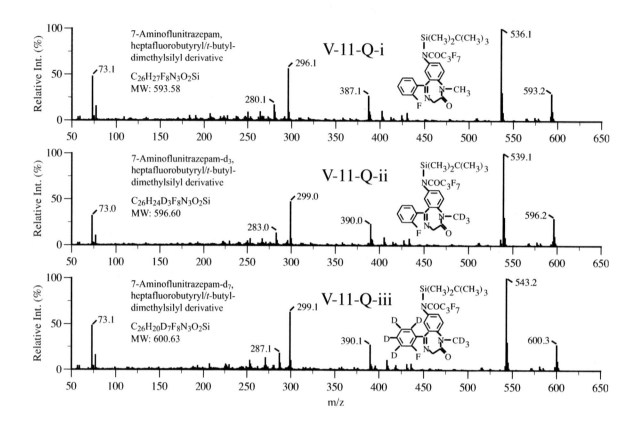

Figure V-12. Mass spectra of *N*-desalkylflurazepam and its deuterated analogs (*N*-desalkylflurazepam-d₄): (A) underivatized; (B) methyl-derivatized; (C) [methyl]₂-derivatized; (D) ethyl-derivatized; (E) propyl-derivatized; (F) butyl-derivatized; (G) acetyl-derivatized; (H) TMS-derivatized; (I) *t*-BDMS-derivatized.

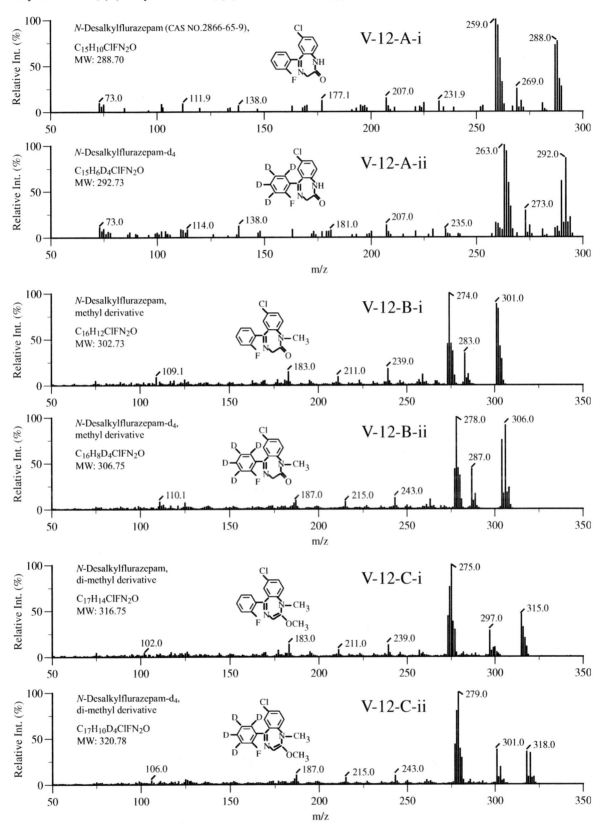

Figure V — Antianxiety Agent

Figure V-12. (Continued)

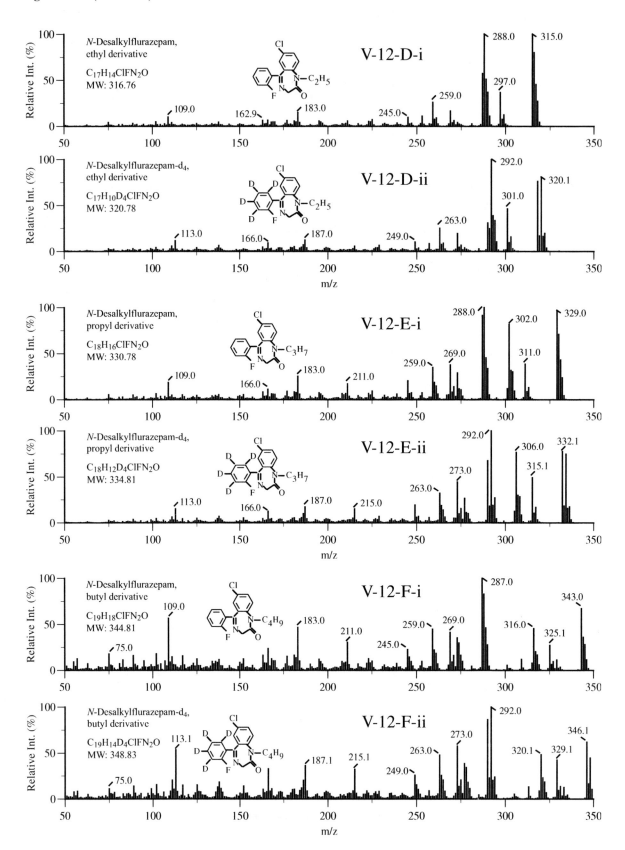

Figure V-12. (Continued)

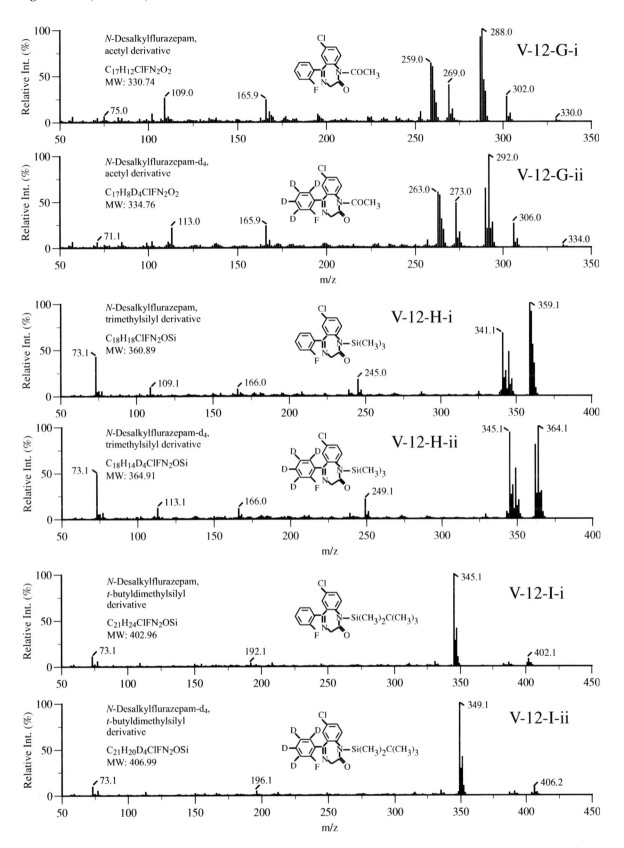

Figure V — Antianxiety Agent

Figure V-13. Mass spectra of *N*-desmethylflunitrazepam and its deuterated analogs (*N*-desmethylflunitrazepam-d₄):
(A) [methyl]₂-derivatized; (B) ethyl-derivatized; (C) propyl-derivatized; (D) butyl-derivatized; (E) acetyl-derivatized;
(F) TMS-derivatized; (G) *t*-BDMS-derivatized.

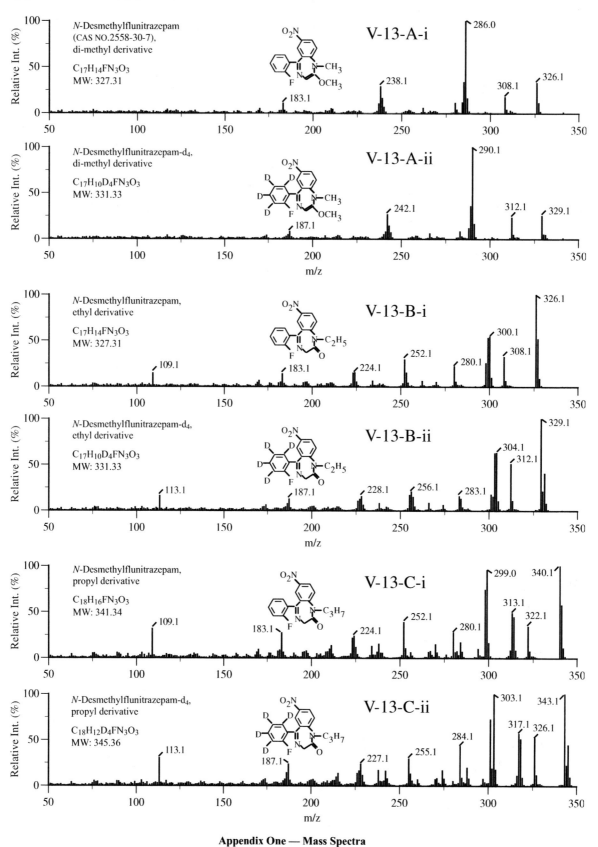

Figure V-13. (Continued)

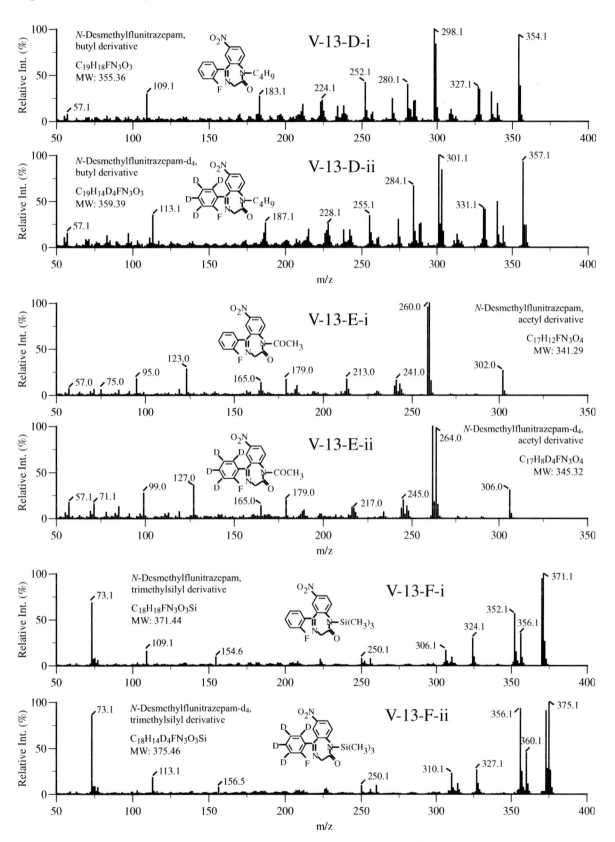

Figure V — Antianxiety Agent

Figure V-13. (Continued)

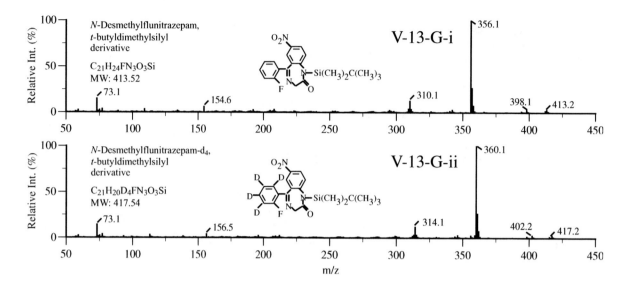

317

Figure V-14. Mass spectra of 2-hydroxyethylflurazepam and its deuterated analogs (2-hydroxyethylflurazepam-d₄): (A) underivatized; (B) butyl-derivatized; (C) TMS-derivatized; (D) *t*-BDMS-derivatized.

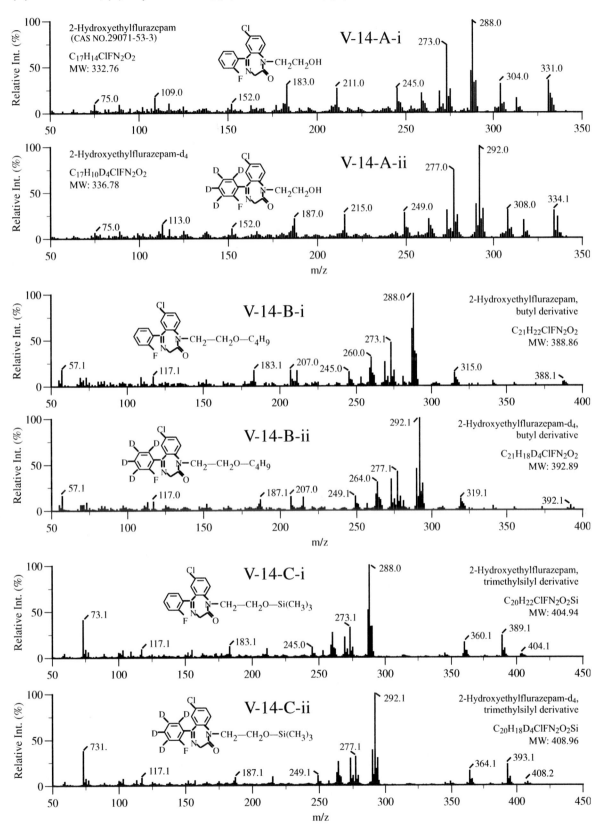

Figure V — Antianxiety Agent

Figure V-14. (Continued)

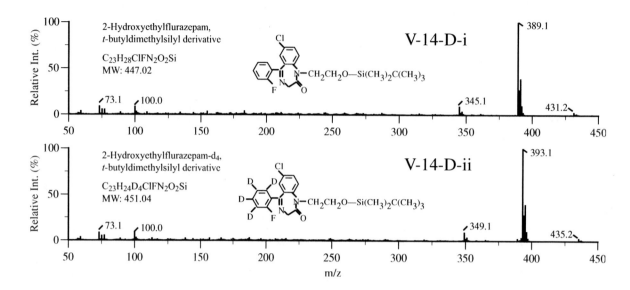

Figure V-15. Mass spectra of estazolam and its deuterated analogs (estazolam-d₅).

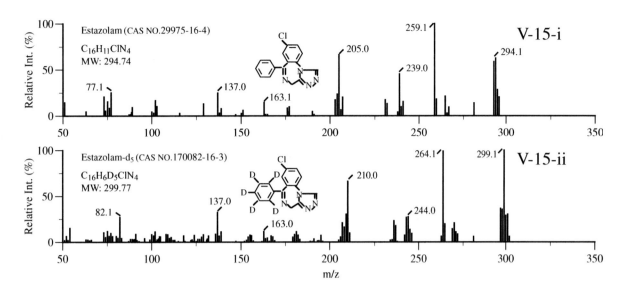

Figure V — Antianxiety Agent

Figure V-16. Mass spectra of alprazolam and its deuterated analogs (alprazolam-d₅).

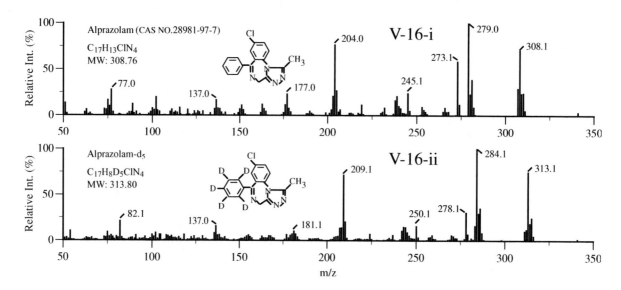

321

Figure V-17. Mass spectra of α-hydroxyalprazolam and its deuterated analogs (α-hydroxyalprazolam-d₅): (A) TMS-derivatized; (B) *t*-BDMS-derivatized.

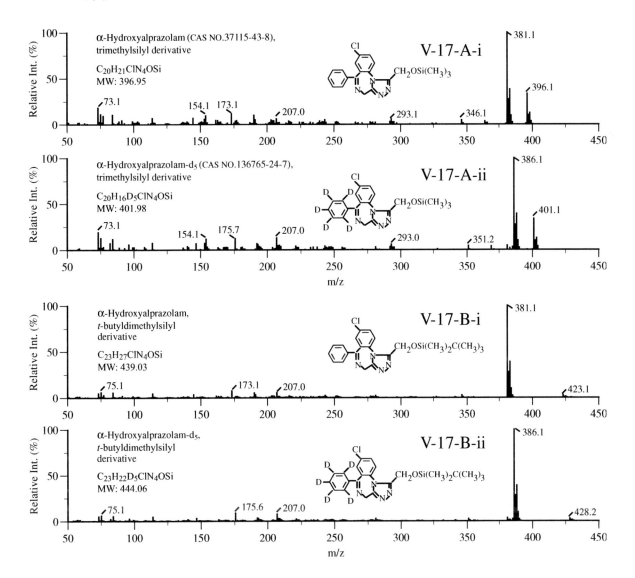

Figure V — Antianxiety Agent

Figure V-18. Mass spectra of α-hydroxytriazolam and its deuterated analogs (α-hydroxytriazolam-d₄): (A) TMS-derivatized; (B) *t*-BDMS-derivatized.

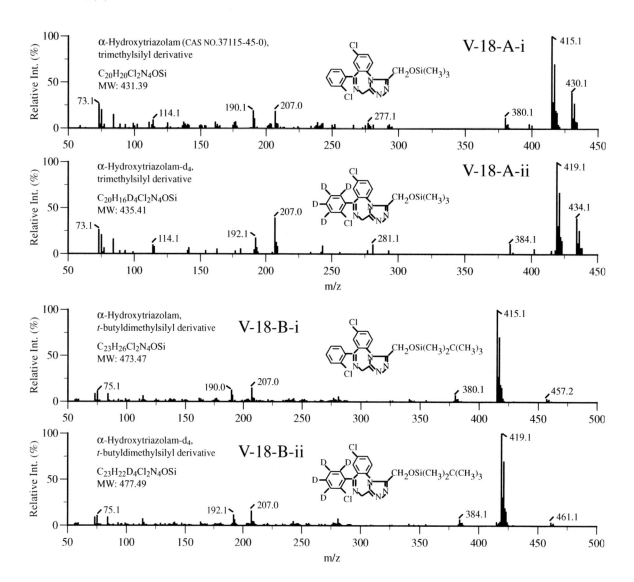

Figure V-19. Mass spectra of mianserin and its deuterated analogs (mianserin-d₃).

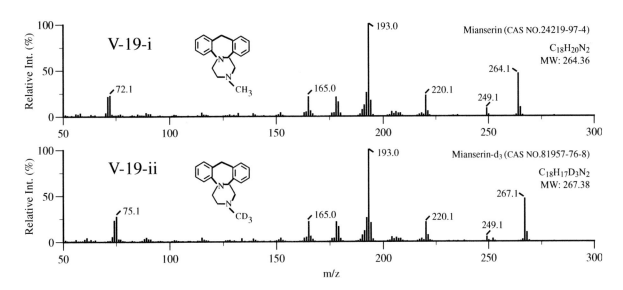

Figure V — Antianxiety Agent

Figure V-20. Mass spectra of methaqualone and its deuterated analogs (methaqualone-d$_7$).

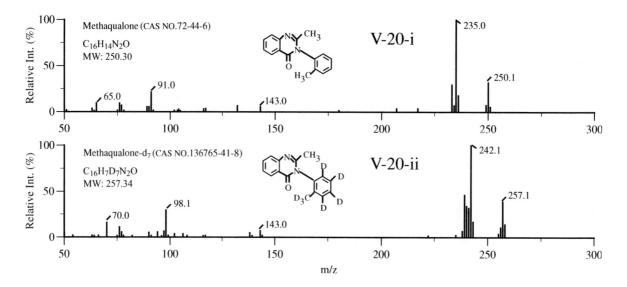

Figure V-21. Mass spectra of haloperidol and its deuterated analogs (haloperidol-d₄): (A) TMS-derivatized.

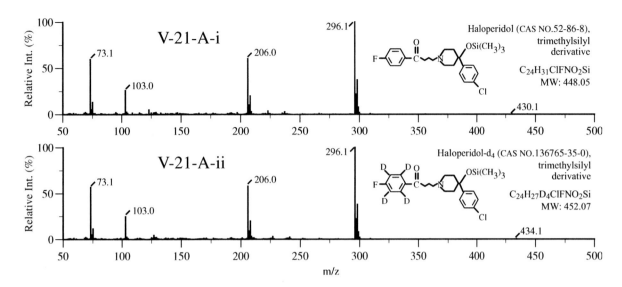

Figure V — Antianxiety Agent

**Summary of Drugs, Isotopic Analogs, and Chemical Derivatization Groups Included in
Figure VI (Antidepressants)**

Compound	Isotopic analog	Chemical derivatization group (no. of spectra)	Figure #
Imipramine	d₃	None (2)	VI-1
Desipramine	d₃	None, acetyl, TCA, TFA, PFP, 4-CB, TMS, *t*-BDMS (16)	VI-2
Trimipramine	d₃	None (2)	VI-3
Clomipramine	d₃	None (2)	VI-4
Nortriptyline	d₃	None, acetyl, TCA, TFA, PFP, HFB, 4-CB, TMS, *t*-BDMS (18)	VI-5
Protriptyline	d₃	None, acetyl, TCA, TFA, PFP, HFB, 4-CB, TMS, *t*-BDMS (18)	VI-6
Doxepin	d₃	None (2)	VI-7
Dothiepin	d₃	None (2)	VI-8
Amitriptyline	d₃	None (2)	VI-9
Maprotiline	d₃	None, acetyl, TCA, TFA, PFP, HFB, 4-CB, TMS, *t*-BDMS (18)	VI-10
		Total no. of mass spectra: 82	

Figure VI — Antidepressant

Appendix One — Figure VI
Mass Spectra of Commonly Abused Drugs and Their Isotopically Labeled Analogs
in Various Derivatization Forms — Antidepressants

Figure VI — Antidepressant

Figure VI-1. Mass spectra of imipramine and its deuterated analogs (imipramine-d₃).

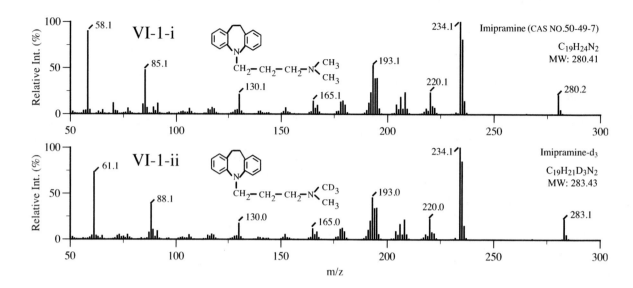

Figure VI-2. Mass spectra of desipramine and its deuterated analogs (desipramine-d₃): (A) underivatized; (B) acetyl-derivatized; (C) TCA-derivatized; (D) TFA-derivatized; (E) PFP-derivatized; (F) 4-CB-derivatized; (G) TMS-derivatized; (H) t-BDMS-derivatized.

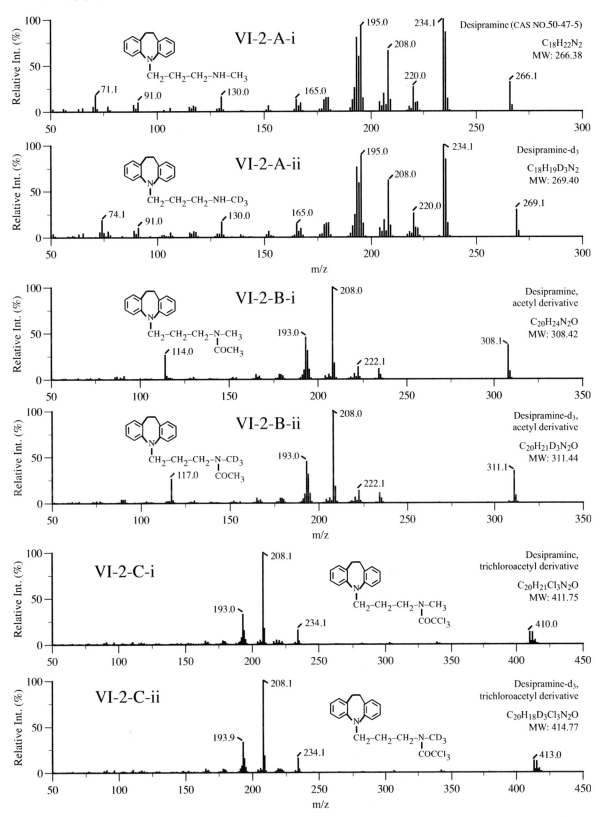

Figure VI — Antidepressant

Figure VI-2. (Continued)

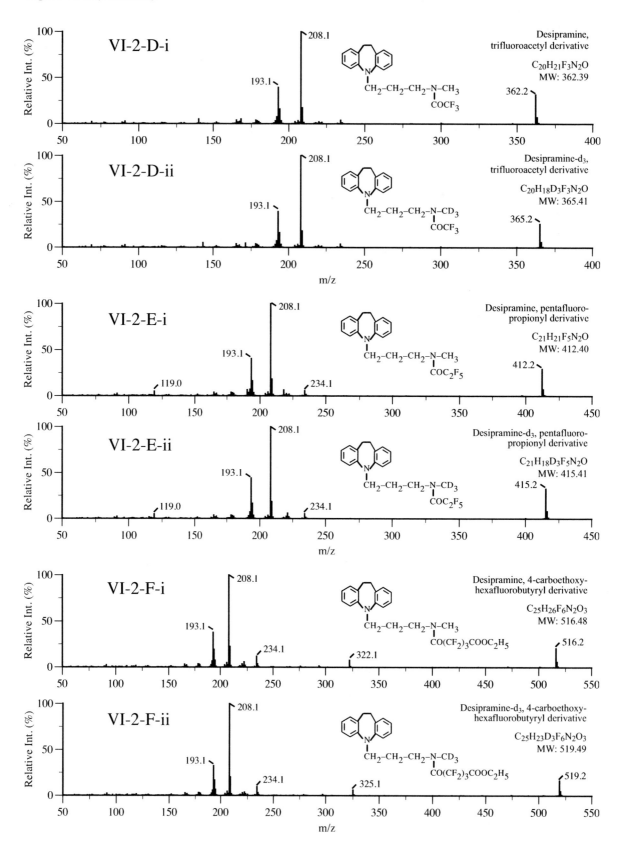

VI-2-D-i

Desipramine,
trifluoroacetyl derivative

$C_{20}H_{21}F_3N_2O$
MW: 362.39

VI-2-D-ii

Desipramine-d_3,
trifluoroacetyl derivative

$C_{20}H_{18}D_3F_3N_2O$
MW: 365.41

VI-2-E-i

Desipramine, pentafluoro-
propionyl derivative

$C_{21}H_{21}F_5N_2O$
MW: 412.40

VI-2-E-ii

Desipramine-d_3, pentafluoropropionyl derivative

$C_{21}H_{18}D_3F_5N_2O$
MW: 415.41

VI-2-F-i

Desipramine, 4-carboethoxy-
hexafluorobutyryl derivative

$C_{25}H_{26}F_6N_2O_3$
MW: 516.48

VI-2-F-ii

Desipramine-d_3, 4-carboethoxy-
hexafluorobutyryl derivative

$C_{25}H_{23}D_3F_6N_2O_3$
MW: 519.49

Appendix One — Mass Spectra

Figure VI-2. (Continued)

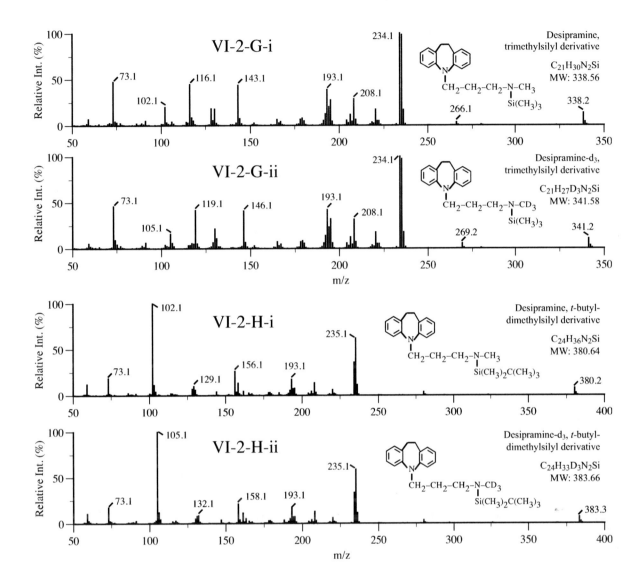

Figure VI — Antidepressant

Figure VI-3. Mass spectra of trimipramine and its deuterated analogs (trimipramine-d₃).

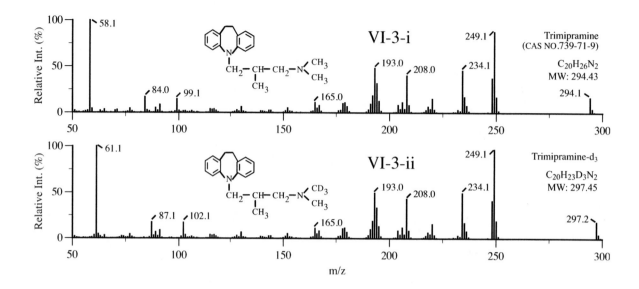

Figure VI-4. Mass spectra of clomipramine and its deuterated analogs (clomipramine-d$_3$).

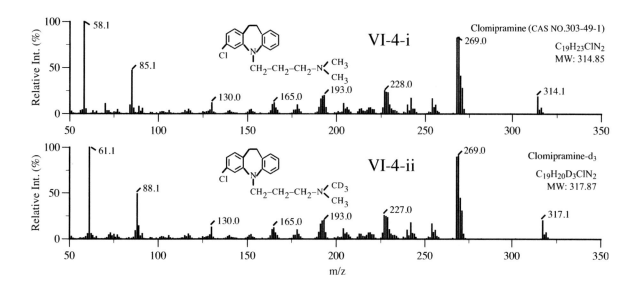

Figure VI — Antidepressant

Figure VI-5. Mass spectra of nortriptyline and its deuterated analogs (nortriptyline-d₃): (A) underivatized; (B) acetyl-derivatized; (C) TCA-derivatized; (D) TFA-derivatized; (E) PFP-derivatized; (F) HFB-derivatized; (G) 4-CB-derivatized; (H) TMS-derivatized; (I) *t*-BDMS-derivatized.

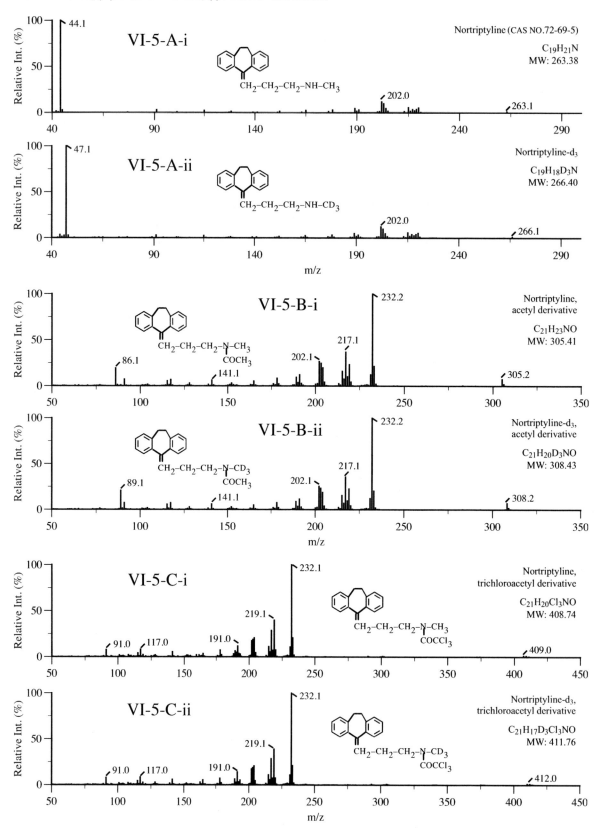

Figure VI-5. (Continued)

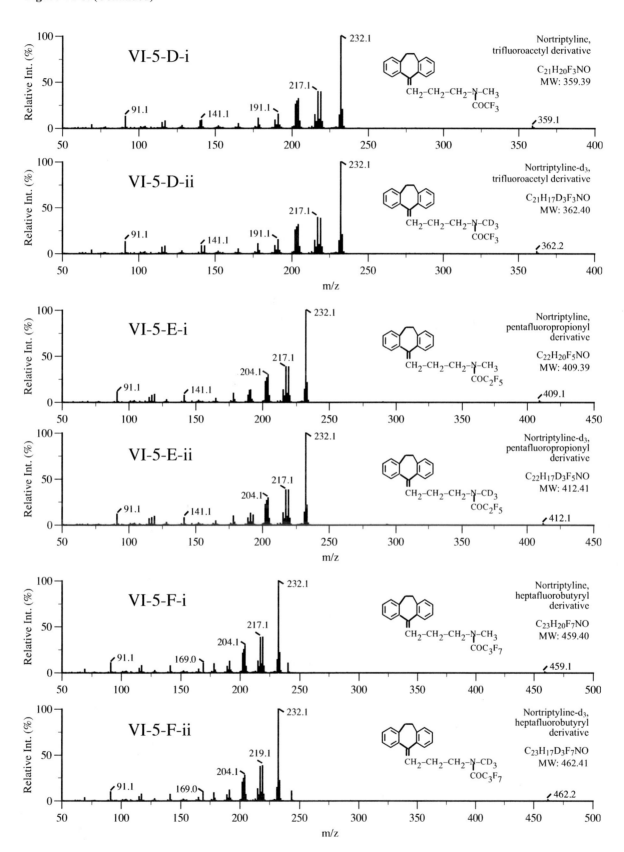

Figure VI — Antidepressant

338

Figure VI-5. (Continued)

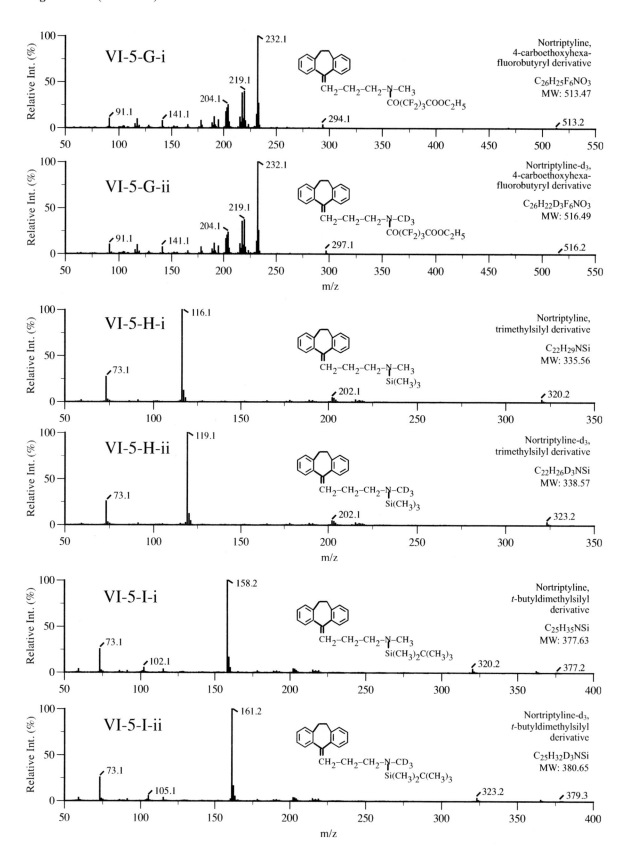

339

Figure VI-6. Mass spectra of protriptyline and its deuterated analogs (protriptyline-d₃): (A) underivatized; (B) acetyl-derivatized; (C) TCA-derivatized; (D) TFA-derivatized; (E) PFP-derivatized; (F) HFB-derivatized; (G) 4-CB-derivatized; (H) TMS-derivatized; (I) *t*-BDMS-derivatized.

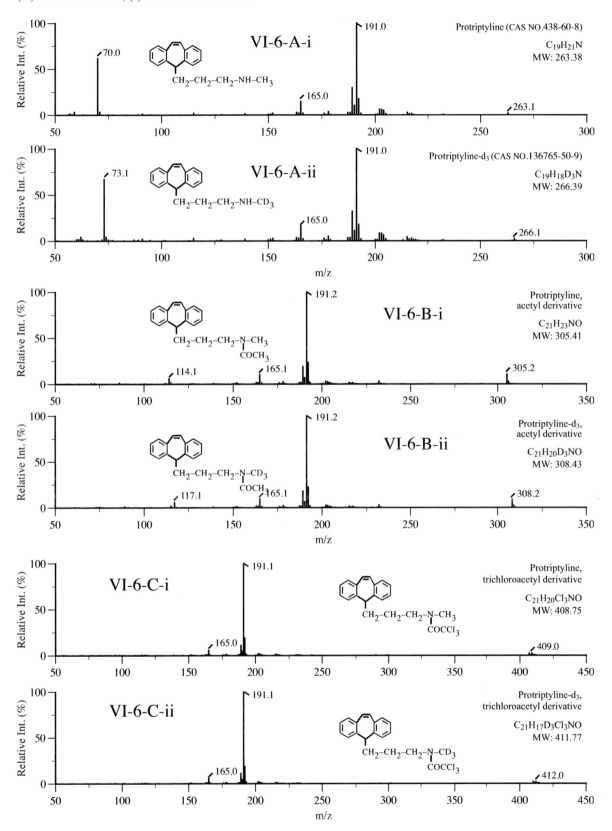

Figure VI — Antidepressant

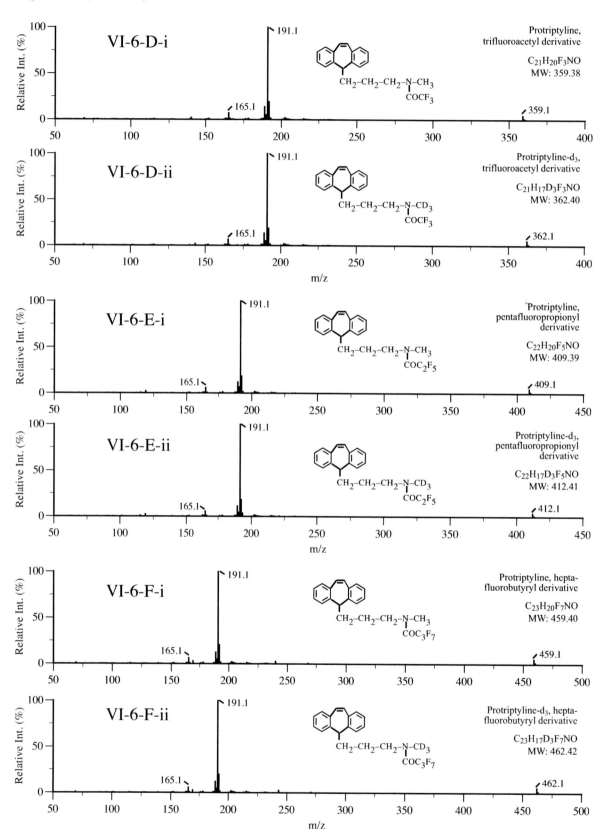

Figure VI-6. (Continued)

Appendix One — Mass Spectra

Figure IV-6. (Continued)

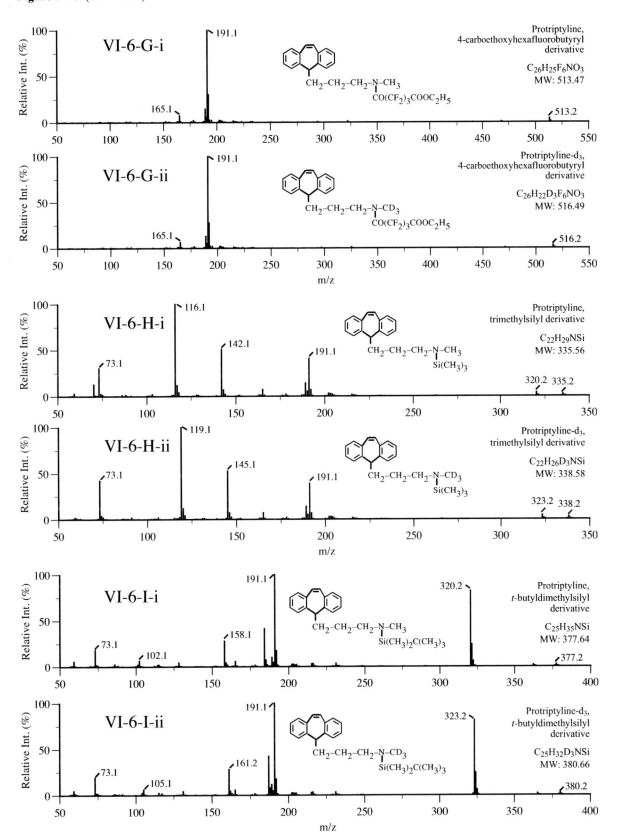

Figure VI — Antidepressant

Figure VI-7. Mass spectra of doxepin and its deuterated analogs (doxepin-d$_3$).

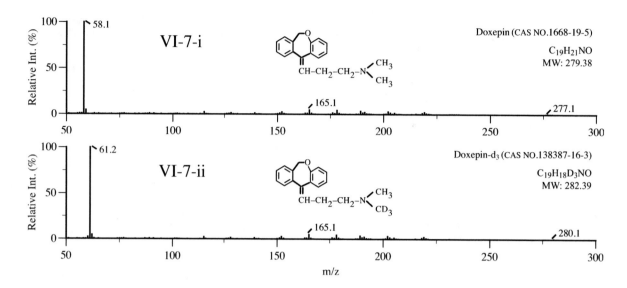

Figure VI-8. Mass spectra of dothiepin and its deuterated analogs (dothiepin-d₃).

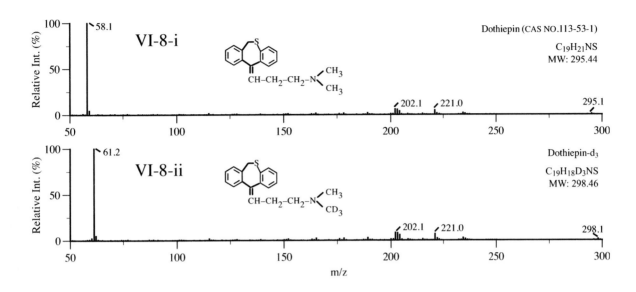

Figure VI — Antidepressant

344

Figure VI-9. Mass spectra of amitriptyline and its deuterated analogs (amitriptyline-d₃).

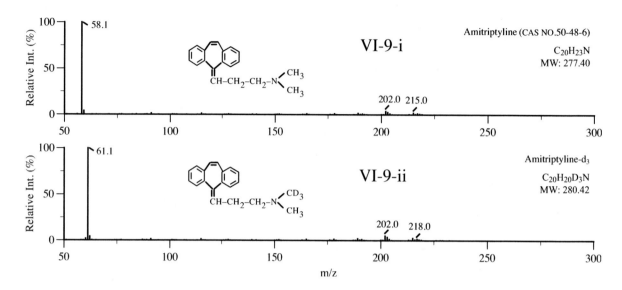

FigureVI-10 Mass spectra of maprotiline and its deuterated analogs (maprotiline-d₃): (A) underivatized; (B) acetyl-derivatized; (C) TCA-derivatized; (D) TFA-derivatized; (E) PFP-derivatized; (F) HFB-derivatized; (G) 4-CB-derivatized; (H) TMS-derivatized; (I) *t*-BDMS-derivatized.

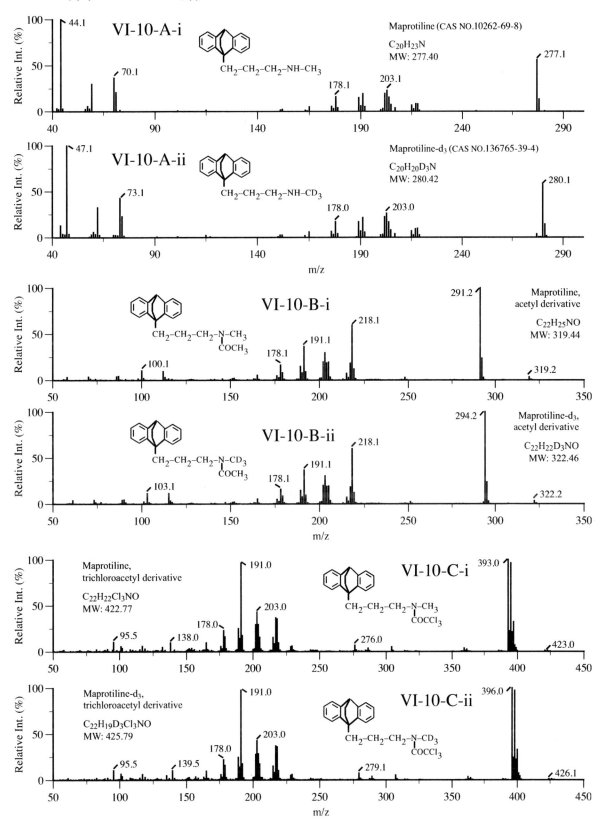

Figure VI — Antidepressant

Figure VI-10. (Continued)

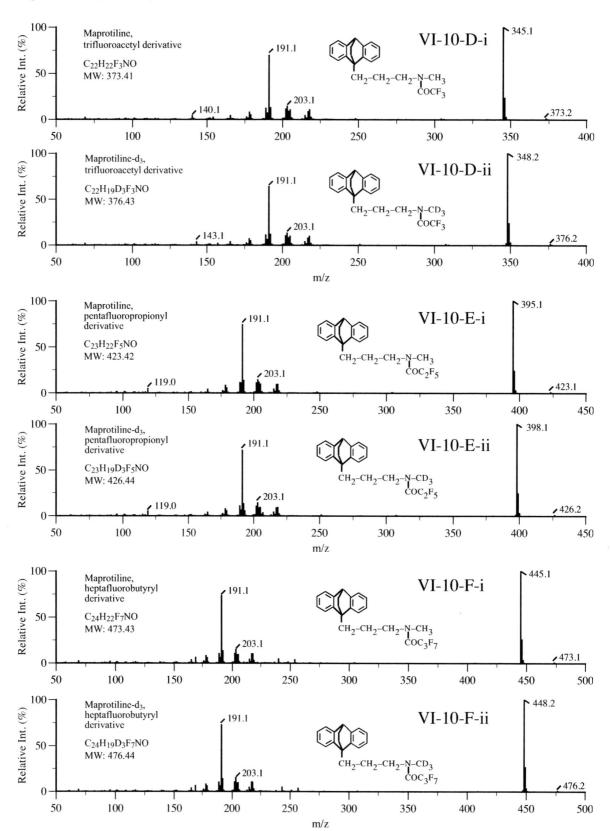

Figure VI-10. (Continued)

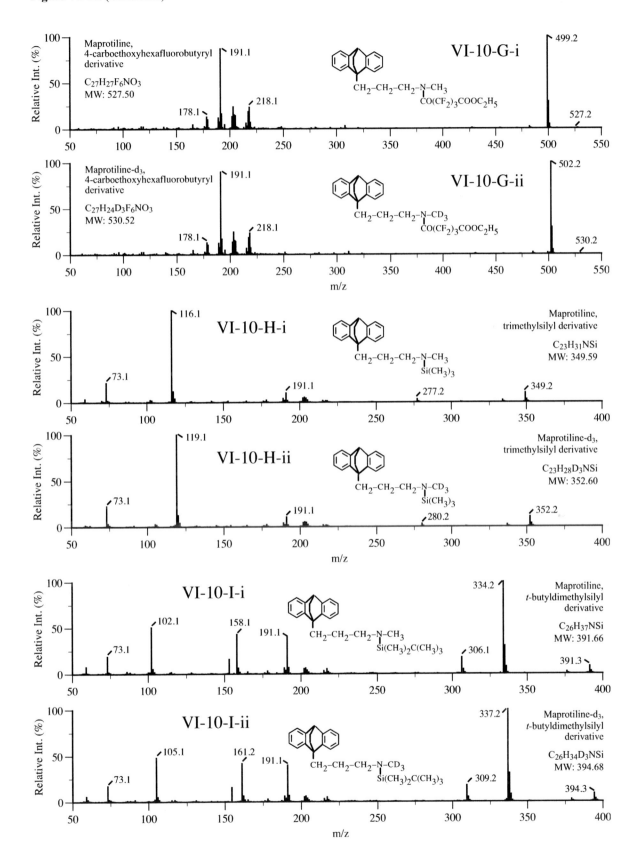

Figure VI — Antidepressant

Summary of Drugs, Isotopic Analogs, and Chemical Derivatization Groups Included in Figure VII (Others)

Compound	Isotopic analog	Chemical derivatization group (no. of spectra)	Figure #
Diphenhydramine	d_3	None (2)	VII-1
Cotinine	d_3	None (2)	VII-2
Nicotine	d_4	None (2)	VII-3
5-α-Estran-3α-ol-17-one	d_3	None, acetyl, TMS (6)	VII-4
5-β-Estran-3α-ol-17-one	d_3	None, acetyl, TMS (6)	VII-5
Stanozolol	d_3	None, acetyl, [TMS]$_2$, t-BDMS (8)	VII-6
3-Hydroxystanozolol	d_3	[TMS]$_2$, [t-BDMS]$_2$ (4)	VII-7
Promethazine	d_3	None (2)	VII-8
Chlorpromazine	d_3	None (2)	VII-9
Acetaminophen	d_4	None, [acetyl]$_2$, TCA, TFA, PFP, HFB, 4-CB, TMS, [TMS]$_2$, t-BDMS, [t-BDMS]$_2$ (22)	VII-10
Clonidine	d_4	None, acetyl, [acetyl]$_2$, TMS, [TMS]$_2$, [t-BDMS]$_2$ (12)	VII-11
Chloramphenicol	d_5	None, [acetyl]$_2$, TMS, [TMS]$_2$ (8)	VII-12
Melatonin	d_7	None, acetyl, TFA, PFP, HFB, TMS (12)	VII-13
		Total no. of mass spectra: 88	

Figure VII — Others

Appendix One — Figure VII
Mass Spectra of Commonly Abused Drugs and Their Isotopically Labeled Analogs
in Various Derivatization Forms — Others

Figure VII — Others

Figure VII-1. Mass spectra of diphenhydramine and its deuterated analogs (diphenhydramine-d₃).

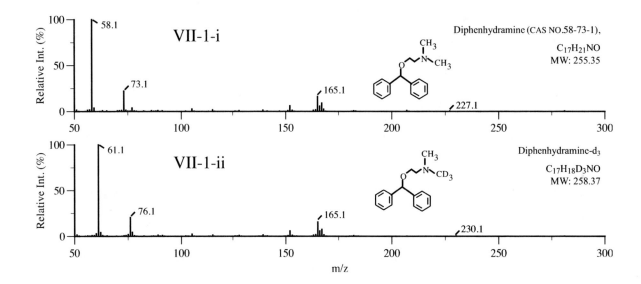

Figure VII-2. Mass spectra of cotinine and its deuterated analogs (cotinine-d₃).

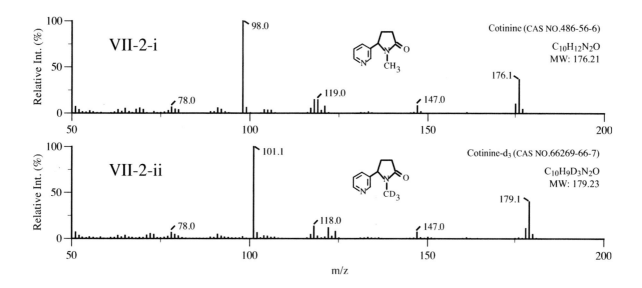

Figure VII — Others

Figure VII-3. Mass spectra of nicotine and its deuterated analogs (nicotine-d₄).

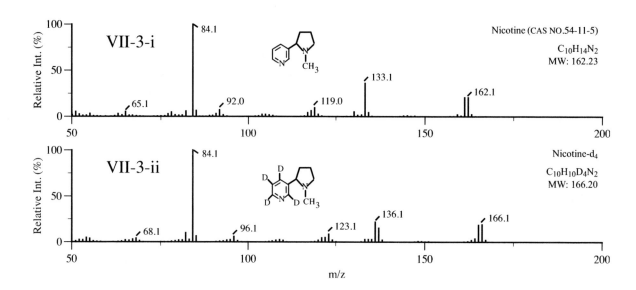

Figure VII-4. Mass spectra of 5-α-estran-3α-ol-17-one and its deuterated analogs (5-α-estran-3α-ol-17-one-d₃): (A) underivatized; (B) acetyl-derivatized; (C) TMS-derivatized.

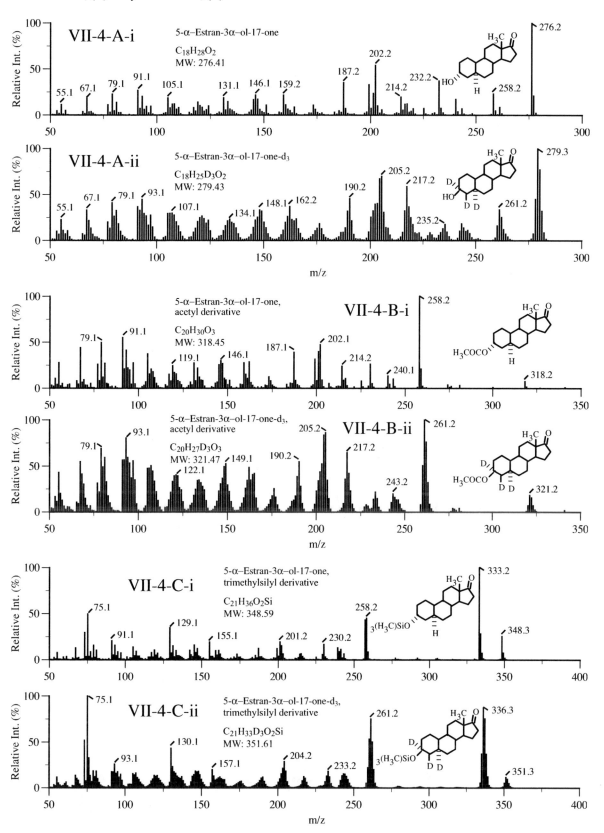

Figure VII — Others

Figure VII-5. Mass spectra of 5-β-estran-3α-ol-17-one and its deuterated analogs (5-β-estran-3α-ol-17-one-d₃): (A) underivatized; (B) acetyl-derivatized; (C) TMS-derivatized.

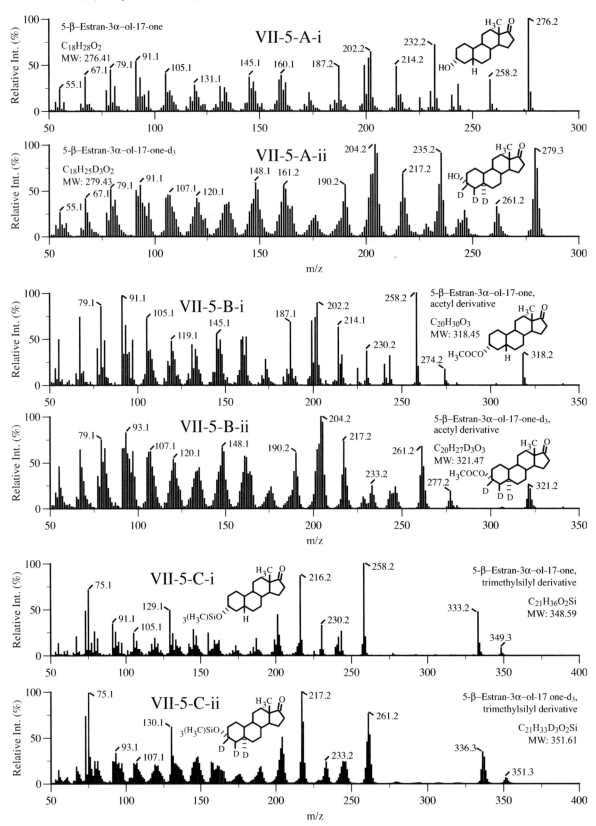

Figure VII-6. Mass spectra of stanozolol and its deuterated analogs (stanozolol-d₃): (A) underivatized; (B) acetyl-derivatized; (C) [TMS]₂-derivatized. (D) *t*-BDMS-derivatized.

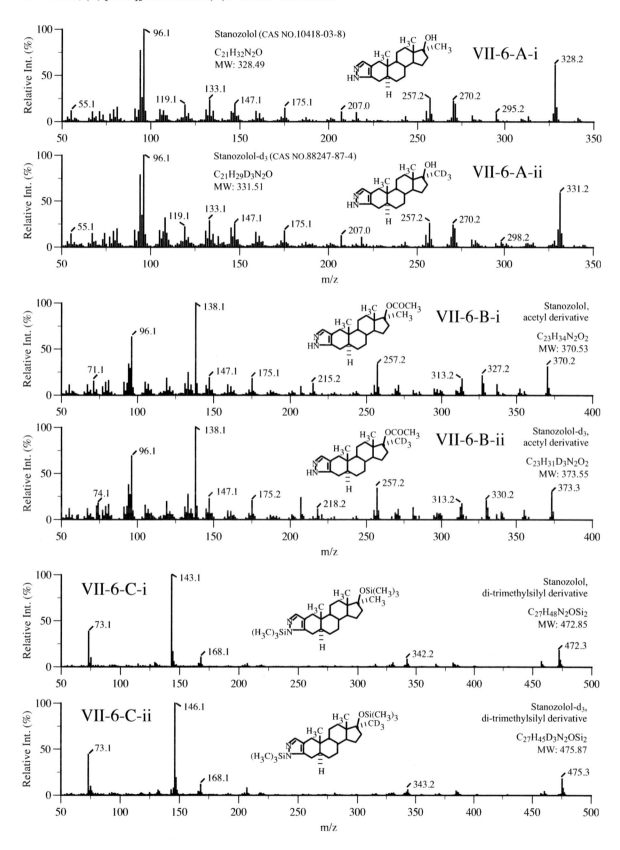

Figure VII — Others

358

Figure VII-6. (Continued)

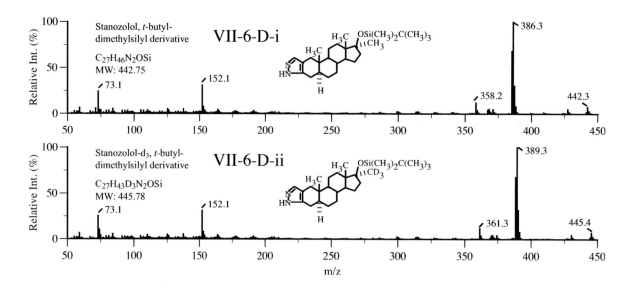

Figure VII-7. Mass spectra of 3-hydroxystanozolol and its deuterated analogs (3-hydroxystanozolol-d₃): (A) [TMS]₂-derivatized. (B) [*t*-BDMS]₂-derivatized.

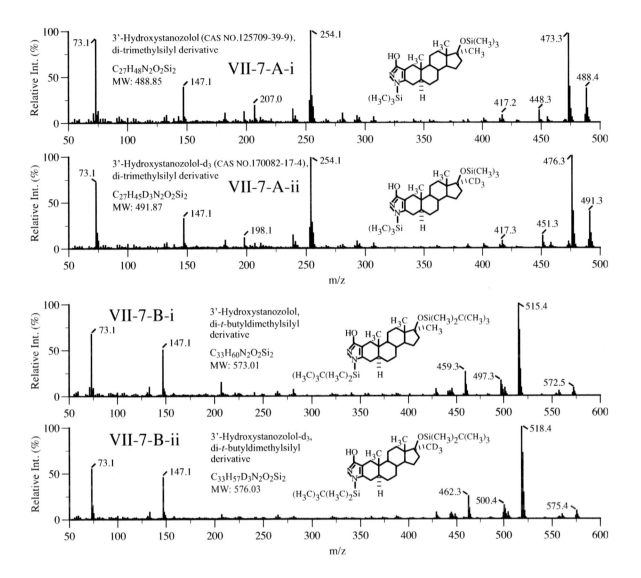

Figure VII — Others

Figure VII-8. Mass spectra of promethazine and its deuterated analogs (promethazine-d$_3$).

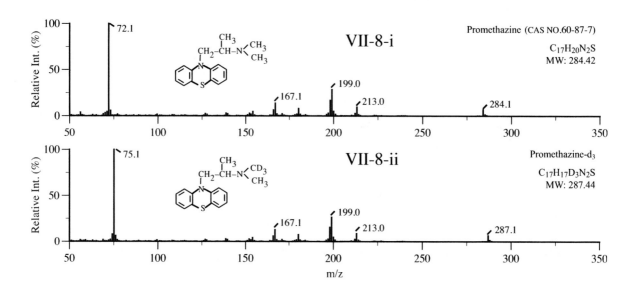

Figure VII-9. Mass spectra of chlorpromazine and its deuterated analogs (chlorpromazine-d₃).

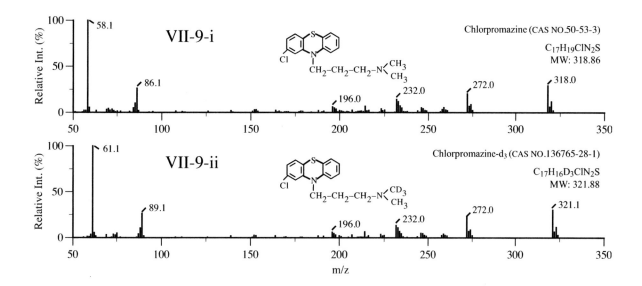

Figure VII — Others

Figure VII-10. Mass spectra of acetaminophen and its deuterated analogs (acetaminophen-d₄): (A) underivatized; (B) [acetyl]₂-derivatized; (C) TCA-derivatized; (D) TFA-derivatized; (E) PFP-derivatized; (F) HFB-derivatized; (G) 4-CB-derivatized; (H) TMS-derivatized; (I) [TMS]₂-derivatized; (J) *t*-BDMS-derivatized; (K) [*t*-BDMS]₂-derivatized.

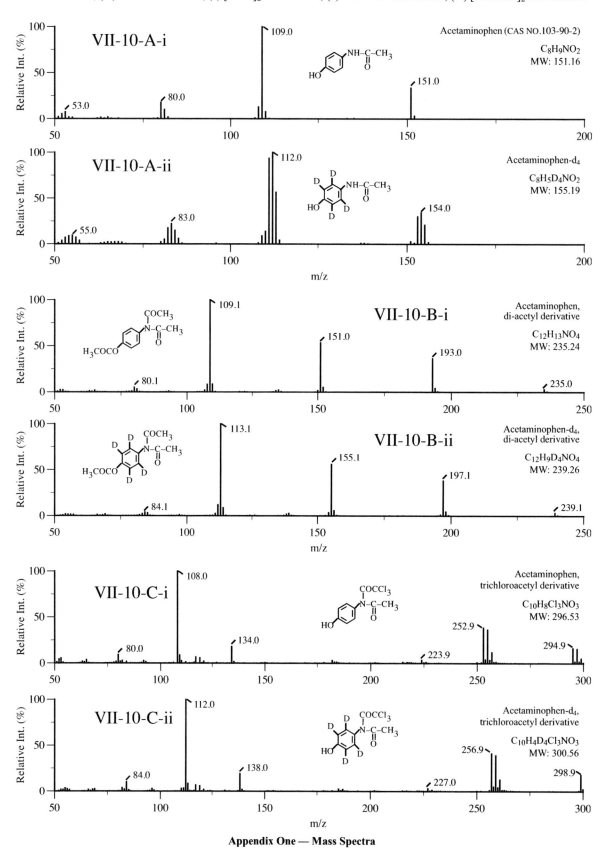

Figure VII-10. (Continued)

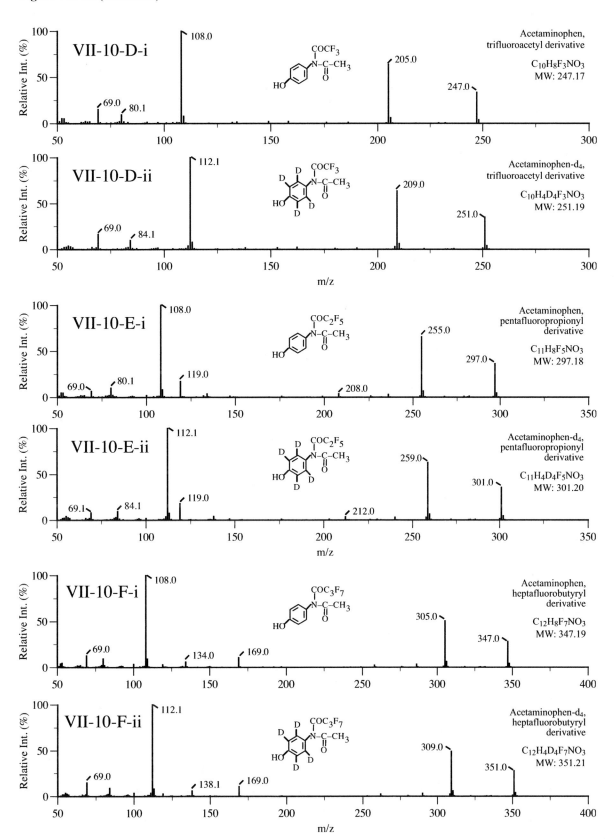

Figure VII — Others

Figure VII-10. (Continued)

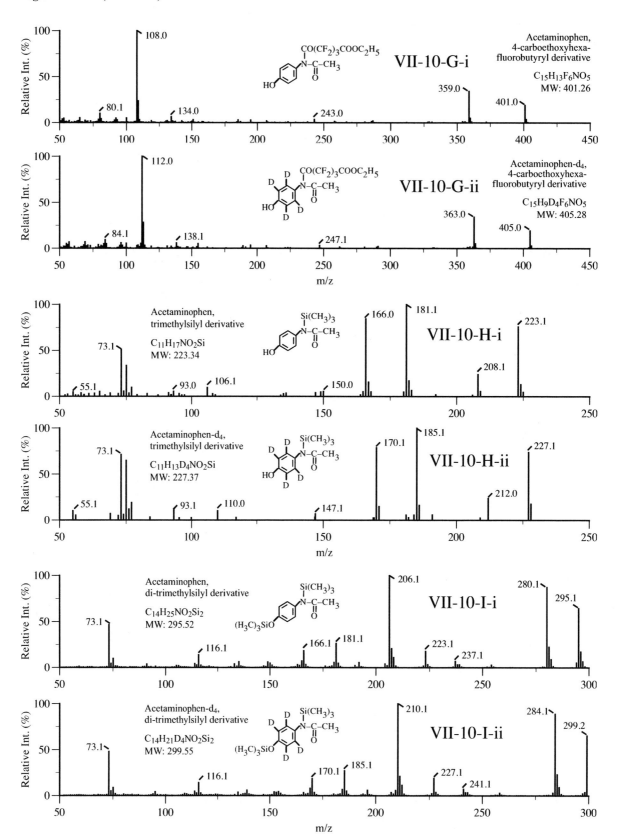

Acetaminophen, 4-carboethoxyhexa-fluorobutyryl derivative
$C_{15}H_{13}F_6NO_5$
MW: 401.26
VII-10-G-i

Acetaminophen-d_4, 4-carboethoxyhexa-fluorobutyryl derivative
$C_{15}H_9D_4F_6NO_5$
MW: 405.28
VII-10-G-ii

Acetaminophen, trimethylsilyl derivative
$C_{11}H_{17}NO_2Si$
MW: 223.34
VII-10-H-i

Acetaminophen-d_4, trimethylsilyl derivative
$C_{11}H_{13}D_4NO_2Si$
MW: 227.37
VII-10-H-ii

Acetaminophen, di-trimethylsilyl derivative
$C_{14}H_{25}NO_2Si_2$
MW: 295.52
VII-10-I-i

Acetaminophen-d_4, di-trimethylsilyl derivative
$C_{14}H_{21}D_4NO_2Si_2$
MW: 299.55
VII-10-I-ii

Figure VII-10. (Continued)

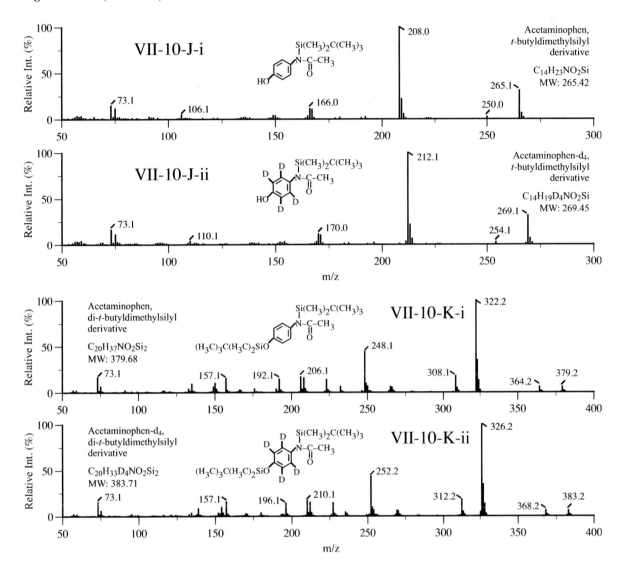

Figure VII — Others

Figure VII-11. Mass spectra of clonidine and its deuterated analogs (clonidine-d₄): (A) underivatized; (B) acetyl-derivatized; (C) [acetyl]₂-derivatized; (D) TMS-derivatized; (E) [TMS]₂-derivatized; (F) [*t*-BDMS]₂-derivatized.

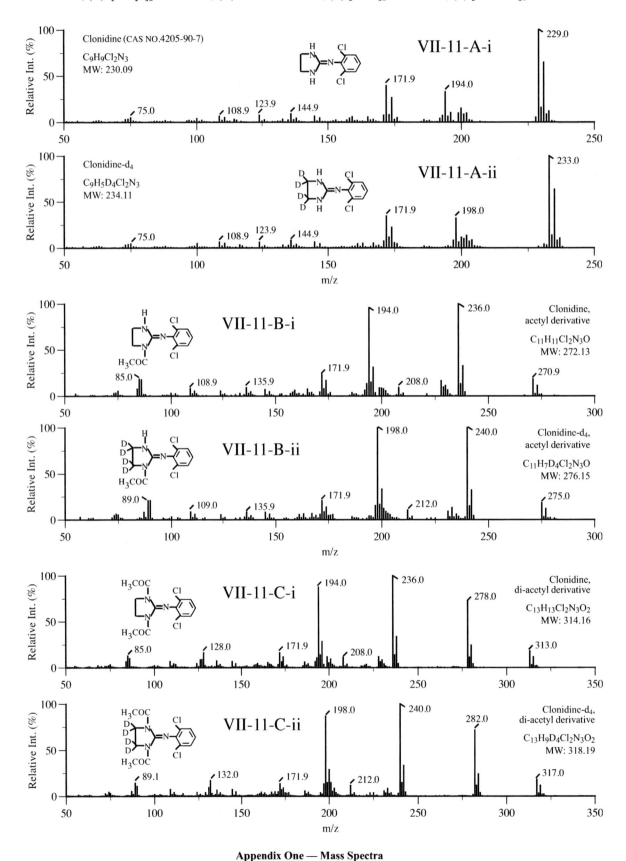

Figure VII-11. (Continued)

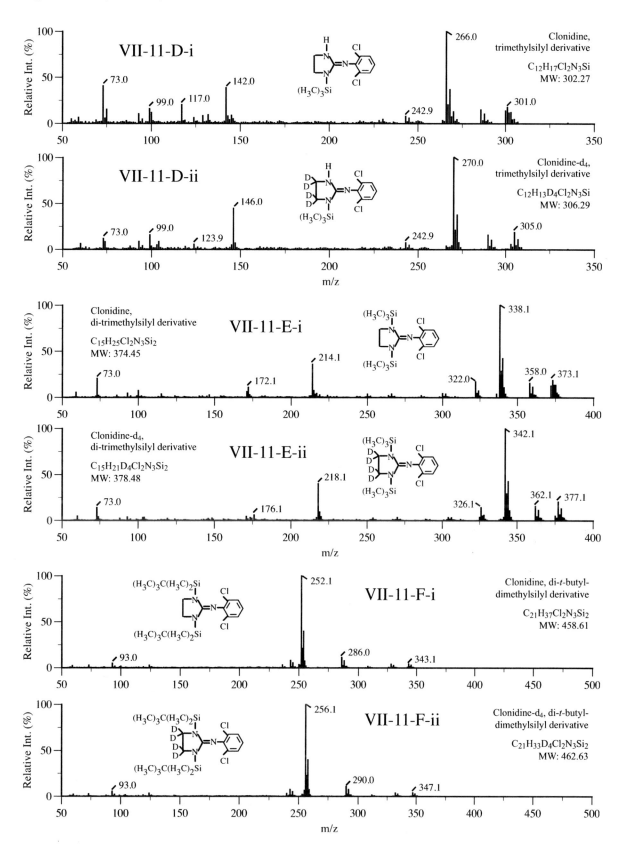

VII-11-D-i
73.0 99.0 117.0 142.0 242.9 266.0 301.0
Clonidine, trimethylsilyl derivative
$C_{12}H_{17}Cl_2N_3Si$
MW: 302.27

VII-11-D-ii
73.0 99.0 123.9 146.0 242.9 270.0 305.0
Clonidine-d_4, trimethylsilyl derivative
$C_{12}H_{13}D_4Cl_2N_3Si$
MW: 306.29

Clonidine, di-trimethylsilyl derivative
$C_{15}H_{25}Cl_2N_3Si_2$
MW: 374.45
VII-11-E-i
73.0 172.1 214.1 322.0 338.1 358.0 373.1

Clonidine-d_4, di-trimethylsilyl derivative
$C_{15}H_{21}D_4Cl_2N_3Si_2$
MW: 378.48
VII-11-E-ii
73.0 176.1 218.1 326.1 342.1 362.1 377.1

VII-11-F-i
93.0 252.1 286.0 343.1
Clonidine, di-t-butyl-dimethylsilyl derivative
$C_{21}H_{37}Cl_2N_3Si_2$
MW: 458.61

VII-11-F-ii
93.0 256.1 290.0 347.1
Clonidine-d_4, di-t-butyl-dimethylsilyl derivative
$C_{21}H_{33}D_4Cl_2N_3Si_2$
MW: 462.63

Figure VII — Others

Figure VII-12. Mass spectra of chloramphenicol and its deuterated analogs (chloramphenicol-d₅): (A) underivatized; (B) [acetyl]₂-derivatized; (C) TMS-derivatized; (D) [TMS]₂-derivatized.

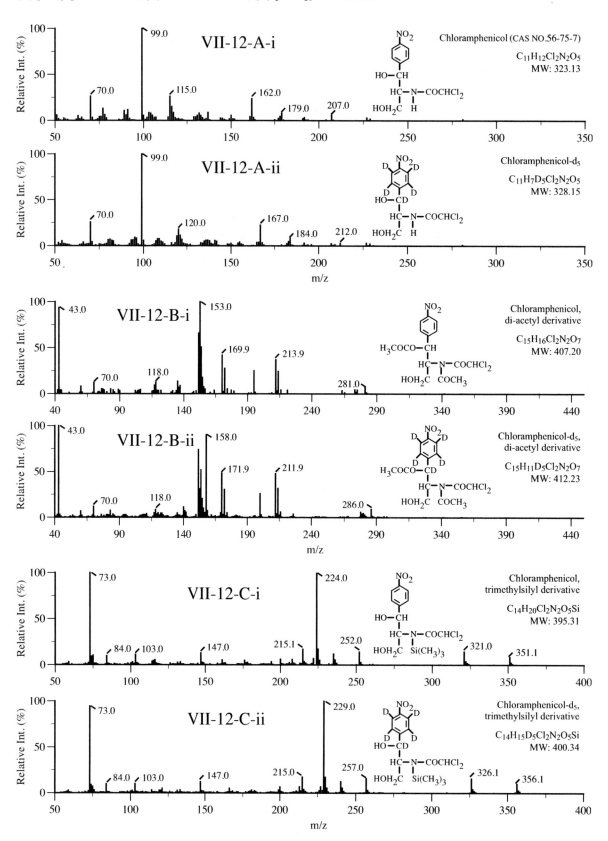

Figure VII-12. (Continued)

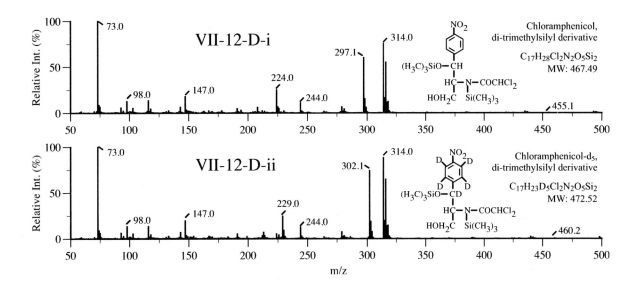

Figure VII — Others

Figure VII-13. Mass spectra of melatonin and its deuterated analogs (melatonin-d₇): (A) underivatized; (B) acetyl-derivatized; (C) TFA-derivatized; (D) PFP-derivatized; (E) HFB-derivatized; (F) TMS-derivatized.

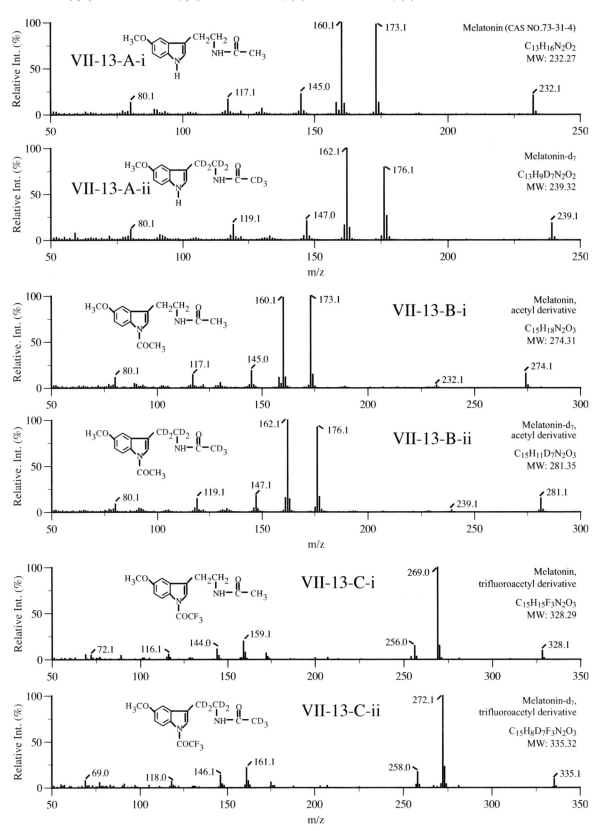

Figure VII-13. (Continued)

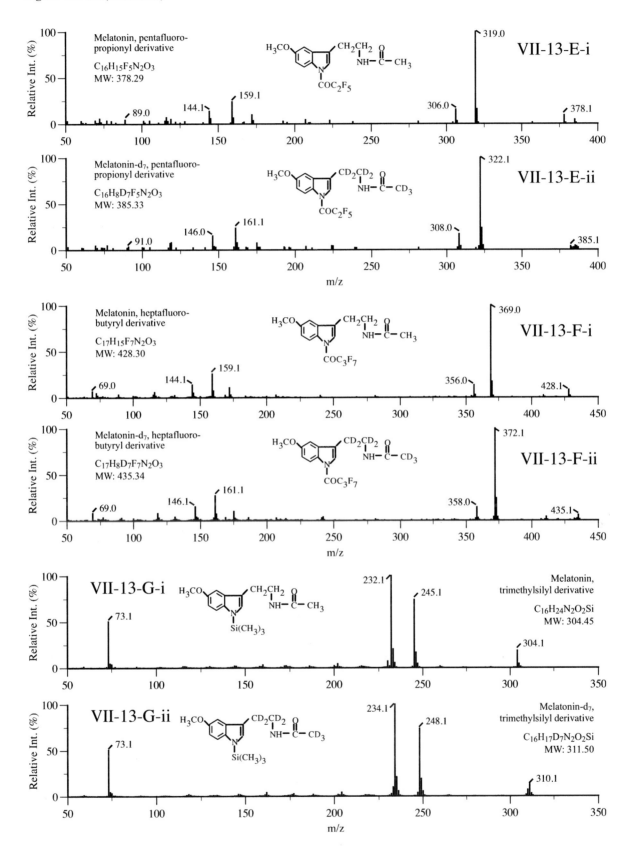

VII-13-E-i
Melatonin, pentafluoro-propionyl derivative
$C_{16}H_{15}F_5N_2O_3$
MW: 378.29

VII-13-E-ii
Melatonin-d7, pentafluoro-propionyl derivative
$C_{16}H_8D_7F_5N_2O_3$
MW: 385.33

VII-13-F-i
Melatonin, heptafluoro-butyryl derivative
$C_{17}H_{15}F_7N_2O_3$
MW: 428.30

VII-13-F-ii
Melatonin-d7, heptafluoro-butyryl derivative
$C_{17}H_8D_7F_7N_2O_3$
MW: 435.34

VII-13-G-i
Melatonin, trimethylsilyl derivative
$C_{16}H_{24}N_2O_2Si$
MW: 304.45

VII-13-G-ii
Melatonin-d7, trimethylsilyl derivative
$C_{16}H_{17}D_7N_2O_2Si$
MW: 311.50

Figure VII — Others

PART THREE

CROSS-CONTRIBUTIONS OF ION INTENSITY BETWEEN ANALYTES AND THEIR ISOTOPICALLY LABELED ANALOGS IN VARIOUS DERIVATIZATION FORMS

Appendix Two

Cross-Contributions between Ions Designating Drugs and Their Isotopically Labeled Analogs in Various Derivatization Forms

Table of Contents for Appendix Two

Table I — Stimulants

Summary of Drugs, Isotopic Analogs, and Chemical Derivatization Groups Included in Table I (Stimulants)

Compound	Isotopic analog	Chemical derivatization group	Table #
Amphetamine	d_5, d_5 (ring), d_6, d_8, d_{10}, d_{11}	None, acetyl, TCA, TFA, PFP, HFB, 4-CB, PFB, propylformyl, *l*-TPC, *d*-TPC, *l*-MTPA, *d*-MTPA, TMS, *t*-BDMS, TFA/*t*-BDMS, PFP/*t*-BDMS, HFB/*t*-BDMS	I-1
Methamphetamine	d_5, d_8, d_9, d_{11}, d_{14}	None, acetyl, TCA, TFA, PFP, HFB, 4-CB, PFB, propylformyl, *l*-TPC, *d*-TPC, *l*-MTPA, *d*-MTPA, TMS, *t*-BDMS	I-2
Ephedrine	d_3	None, acetyl, TCA, [TFA]$_2$, [PFP]$_2$, [HFB]$_2$, 4-CB, PFB, propylformyl, *d*-TPC, *d*-MTPA, [TMS]$_2$	I-3
Phenylpropanolamine	d_3	None, acetyl, TCA, [TFA]$_2$, [PFP]$_2$, [HFB]$_2$, 4-CB, PFB, *l*-TPC, *d*-TPC, *l*-MTPA, *d*-MTPA, [TMS]$_2$, *t*-BDMS, [*t*-BDMS]$_2$, TFA/[*t*-BDMS]$_2$, PFP/[*t*-BDMS]$_2$, HFB/[*t*-BDMS]$_2$	I-4
MDA	d_5	None, acetyl, TCA, TFA, PFP, HFB, 4-CB, PFB, propylformyl, *l*-TPC, *d*-TPC, *l*-MTPA, *d*-MTPA, TMS, TFA/*t*-BDMS, PFP/*t*-BDMS, HFB/*t*-BDMS	I-5
MDMA	d_5	None, acetyl, TCA, TFA, PFP, HFB, 4-CB, PFB, propylformyl, *l*-TPC, *d*-TPC, *l*-MTPA, *d*-MTPA, TMS	I-6
MDEA	d_5, d_6	None, acetyl, TCA, TFA, PFP, HFB, 4-CB, PFB, propylformyl, *l*-TPC, *d*-TPC, *l*-MTPA, *d*-MTPA, TMS	I-7
MBDB	d_5	None, acetyl, TCA, TFA, PFP, HFB, 4-CB, PFB, propylformyl, *l*-TPC, *d*-TPC, *l*-MTPA *d*-MTPA, TMS	I-8
Selegiline	d_8	None	I-9
N-Desmethylselegiline	d_{11}	None, acetyl, TCA, TFA, PFP, HFB, 4-CB, TMS	I-10
Fenfluramine	d_{10}	None, acetyl, TCA, TFA, PFP, HFB, 4-CB	I-11
Norcocaine	d_3	None, TFA, PFP, HFB, TMS	I-12
Cocaine	d_3	None	I-13
Cocaethylene	d_3, d_8	None	I-14
Ecgonine methyl ester	d_3	None, TFA, PFP, HFB, TMS, *t*-BDMS	I-15
Benzoylecgonine	d_3, d_8	Methyl, ethyl, propyl, butyl, PFPoxy, HFPoxy, TMS, *t*-BDMS	I-16
Ecgonine	d_3	[TMS]$_2$, [*t*-BDMS]$_2$, HFPoxy/TFA, PFPoxy/PFP, HFPoxy/HFB	I-17
Anhydroecgonine methyl ester	d_3	None	I-18
Caffeine	$^{13}C_3$	None	I-19
Methylphenidate	d_3	None, TFA, PFP, HFB, 4-CB, TMS	I-20
Ritalinic acid	d_5	4-CB, [TMS]$_2$, *t*-BDMS	I-21

Table I — Stimulants

Appendix Two — Table I
Cross-Contributions Between ions Designating the Drugs and Their Isotopically Labeled Analogs in Various Derivatization Forms — Stimulants

Table I — Stimulants

Table I-1a. Relative intensity and cross-contribution data[a] of ions[b] with potential for designating the analyte and the adapted internal standard — Amphetamine/amphetamine-d$_5$

CD Group[c]	Amphetamine			Amphetamine-d$_5$		
	Ion (m/z)[d]	Rel. int.	Analog's cont.	Ion (m/z)[d]	Rel. int.	Analog's cont.
None	<u>44</u>	100	3.76	<u>48</u>	100	0.00
Acetyl	<u>44</u>	100	—	<u>48</u>	100	—
	86	62.4	0.60	90	60.9	0.21
TCA	<u>118</u>	100	4.09	<u>123</u>	74.9	0.29
	190	66.2	1.02	194	94.8	1.96
TFA	<u>140</u>	100	0.23	<u>144</u>	100	0.85
PFP	118	73.0	3.48	123	44.3	0.19
	<u>190</u>	100	0.10	<u>194</u>	100	0.14
HFB	118	57.3	3.77	123	33.9	0.38
	223	3.35	0.69	227	5.11	0.50
	<u>240</u>	100	0.09	<u>244</u>	100	0.01
4-CB	<u>118</u>	100	3.14	<u>123</u>	100	0.14
	248	35.0	0.29	251	33.4	0.94
	266	41.8	0.19	270	68.2	0.08
	294	43.6	0.19	298	77.3	0.06
PFB	118	26.9	3.92	123	14.4	0.10
	<u>238</u>	48.8	0.27	<u>242</u>	44.0	0.02
Propylformyl	<u>130</u>	100	0.03	<u>134</u>	100	0.96
	162	3.40	0.27	167	3.31	0.27
l-TPC (l-l)	118	16.4	4.28	123	10.2	0.74
	<u>237</u>	39.4	0.07	<u>241</u>	40.2	0.04
l-TPC (d-d)	118	17.3	4.10	123	10.2	0.68
	<u>237</u>	40.3	0.07	<u>241</u>	42.6	0.04
l-MTPA (l-l)	<u>91</u>	65.8	—	<u>93</u>	33.2	—
	162	14.7	2.51	167	16.7	2.45
	260	55.7	1.70	264	55.5	0.11
l-MTPA (d-d)	<u>91</u>	71.9	—	<u>93</u>	37.1	—
	162	17.3	2.75	167	19.5	4.90
	260	57.9	1.92	264	56.2	0.11
TMS	<u>116</u>	100	0.54	<u>120</u>	100	0.73
	192	7.80	1.85	197	7.77	0.20
t-BDMS	100	16.0	2.02	104	15.4	1.30
	<u>158</u>	100	0.20	<u>162</u>	100	0.37
	192	20.0	0.27	197	21.0	0.00
	234	3.66	0.95	239	2.68	0.00
TFA/	91	37.1	2.59	93	35.8	0.91
t-BDMS	119	19.8	4.60	124	32.9	0.46
	<u>254</u>	100	0.08	<u>258</u>	100	0.12
	288	24.8	0.17	293	25.2	0.16
	289	4.76	0.63	294	5.18	0.24
PFP/	91	48.0	2.39	93	28.0	0.65
t-BDMS	<u>304</u>	100	0.12	<u>308</u>	100	0.05
	338	14.4	0.15	343	15.9	0.00
	339	3.17	0.56	344	3.48	0.03
HFB/	91	47.8	2.32	93	24.3	0.79
t-BDMS	<u>354</u>	100	0.14	<u>358</u>	100	0.05
	388	13.1	0.14	393	15.2	0.00
	389	3.09	0.43	394	3.48	0.00

[a] Relative intensities are based on full-scan data (see corresponding mass spectra in Appendix One), while cross-contributions (CC) are derived from selected ion monitoring (SIM) data.
[b] Ion-pairs with 5% (or higher) CC by the analog are not listed.

Table I — Stimulants

Table I-1a. (Continued)

[c] *See* Table 2 in Chapter 2 for the abbreviations for the derivatization groups.

[d] The SIM intensities of the underlined ion-pair observed in two separate runs (for the analyte and its analog) were used to normalize the intensties of all ions for the calculations of CC data listed in the table.

Table I-1b. Relative intensity and cross-contribution data[a] of ions[b] with potential for designating the analyte and the adapted internal standard — Amphetamine/amphetamine-d$_5$ (ring)

CD Group[c]	Amphetamine			Amphetamine-d$_5$		
	Ion (m/z)[d]	Rel. int.	Analog's cont.	Ion (m/z)[d]	Rel. int.	Analog's cont.
None	<u>91</u>	16.7	2.37	<u>96</u>	13.6	0.00
Acetyl	91	17.6	0.77	96	31.2	0.00
	117	11.4	1.46	122	13.6	0.00
	<u>118</u>	26.4	3.75	<u>123</u>	47.6	0.00
TCA	91	63.8	1.44	96	53.2	0.44
	<u>118</u>	100	3.02	<u>123</u>	100	0.40
TFA	91	36.3	0.49	96	37.2	1.21
	117	16.1	1.89	122	13.5	0.54
	<u>118</u>	83.9	1.60	<u>123</u>	88.9	0.06
PFP	91	29.9	0.54	96	29.9	0.64
	117	12.4	2.12	122	10.6	1.08
	<u>118</u>	68.1	1.75	<u>123</u>	73.6	0.08
HFB	91	27.7	0.49	96	25.8	0.13
	117	11.0	2.17	122	8.28	1.19
	<u>118</u>	60.7	1.76	<u>123</u>	60.7	0.20
4-CB	91	49.9	0.54	96	43.3	0.45
	<u>118</u>	100	1.36	<u>123</u>	100	0.17
PFB	91	11.3	1.02	96	10.4	0.16
	<u>118</u>	26.9	2.42	<u>123</u>	27.9	0.05
Propylformyl	<u>91</u>	63.3	0.26	<u>96</u>	56.7	0.22
l-TPC (*l-l*)	<u>91</u>	21.8	—	<u>96</u>	29.6	—
	117	3.87	3.26	122	3.35	1.99
	118	16.4	3.40	123	15.1	0.45
l-TPC (*d-d*)	<u>91</u>	22.2	—	<u>96</u>	27.0	—
	117	4.09	3.22	122	3.20	3.48
	118	17.3	3.26	123	14.8	0.44
l-MTPA (*l-l*)	<u>91</u>	65.9	—	<u>96</u>	53.7	—
	118	10.8	3.33	123	13.8	4.53
	162	14.7	0.72	167	12.9	4.82
l-MTPA (*d-d*)	<u>91</u>	71.9	—	<u>96</u>	55.2	—
TMS	<u>91</u>	9.28	0.92	<u>96</u>	9.61	1.54
	192	8.43	0.00	197	8.42	0.65
t-BDMS	91	7.29	1.03	96	8.35	2.93
	<u>192</u>	20.0	0.35	<u>197</u>	19.4	0.13
	234	3.66	0.70	239	3.50	0.25
TFA/	<u>91</u>	37.1	0.41	<u>96</u>	66.2	0.35
t-BDMS	288	24.8	0.11	293	21.7	0.02
	289	4.76	0.28	294	4.46	0.19
PFP/	<u>91</u>	48.0	0.38	<u>96</u>	43.9	0.85
t-BDMS	338	14.4	0.06	343	14.3	0.01
	339	3.17	0.62	344	3.13	0.02
HFB/	<u>91</u>	47.8	0.54	<u>96</u>	43.6	0.09
t-BDMS	388	13.1	0.04	393	12.3	0.00
	389	3.09	0.65	394	3.03	0.00

[a–d] See the corresponding footnotes in Table I-1a.

Appendix Two — Ion Intensity Cross-Contribution Data

Table I-1c. Relative intensity and cross-contribution data[a] of ions[b] with potential for designating the analyte and the adapted internal standard — Amphetamine/amphetamine-d_6

CD Group[c]	Amphetamine			Amphetamine-d_6		
	Ion (m/z)[d]	Rel. int.	Analog's cont.	Ion (m/z)[d]	Rel. int.	Analog's cont.
None	<u>44</u>	100	3.50	<u>48</u>	100	2.28
Acetyl	<u>44</u>	100	0.96	<u>48</u>	100	0.69
	86	62.4	0.68	90	58.5	1.00
	118	45.4	2.08	123	43.9	0.13
TCA	91	63.8	4.47	93	59.2	1.45
	<u>118</u>	100	1.65	<u>123</u>	100	0.30
	190	66.2	0.55	194	72.4	3.15
TFA	<u>140</u>	100	0.26	<u>144</u>	100	0.87
PFP	91	36.0	4.69	93	30.4	2.47
	118	73.0	2.42	123	69.8	0.12
	<u>190</u>	100	0.09	<u>194</u>	100	0.19
HFB	91	27.0	4.27	93	33.7	3.28
	118	57.3	2.46	123	76.4	0.23
	223	3.35	0.57	227	5.52	0.64
	<u>240</u>	100	0.06	<u>244</u>	100	0.01
4-CB	91	42.1	3.01	93	39.6	3.74
	<u>118</u>	100	1.21	<u>123</u>	100	0.20
	248	35.0	0.12	251	23.2	1.48
	266	41.8	0.12	270	48.1	0.17
	294	43.6	0.11	298	56.0	0.13
PFB	118	26.9	2.75	123	24.4	0.13
	<u>238</u>	48.8	0.17	<u>242</u>	47.9	0.12
Propylformyl	91	41.7	3.78	93	37.1	1.20
	<u>130</u>	100	0.00	<u>134</u>	100	1.00
	162	3.40	0.10	168	3.50	0.02
l-TPC (l-l)	118	16.4	2.78	123	16.2	0.45
	<u>237</u>	39.4	0.13	<u>241</u>	43.0	0.02
l-TPC (d-d)	118	17.3	2.69	123	17.1	0.39
	<u>237</u>	40.3	0.14	<u>241</u>	44.0	0.01
l-MTPA (l-l)	<u>91</u>	65.8	—	<u>93</u>	57.0	—
	162	14.7	0.21	168	17.7	0.84
	234	15.3	2.12	236	9.78	1.54
	260	55.7	0.23	264	53.9	0.18
l-MTPA (d-d)	<u>91</u>	71.9	—	<u>93</u>	56.2	—
	162	17.3	0.32	168	19.1	1.91
	234	16.1	3.12	236	10.3	1.59
	260	57.9	0.30	264	58.6	0.18
TMS	<u>116</u>	100	0.30	<u>120</u>	100	0.47
	192	7.80	1.74	198	7.82	1.04
t-BDMS	100	16.0	0.97	104	15.4	1.07
	<u>158</u>	100	0.19	<u>162</u>	100	0.39
	192	20.0	0.21	198	22.5	0.02
	234	3.66	0.87	240	2.79	0.00
TFA/ t-BDMS	91	37.1	1.51	93	49.6	0.58
	119	19.8	4.28	125	28.8	0.28
	<u>254</u>	100	0.11	<u>258</u>	100	0.10
	288	24.8	0.18	294	26.9	0.03
	289	4.76	0.16	295	5.40	0.12
PFP/ t-BDMS	91	48.0	1.27	93	46.1	0.44
	<u>304</u>	100	0.10	<u>308</u>	100	0.05
	338	14.4	0.07	344	16.7	0.00
	339	3.17	0.20	345	3.63	0.07

Table I — Stimulants

Table I-1c. (Continued)

CD Group[c]	Amphetamine			Amphetamine-d$_6$		
	Ion (m/z)[d]	Rel. int.	Analog's cont.	Ion (m/z)[d]	Rel. int.	Analog's cont.
HFB/	91	47.8	1.22	93	45.0	0.53
t-BDMS	354	100	0.18	358	100	0.06
	388	13.1	0.09	394	14.6	0.00
	389	3.09	0.00	395	3.43	0.00

[a–d] See the corresponding footnotes in Table I-1a.

Table I-1d. Relative intensity and cross-contribution data[a] of ions[b] with potential for designating the analyte and the adapted internal standard — Amphetamine/amphetamine-d$_8$

CD Group[c]	Amphetamine			Amphetamine-d$_8$		
	Ion (m/z)[d]	Rel. int.	Analog's cont.	Ion (m/z)[d]	Rel. int.	Analog's cont.
None	44	100	—	47	100	—
	91	15.6	4.41	96	11.8	0.16
Acetyl	44	100	2.10	47	100	0.01
	86	62.4	1.97	89	62.2	1.71
	91	31.6	0.89	96	24.3	0.02
	118	45.4	0.41	126	40.7	0.03
TCA	91	63.8	0.84	96	38.9	0.23
	118	100	—	126	100	—
	188	70.0	4.48	191	73.1	3.54
	190	66.2	4.05	193	69.6	1.59
TFA	91	43.4	0.52	96	39.6	3.40
	118	89.3	0.25	126	100	0.15
	140	100	4.05	143	98.9	0.14
PFP	91	36.0	0.40	96	29.3	0.98
	118	73.0	0.28	126	78.2	0.33
	173	1.07	3.54	176	6.36	0.93
	190	100	1.99	193	100	0.08
HFB	91	27.0	0.43	96	27.4	0.24
	118	57.3	0.20	126	77.0	0.08
	192	3.81	2.40	195	4.12	1.43
	240	100	1.97	243	100	0.03
4-CB	91	42.1	0.32	96	26.8	0.57
	118	100	0.16	126	100	0.05
	220	12.4	2.37	223	7.25	2.44
	248	35.0	1.96	250	19.4	2.19
	266	41.8	1.75	269	41.1	0.55
	294	43.6	1.78	297	46.5	0.15
PFB	91	11.3	0.32	96	10.8	0.05
	118	26.9	1.51	126	25.5	0.07
	238	48.8	1.89	241	47.7	0.06
	239	5.04	4.58	242	5.27	1.08
Propylformyl	118	6.96	1.79	126	6.21	0.32
	130	100	1.96	133	100	0.51
	162	3.40	0.28	170	2.35	0.80
l-TPC (l-l)	237	39.4	1.84	240	43.0	0.10
l-TPC (d-d)	237	40.3	1.83	240	45.2	0.10
l-MTPA (l-l)	91	65.8	—	97	36.2	—
	118	10.8	2.87	126	14.5	0.26
	260	55.7	4.43	263	58.0	0.06

Table I-1d. (Continued)

CD Group[c]	Amphetamine			Amphetamine-d[8]		
	Ion (m/z)[d]	Rel. int.	Analog's cont.	Ion (m/z)[d]	Rel. int.	Analog's cont.
l-MTPA (d-d)	91	71.9	—	97	39.9	—
	118	12.1	3.00	126	15.3	0.21
	260	57.9	4.88	263	57.9	0.06
TMS	91	11.5	1.71	96	9.97	2.37
	116	100	2.18	119	100	0.74
	192	7.80	2.05	200	6.64	2.58
t-BDMS	100	16.0	4.47	103	17.6	3.33
	158	100	1.85	161	100	0.52
	159	15.5	2.77	162	14.8	2.79
	192	20.0	0.13	200	19.8	0.00
	234	3.66	0.80	242	2.67	0.80
TFA/	91	37.1	0.29	97	26.8	0.58
t-BDMS	119	19.8	1.59	127	23.4	0.94
	254	100	1.83	257	100	0.54
	255	15.9	2.03	258	16.7	0.59
	288	24.8	0.08	296	25.2	0.01
	289	4.76	0.18	297	5.04	0.00
PFP/	91	48.0	0.30	97	31.8	0.48
t-BDMS	119	21.9	3.76	127	22.4	0.71
	304	100	1.92	307	100	0.54
	305	18.1	2.17	308	18.1	0.29
	338	14.4	0.11	346	14.6	0.00
	339	3.17	0.20	347	3.21	0.00
HFB/	91	47.8	0.45	97	29.0	0.40
t-BDMS	119	21.7	3.73	127	21.0	0.06
	354	100	1.91	357	100	0.59
	355	19.7	2.13	358	19.8	0.29
	388	13.1	0.06	396	13.1	0.00
	389	3.09	0.00	397	3.01	4.15

[a–d] See the corresponding footnotes in Table I-1a.

Table I-1e. Relative intensity and cross-contribution data[a] of ions[b] with potential for designating the analyte and the adapted internal standard — Amphetamine/amphetamine-d[10]

CD Group[c]	Amphetamine			Amphetamine-d[10]		
	Ion (m/z)[d]	Rel. int.	Analog's cont.	Ion (m/z)[d]	Rel. int.	Analog's cont.
None	44	100	1.60	48	100	0.38
	91	15.6	0.14	97	11.5	0.65
Acetyl	44	100	0.81	48	100	0.06
	86	62.4	0.54	90	61.0	0.73
	118	45.4	0.21	127	27.1	0.17
TCA	91	63.8	1.13	97	63.1	3.89
	92	8.81	2.53	98	37.4	3.43
	118	100	0.40	128	89.5	1.35
TFA	91	43.4	0.30	97	37.7	3.14
	118	89.3	0.17	128	51.8	1.02
	140	100	0.44	144	100	0.85
PFP	91	36.0	0.21	97	32.7	0.23
	118	73.0	0.20	128	47.9	1.03
	190	100	0.08	194	100	0.22
HFB	91	27.0	0.27	97	31.3	0.17
	118	57.3	0.09	128	47.3	0.72
	223	3.35	0.49	227	7.98	0.43

Table I — Stimulants

Table I-1e. (Continued)

CD Group[c]	Amphetamine			Amphetamine-d$_{10}$		
	Ion (m/z)[d]	Rel. int.	Analog's cont.	Ion (m/z)[d]	Rel. int.	Analog's cont.
4-CB	91	42.1	0.45	97	52.8	0.16
	<u>118</u>	100	0.19	<u>128</u>	100	0.42
	248	35.0	0.27	251	33.3	1.01
	266	41.8	0.18	270	69.2	0.06
	294	43.6	0.15	298	82.9	0.04
PFB	91	11.3	0.32	97	11.3	0.02
	118	26.9	1.42	128	15.0	0.11
	<u>238</u>	48.8	0.15	<u>242</u>	48.8	0.04
	239	5.04	2.01	243	4.98	0.17
Propylformyl	91	41.7	0.54	97	34.5	0.29
	118	6.96	2.55	126	2.75	0.82
	<u>130</u>	100	0.93	<u>134</u>	100	1.08
	162	3.40	0.89	172	2.47	0.00
l-TPC (l-l)	<u>237</u>	39.4	0.05	<u>241</u>	44.9	0.05
l-TPC (d-d)	91	22.2	4.64	97	16.2	4.93
	<u>237</u>	40.3	0.06	<u>241</u>	46.4	0.02
l-MTPA (l-l)	<u>91</u>	65.8	—	<u>97</u>	38.1	—
	118	10.8	2.31	128	10.2	2.17
	162	14.7	0.30	172	13.8	0.14
	234	15.3	4.11	236	5.75	1.48
	260	55.7	0.17	264	56.0	0.10
	261	7.73	2.29	265	7.49	0.14
l-MTPA (d-d)	<u>91</u>	71.9	—	<u>97</u>	41.6	—
	118	12.1	2.83	128	11.1	2.00
	162	17.3	0.48	172	16.3	0.16
	260	57.9	0.33	264	59.3	0.10
	261	7.93	2.64	265	8.01	0.16
TMS	91	11.5	2.50	97	10.3	1.05
	<u>116</u>	100	0.30	<u>120</u>	100	0.51
	192	7.80	2.11	202	7.20	1.04
t-BDMS	91	7.29	3.40	97	6.19	1.35
	100	16.0	1.72	104	18.2	0.95
	<u>158</u>	100	0.03	<u>162</u>	100	0.40
	159	15.5	1.40	163	14.6	0.63
	192	20.0	0.40	202	20.9	0.00
	234	3.66	0.46	244	2.48	0.00
TFA/	91	37.1	0.28	97	31.3	0.55
t-BDMS	119	19.8	0.81	129	25.6	0.34
	<u>254</u>	100	0.06	<u>258</u>	100	0.10
	255	15.9	1.06	259	16.8	0.31
	288	24.8	0.09	298	25.9	0.00
	289	4.76	0.18	299	5.07	0.00
PFP/	91	48.0	0.28	98	36.4	0.25
t-BDMS	119	21.9	2.81	129	26.9	0.36
	<u>304</u>	100	0.05	<u>308</u>	100	0.05
	305	18.1	1.13	309	18.5	0.08
	338	14.4	0.06	348	15.4	0.02
	349	3.17	0.25	349	3.37	0.00
HFB/	91	47.8	0.50	98	39.0	0.29
t-BDMS	119	21.7	4.12	129	28.8	0.94
	<u>354</u>	100	0.07	<u>358</u>	100	0.05
	355	19.7	1.86	359	18.4	0.09
	388	13.1	0.00	398	12.9	0.38
	389	3.09	0.00	399	3.01	1.15

[a–d] See the corresponding footnotes in Table I-1a.

Table I-1f. Relative intensity and cross-contribution data[a] of ions[b] with potential for designating the analyte and the adapted internal standard — Amphetamine/amphetamine-d$_{11}$

CD Group[c]	Amphetamine			Amphetamine-d$_{11}$		
	Ion (m/z)[d]	Rel. int.	Analog's cont.	Ion (m/z)[d]	Rel. int.	Analog's cont.
None	<u>44</u>	100	4.96	<u>48</u>	100	2.68
	91	15.6	3.80	98	12.0	3.88
	120	3.54	3.98	128	2.99	2.31
Acetyl	<u>44</u>	100	0.19	<u>48</u>	100	0.43
	86	62.4	0.06	90	61.1	1.26
	91	31.6	3.51	98	34.4	0.16
	118	45.4	0.18	128	42.9	0.28
TCA	91	63.8	0.82	98	66.8	2.18
	<u>118</u>	100	0.27	<u>128</u>	100	1.38
	190	66.2	2.45	194	68.3	3.41
TFA	91	43.4	0.30	98	41.6	0.35
	118	89.3	0.17	128	82.3	0.64
	<u>140</u>	100	0.38	<u>144</u>	100	0.86
PFP	91	36.0	0.15	98	36.1	0.13
	118	73.0	0.20	128	72.7	0.63
	<u>190</u>	100	0.09	<u>194</u>	100	0.00
HFB	91	27.0	0.19	98	38.5	0.08
	118	57.3	0.09	128	74.8	0.45
	223	3.35	0.53	227	7.08	0.47
	<u>240</u>	100	0.06	<u>244</u>	100	0.01
4-CB	91	42.1	0.28	98	46.3	0.10
	<u>118</u>	100	0.12	<u>128</u>	100	0.46
	248	35.0	0.13	251	21.9	1.53
	266	41.8	0.12	270	46.3	0.15
	294	43.6	0.11	298	56.6	0.12
PFB	118	26.9	1.42	128	24.4	0.10
	<u>238</u>	48.8	0.18	<u>242</u>	29.9	0.09
Propylformyl	91	41.7	0.32	98	46.8	0.01
	<u>130</u>	100	—	<u>134</u>	100	—
	118	6.96	0.47	126	3.84	0.72
	162	3.40	0.00	173	2.19	0.68
l-TPC (l-l)	<u>237</u>	39.4	0.04	<u>241</u>	45.0	0.03
l-TPC (d-d)	<u>237</u>	40.3	0.05	<u>241</u>	45.7	0.02
l-MTPA (l-l)	<u>91</u>	65.8	—	<u>98</u>	71.1	—
	118	10.8	1.21	128	10.7	3.49
	162	14.7	0.11	173	14.9	0.77
	234	15.3	1.83	236	10.4	1.20
	260	55.7	0.06	264	55.9	0.17
	261	7.73	4.77	265	7.43	0.32
l-MTPA (d-d)	<u>91</u>	71.9	—	<u>98</u>	70.5	—
	118	12.1	1.36	128	11.0	3.54
	162	17.3	0.21	173	16.3	0.77
	234	16.1	2.73	236	11.1	1.26
	260	57.9	0.13	264	59.7	0.14
TMS	91	11.5	2.70	98	12.5	2.85
	<u>116</u>	100	0.09	<u>120</u>	100	0.54
	192	7.80	0.11	203	6.51	1.44
t-BDMS	100	16.0	0.58	104	19.0	0.84
	<u>158</u>	100	0.21	<u>162</u>	100	0.40
	159	15.5	4.77	163	14.4	0.56
	192	20.0	0.29	203	21.5	0.00
	234	3.66	0.76	245	2.73	1.51
TFA/ t-BDMS	91	37.1	0.24	98	56.1	0.12
	119	19.8	0.70	130	25.9	0.47
	<u>254</u>	100	0.04	<u>258</u>	100	0.09
	255	15.9	4.04	259	16.9	0.32
	288	24.8	0.19	299	26.5	0.00
	289	4.76	0.34	300	5.18	0.57

Table I — Stimulants

Table I-1f. (Continued)

CD Group[c]	Amphetamine			Amphetamine-d$_{11}$		
	Ion (m/z)[d]	Rel. int.	Analog's cont.	Ion (m/z)[d]	Rel. int.	Analog's cont.
PFP/	91	48.0	0.29	98	65.9	0.17
t-BDMS	119	21.9	2.83	130	26.6	0.55
	<u>304</u>	100	0.07	<u>308</u>	100	0.05
	305	18.1	3.96	309	18.5	0.03
	338	14.4	0.08	349	15.8	0.00
	339	3.17	0.27	350	3.22	0.00
HFB/	91	47.8	0.35	98	59.4	0.16
t-BDMS	119	21.7	3.16	130	23.7	1.54
	<u>354</u>	100	0.05	<u>358</u>	100	0.05
	355	19.7	4.10	359	19.3	0.11
	388	14.4	0.00	399	14.0	0.21
	389	3.17	0.00	400	3.22	0.23

[a–d] See the corresponding footnotes in Table I-1a.

Table I-2a. Relative intensity and cross-contribution data[a] of ions[b] with potential for designating the analyte and the adapted internal standard — Methamphetamine/methamphetamine-d$_5$

CD Group[c]	Methamphetamine			Methamphetamine-d$_5$		
	Ion (m/z)[d]	Rel. int.	Analog's cont.	Ion (m/z)[d]	Rel. int.	Analog's cont.
None	<u>58</u>	100	3.40	<u>62</u>	100	0.60
Acetyl	<u>58</u>	100	0.54	<u>62</u>	100	0.18
	100	76.1	0.25	104	73.7	0.39
TCA	<u>204</u>	95.6	0.07	<u>208</u>	95.5	3.52
TFA	110	24.5	0.63	113	23.6	0.49
	<u>154</u>	100	0.14	<u>158</u>	100	0.57
PFP	160	28.7	0.49	163	26.7	0.12
	<u>204</u>	100	0.05	<u>208</u>	100	0.03
HFB	210	29.7	0.50	213	26.7	0.05
	<u>254</u>	100	0.04	<u>258</u>	100	0.01
4-CB	262	11.3	0.06	266	10.6	0.17
	280	13.2	0.01	284	12.9	0.88
	<u>308</u>	100	0.06	<u>312</u>	100	0.04
	309	11.7	0.81	313	11.7	0.07
PFB	<u>252</u>	83.7	0.23	<u>256</u>	85.4	0.01
	253	9.69	0.96	257	9.63	0.44
Propylformyl	58	41.7	0.95	62	43.4	0.39
	102	30.5	0.50	106	30.3	0.39
	<u>144</u>	100	0.04	<u>148</u>	100	1.29
	176	3.92	0.00	181	4.07	0.13
l-TPC(l-l)	58	46.8	0.35	62	50.2	0.22
	176	2.01	1.70	181	2.16	0.78
	<u>251</u>	45.8	0.04	<u>255</u>	49.6	0.03
l-TPC(d-d)	58	45.1	0.42	62	48.7	0.19
	176	1.84	1.75	181	2.04	0.65
	<u>251</u>	45.3	0.06	<u>255</u>	48.2	0.03
l-MTPA(l-l)	176	4.58	1.48	181	4.30	0.22
	200	7.02	1.52	204	4.68	0.61
	<u>274</u>	57.1	3.18	<u>278</u>	54.2	0.01
	275	8.50	2.75	279	8.00	0.11
l-MTPA(d-d)	176	4.16	2.31	181	4.20	0.34
	200	7.26	4.31	204	5.20	0.12
	<u>274</u>	60.8	0.78	<u>278</u>	61.2	0.01
	275	8.59	0.34	279	9.32	0.11

Appendix Two — Ion Intensity Cross-Contribution Data

Table I-2a. (Continued)

CD Group[c]	Methamphetamine			Methamphetamine-d$_5$		
	Ion (m/z)[d]	Rel. int.	Analog's cont.	Ion (m/z)[d]	Rel. int.	Analog's cont.
TMS	<u>130</u>	100	0.13	<u>134</u>	100	0.62
	206	7.05	0.37	211	6.28	0.17
t-BDMS	<u>172</u>	100	0.26	<u>176</u>	100	1.20
	206	16.2	0.32	211	14.1	1.42
	248	2.58	0.00	253	3.06	0.00

[a–d] See the corresponding footnotes in Table I-1a.

Table I-2b. Relative intensity and cross-contribution data[a] of ions[b] with potential for designating the analyte and the adapted internal standard — Methamphetamine/methamphetamine-d$_8$

CD Group[c]	Methamphetamine			Methamphetamine-d$_8$		
	Ion (m/z)[d]	Rel. int.	Analog's cont.	Ion (m/z)[d]	Rel. int.	Analog's cont.
None	<u>58</u>	100	—	<u>65</u>	100	—
	134	2.79	3.24	139	2.05	4.40
Acetyl	<u>58</u>	100	0.39	<u>62</u>	100	3.10
	100	76.1	0.25	107	72.9	0.02
TCA	<u>202</u>	100	0.96	<u>209</u>	100	0.21
	204	95.6	0.38	211	96.4	0.08
	206	30.6	2.13	213	31.3	0.07
TFA	110	24.5	0.41	113	23.1	0.54
	<u>154</u>	100	0.22	<u>161</u>	100	0.09
PFP	160	28.7	0.12	163	27.3	0.08
	<u>204</u>	100	0.04	<u>211</u>	100	0.01
HFB	210	29.7	0.17	213	23.3	0.07
	<u>254</u>	100	0.05	<u>261</u>	100	0.01
4-CB	262	11.3	0.05	268	7.56	0.21
	280	13.2	0.09	287	11.4	0.00
	<u>308</u>	100	0.07	<u>315</u>	100	0.01
	309	11.7	0.00	316	11.6	0.02
PFB	<u>252</u>	83.7	0.21	<u>259</u>	84.2	0.01
	253	9.69	0.93	260	9.68	0.43
Propylformyl	102	30.5	0.31	109	31.0	0.13
	<u>144</u>	100	0.04	<u>151</u>	100	0.02
	176	3.92	0.12	184	4.19	0.11
l-TPC (l-l)	58	46.8	0.29	65	53.6	4.61
	176	2.01	1.66	184	2.25	0.30
	<u>251</u>	45.8	0.04	<u>258</u>	50.6	0.56
l-TPC (d-d)	58	45.1	0.31	65	51.7	4.65
	176	1.84	1.66	184	2.13	0.38
	<u>251</u>	45.3	0.04	<u>258</u>	50.5	0.60
l-MTPA (l-l)	176	4.58	0.37	184	4.23	0.21
	200	7.02	1.22	204	4.88	0.09
	<u>274</u>	57.1	3.31	<u>281</u>	57.2	0.00
	275	8.50	3.28	282	8.18	0.23
l-MTPA (d-d)	176	4.58	1.27	184	4.09	0.48
	200	7.26	2.64	204	5.41	0.13
	<u>274</u>	60.8	2.11	<u>281</u>	63.8	0.00
	275	8.59	2.48	282	9.35	0.31
TMS	<u>130</u>	100	0.00	<u>137</u>	100	0.03
	206	7.05	0.07	214	6.26	1.54

Table I — Stimulants

Table I-2b. (Continued)

CD Group[c]	Methamphetamine			Methamphetamine-d8		
	Ion (m/z)[d]	Rel. int.	Analog's cont.	Ion (m/z)[d]	Rel. int.	Analog's cont.
t-BDMS	172	100	0.83	179	100	0.00
	206	16.2	0.96	214	16.3	0.13

[a–d] See the corresponding footnotes in Table I-1a.

Table I-2c. Relative intensity and cross-contribution data[a] of ions[b] with potential for designating the analyte and the adapted internal standard — Methamphetamine/methamphetamine-d9

CD Group[c]	Methamphetamine			Methamphetamine-d9		
	Ion (m/z)[d]	Rel. int.	Analog's cont.	Ion (m/z)[d]	Rel. int.	Analog's cont.
None	58	100	—	65	100	—
Acetyl	58	100	0.43	65	100	4.00
	100	76.1	0.26	107	70.4	0.82
TCA	91	42.8	4.14	93	32.6	1.54
	202	100	0.10	209	100	0.24
	204	95.6	0.19	211	96.2	0.16
TFA	110	24.5	0.72	113	23.1	0.51
	118	29.4	4.87	123	26.2	0.62
	154	100	0.29	161	100	0.12
PFP	160	28.7	0.12	163	26.6	0.10
	204	100	0.05	211	100	0.01
HFB	91	16.8	4.89	93	10.3	2.14
	118	23.9	4.97	123	18.0	0.15
	210	29.7	0.17	213	25.3	0.06
	254	100	0.06	261	100	0.01
4-CB	91	19.5	2.47	93	16.4	0.77
	118	14.8	3.89	123	11.6	0.04
	262	11.3	0.06	268	7.63	0.17
	280	13.2	0.09	287	12.1	0.09
	308	100	0.08	315	100	0.07
	309	11.7	0.14	316	11.7	0.09
PFB	252	83.7	0.15	259	82.4	0.08
	253	9.69	0.74	260	9.12	0.52
Propylformyl	102	30.5	0.00	109	32.5	0.14
	144	100	0.00	151	100	0.03
	176	3.92	0.23	185	3.98	0.21
l-TPC (l-l)	58	46.8	0.34	65	51.3	4.56
	176	2.01	1.74	185	2.30	1.25
	251	45.8	0.04	258	50.6	0.57
l-TPC (d-d)	58	45.1	0.35	65	49.1	4.62
	176	1.84	1.74	185	2.19	1.53
	251	45.3	0.04	258	50.1	0.61
l-MTPA (l-l)	176	4.58	0.82	185	4.35	1.28
	200	7.02	0.50	204	4.87	0.49
	274	57.1	0.08	281	57.8	0.00
	275	8.50	0.22	282	8.50	0.00
l-MTPA (d-d)	176	4.58	0.96	185	4.17	2.02
	200	7.26	1.43	204	5.57	0.32
	274	60.8	0.11	281	63.9	0.00
	275	8.59	0.14	282	9.09	0.00
TMS	130	100	0.00	137	100	0.02
	206	7.05	0.00	215	6.35	0.00

Appendix Two — Ion Intensity Cross-Contribution Data

Table I-2c. (Continued)

CD Group[c]	Methamphetamine			Methamphetamine-d[9]		
	Ion (m/z)[d]	Rel. int.	Analog's cont.	Ion (m/z)[d]	Rel. int.	Analog's cont.
t-BDMS	<u>172</u>	100	0.72	<u>179</u>	100	0.17
	206	16.2	1.01	215	17.9	0.34
	248	2.58	0.00	257	2.09	4.82

[a–d] See the corresponding footnotes in Table I-1a.

Table I-2d. Relative intensity and cross-contribution data[a] of ions[b] with potential for designating the analyte and the adapted internal standard — Methamphetamine/methamphetamine-d[11]

CD Group[c]	Methamphetamine			Methamphetamine-d[11]		
	Ion (m/z)[d]	Rel. int.	Analog's cont.	Ion (m/z)[d]	Rel. int.	Analog's cont.
None	<u>58</u>	100	3.49	<u>64</u>	100	0.63
	91	11.1	4.00	96	7.21	0.11
	134	2.79	3.72	142	2.07	0.19
Acetyl	<u>58</u>	100	—	<u>64</u>	100	—
	91	18.2	1.40	96	9.80	0.00
	100	76.1	0.40	106	74.1	0.11
	118	9.50	0.86	126	4.88	0.00
TCA	91	42.8	1.16	96	23.0	1.60
	<u>202</u>	100	0.05	<u>208</u>	100	3.21
	204	95.6	0.67	210	95.3	0.07
	206	30.6	1.20	212	30.6	0.14
TFA	91	14.5	1.16	96	9.69	2.98
	118	29.4	0.27	126	30.2	0.87
	<u>154</u>	100	0.12	<u>160</u>	100	0.18
PFP	91	13.1	0.63	96	7.75	0.66
	118	22.0	0.28	126	21.8	0.56
	160	28.7	1.36	163	25.8	0.10
	<u>204</u>	100	0.06	<u>210</u>	100	0.00
HFB	91	16.8	0.68	96	8.25	0.47
	118	23.9	0.25	126	21.3	0.22
	210	29.7	0.28	213	26.9	0.05
	<u>254</u>	100	0.05	<u>260</u>	100	0.00
4-CB	91	19.5	0.53	97	10.4	0.42
	118	14.8	0.25	126	16.3	0.23
	262	11.3	0.08	267	7.80	0.20
	280	13.2	0.01	286	11.9	0.24
	<u>308</u>	100	0.07	<u>314</u>	100	0.01
	309	11.7	0.53	315	11.6	0.47
PFB	91	7.97	0.73	96	6.85	0.28
	118	6.62	4.45	126	6.18	0.49
	<u>252</u>	83.7	0.25	<u>258</u>	83.1	0.02
	253	9.69	0.97	259	9.80	0.89
Propylformyl	58	41.7	1.37	64	45.9	0.60
	91	26.2	1.83	96	16.0	0.20
	102	30.5	0.51	108	30.5	0.59
	<u>144</u>	100	0.10	<u>150</u>	100	0.13
l-TPC (l-l)	58	46.8	0.38	65	50.3	0.44
	119	7.35	2.49	127	8.76	1.92
	<u>251</u>	45.8	0.04	<u>257</u>	50.8	2.96
l-TPC (d-d)	58	45.1	0.39	65	47.3	0.45
	119	7.59	2.35	127	9.09	1.93
	<u>251</u>	45.3	0.04	<u>257</u>	52.0	3.13

Table I — Stimulants

Table I-2d. (Continued)

CD Group[c]	Methamphetamine			Methamphetamine-d[11]		
	Ion (m/z)[d]	Rel. int.	Analog's cont.	Ion (m/z)[d]	Rel. int.	Analog's cont.
l-MTPA (l-l)	176	4.58	0.49	187	3.71	4.68
	200	7.02	1.19	203	6.65	1.05
	274	57.1	0.31	280	57.2	0.02
	275	8.50	0.11	281	8.86	0.00
l-MTPA (d-d)	176	4.58	0.84	187	3.77	1.34
	200	7.26	2.07	203	7.54	2.03
	274	60.8	0.43	280	61.5	0.01
	275	8.59	0.16	281	9.64	0.00
TMS	91	8.68	2.32	96	6.79	0.44
	130	100	0.00	136	100	0.09
	206	7.05	0.00	217	7.03	0.15
t-BDMS	172	100	1.37	178	100	0.24
	206	16.2	2.71	217	18.2	0.00

[a–d] See the corresponding footnotes in Table I-1a.

Table I-2e. Relative intensity and cross-contribution data[a] of ions[b] with potential for designating the analyte and the adapted internal standard — Methamphetamine/methamphetamine-d[14]

CD Group[c]	Methamphetamine			Methamphetamine-d[14]		
	Ion (m/z)[d]	Rel. int.	Analog's cont.	Ion (m/z)[d]	Rel. int.	Analog's cont.
None	58	100	—	65	100	—
	91	11.1	0.97	98	8.75	1.87
	134	2.79	1.25	145	2.17	3.23
Acetyl	58	100	0.28	65	100	3.29
	91	18.2	0.80	98	16.9	0.18
	100	76.1	0.37	107	70.7	0.11
	118	9.50	0.07	128	7.93	0.63
TCA	91	42.8	0.47	98	41.9	0.05
	202	100	0.24	209	100	0.30
	204	95.6	0.08	211	95.6	0.02
TFA	91	14.5	1.60	98	13.6	1.99
	118	29.4	0.23	128	25.7	3.31
	154	100	0.28	161	100	0.12
PFP	91	13.1	0.32	98	12.8	0.71
	118	22.0	0.21	128	20.2	2.56
	160	28.7	0.56	163	26.7	0.11
	204	100	0.13	211	100	0.00
HFB	91	16.8	0.29	98	13.7	0.33
	118	23.9	0.15	128	17.5	2.34
	210	29.7	0.16	213	24.8	0.06
	254	100	0.03	261	100	0.00
4-CB	91	19.5	0.36	98	19.7	0.16
	118	14.8	0.17	128	11.7	0.64
	262	11.3	0.10	268	7.63	0.16
	280	13.2	0.09	287	12.2	0.04
	308	100	0.08	315	100	0.02
	309	11.7	0.00	316	12.0	0.00
PFB	91	7.97	0.24	98	8.98	0.08
	118	6.62	4.50	128	5.27	0.72
	252	83.7	0.02	259	80.6	0.03
	253	9.69	0.58	260	9.17	0.44

Table I-2e. (Continued)

CD Group[c]	Methamphetamine Ion (m/z)[d]	Rel. int.	Analog's cont.	Methamphetamine-d₁₄ Ion (m/z)[d]	Rel. int.	Analog's cont.
Propylformyl	91	26.2	1.52	98	30.9	0.12
	102	30.5	0.23	109	32.5	0.18
	144	100	0.41	151	100	0.03
	176	3.92	0.45	190	3.09	2.71
l-TPC (l-l)	58	46.8	0.46	65	48.9	4.81
	119	7.35	4.51	130	8.23	1.81
	251	45.8	0.02	258	53.2	0.56
l-TPC (d-d)	58	45.1	0.46	65	48.3	4.74
	119	7.59	4.73	130	8.44	1.72
	251	45.3	0.02	258	51.0	0.58
l-MTPA (l-l)	200	7.02	0.57	204	5.11	0.90
	274	57.1	0.02	281	59.9	0.00
	275	8.50	0.07	282	8.82	0.00
l-MTPA (d-d)	200	7.26	0.46	204	5.15	0.24
	274	60.8	0.25	281	64.1	0.00
	275	8.59	0.31	282	9.78	0.00
TMS	91	8.68	1.89	98	9.26	0.82
	130	100	0.50	137	100	0.02
t-BDMS	172	100	0.64	179	100	0.00
	206	16.2	1.27	220	18.3	0.00

[a–d] See the corresponding footnotes in Table I-1a.

Table I-3. Relative intensity and cross-contribution data[a] of ions[b] with potential for designating the analyte and the adapted internal standard — Ephedrine/ephedrine-d₃

CD Group[c]	Ephedrine Ion (m/z)[d]	Rel. int.	Analog's cont.	Ephedrine-d₃ Ion (m/z)[d]	Rel. int.	Analog's cont.
None	58	100	2.49	61	100	0.05
Acetyl	58	100	2.22	61	100	0.18
	100	69.4	0.60	103	65.8	1.38
	101	27.0	1.14	104	26.3	1.80
	189	2.07	2.05	192	2.18	2.98
TCA	42	10.2	1.52	45	9.69	3.18
	202	100	0.74	205	100	2.14
	204	95.9	3.37	207	95.2	0.20
[TFA]₂	110	20.0	0.55	113	19.4	0.42
	154	100	0.06	157	100	0.11
	244	3.34	2.86	247	4.15	0.17
[PFP]₂	160	21.8	0.57	163	20.8	0.23
	204	100	0.22	207	100	0.17
	294	3.72	2.59	297	4.84	0.21
[HFB]₂	210	19.4	0.50	213	22.6	0.41
	254	100	0.30	257	100	0.46
	344	5.40	2.10	347	3.81	0.43
4-CB	166	9.88	3.64	169	10.3	2.58
	262	18.0	1.22	265	17.2	2.35
	280	21.1	1.35	283	20.3	0.36
	308	100	1.16	311	100	1.07
	309	79.2	1.50	312	78.9	0.42
PFB	252	90.8	0.41	255	94.8	0.11
	253	10.9	4.10	256	11.3	0.26

Table I — Stimulants

Table I-3. (Continued)

CD Group[c]	Ephedrine Ion (m/z)[d]	Rel. int.	Analog's cont.	Ephedrine-d₃ Ion (m/z)[d]	Rel. int.	Analog's cont.
Propylformyl	58	50.9	2.31	61	50.1	0.20
	_144_	100	0.08	_147_	100	0.50
l-TPC (d-d)	_58_	20.1	1.45	_61_	31.4	0.00
	251	13.9	1.05	254	22.6	0.00
l-MTPA (d-d)	200	6.96	2.26	203	6.90	0.16
	_274_	52.9	1.34	_277_	52.7	0.23
	275	16.4	1.64	278	16.5	0.07
[TMS]₂	_130_	100	0.31	_133_	100	1.55
	294	2.31	0.72	297	2.29	1.68

[a–d] See the corresponding footnotes in Table I-1a.

Table I-4. Relative intensity and cross-contribution data[a] of ions[b] with potential for designating the analyte and the adapted internal standard — Phenylpropanolamine/phenylpropanolamine-d₃

CD Group[c]	Phenylpropanolamine Ion (m/z)[d]	Rel. int.	Analog's cont.	Phenylpropanolamine-d₃ Ion (m/z)[d]	Rel. int.	Analog's cont.
None	_44_	100	2.76	_47_	100	0.01
Acetyl	86	100	0.36	89	93.1	3.83
	176	6.35	4.69	179	5.42	0.68
TCA	160	18.0	3.63	163	18.6	3.67
	_188_	100	1.64	_191_	100	4.86
	190	94.8	2.22	193	95.0	1.71
[TFA]₂	140	100	3.42	143	100	0.01
[PFP]₂	190	100	1.66	193	100	0.12
	280	17.6	4.97	283	43.4	0.11
[HFB]₂	240	100	1.76	243	100	0.05
4-CB	248	19.3	0.28	250	9.57	4.43
	266	30.6	0.23	269	27.8	0.18
	_294_	100	0.31	_297_	100	0.09
	295	11.0	3.40	298	10.7	0.09
	338	6.09	0.40	341	6.22	1.97
	384	15.8	0.33	387	15.4	0.21
PFB	_238_	69.3	0.40	_241_	69.7	0.13
	239	7.32	2.17	242	7.53	0.00
l-TPC (l-l)	220	5.97	3.91	223	5.69	0.00
	_237_	25.4	3.80	_240_	23.8	0.13
l-MTPA (l-l)	229	17.6	1.10	232	16.6	0.74
	260	20.4	0.88	263	19.6	3.53
	_261_	56.6	1.00	_264_	55.5	0.25
l-MTPA (d-d)	229	19.2	1.40	232	17.5	0.73
	260	20.2	0.83	263	19.0	3.50
	_261_	58.6	0.93	_264_	56.6	0.17
[TMS]₂	116	100	3.71	119	100	1.01
t-BDMS	158	100	4.24	161	100	0.71
[t-BDMS]₂	None (No ion pair meets the selection criteria)					

Table I-4. (Continued)

CD Group[c]	Phenylpropanolamine			Phenylpropanolamine-d₃		
	Ion (m/z)[d]	Rel. int.	Analog's cont.	Ion (m/z)[d]	Rel. int.	Analog's cont.
TFA/	191	7.33	4.14	194	6.35	1.90
[t-BDMS]₂	254	5.50	1.75	257	5.68	0.84
	<u>418</u>	35.8	0.24	<u>421</u>	41.9	2.35
	419	11.7	0.45	422	13.6	1.30
PFP/	304	6.07	1.39	307	6.14	1.16
[t-BDMS]₂	<u>468</u>	36.9	0.25	<u>471</u>	33.5	2.51
	469	12.2	0.96	472	10.5	1.40
HFB/	191	4.41	4.68	194	4.33	4.18
[t-BDMS]₂	354	6.99	1.51	357	7.41	3.78
	<u>518</u>	19.6	0.25	<u>521</u>	25.0	2.50
	519	6.60	0.53	522	8.67	1.35

[a–d] See the corresponding footnotes in Table I-1a.

Table I-5. Relative intensity and cross-contribution data[a] of ions[b] with potential for designating the analyte and the adapted internal standard — MDA/MDA-d₅

CD Group[c]	MDA			MDA-d₅		
	Ion (m/z)[d]	Rel. int.	Analog's cont.	Ion (m/z)[d]	Rel. int.	Analog's cont.
None	<u>44</u>	100	—	<u>48</u>	100	—
Acetyl	44	56.6	2.71	48	95.6	0.01
	86	12.4	2.82	90	17.2	1.86
	<u>162</u>	100	2.98	<u>166</u>	95.6	0.02
	221	6.31	4.02	226	11.9	0.24
TCA	<u>162</u>	100	1.06	<u>167</u>	68.6	0.07
	190	13.5	1.75	194	16.2	2.07
	323	4.21	1.25	328	6.03	2.39
TFA	<u>135</u>	100	—	<u>136</u>	100	—
	162	45.6	0.47	167	27.5	0.03
	275	15.7	0.26	280	16.9	0.00
PFP	<u>135</u>	100	—	<u>136</u>	100	—
	162	48.7	0.46	167	29.6	0.04
	190	8.11	0.64	194	8.75	0.24
	325	12.7	0.25	330	13.8	0.00
HFB	<u>135</u>	100	—	<u>136</u>	100	—
	162	54.6	0.45	167	33.5	0.09
	240	9.69	0.65	244	10.7	0.03
	375	12.8	0.25	380	13.8	0.00
4-CB	<u>162</u>	100	1.30	<u>166</u>	74.4	0.09
	248	8.86	2.35	251	7.37	0.57
	266	9.08	2.64	270	12.6	0.05
	429	4.99	1.23	434	7.79	0.00
PFB	<u>162</u>	77.6	0.78	<u>166</u>	35.9	0.09
	238	16.5	1.44	232	15.9	0.12
	373	4.82	2.86	378	5.96	0.00
Propylformyl	<u>130</u>	100	0.34	<u>134</u>	100	2.04
	162	37.8	0.70	167	24.3	1.28
	206	8.21	0.52	211	9.69	0.10
	265	25.5	0.40	270	28.6	0.00
l-TPC (l-l)	<u>135</u>	19.2	—	<u>136</u>	12.1	—
	237	7.51	0.51	241	4.73	0.10
	372	7.65	0.35	377	5.31	0.02

Table I — Stimulants

Table I-5. (Continued)

CD Group[c]	MDA			MDA-d₅		
	Ion (m/z)[d]	Rel. int.	Analog's cont.	Ion (m/z)[d]	Rel. int.	Analog's cont.
l-TPC (d-d)	<u>135</u>	19.1	—	<u>136</u>	11.7	—
	237	8.92	0.53	241	5.67	0.00
	372	6.77	0.33	377	5.27	0.07
l-MTPA (l-l)	<u>162</u>	100	0.20	<u>166</u>	74.8	0.07
	163	22.6	4.98	167	86.9	0.30
	206	5.03	0.85	211	8.04	0.59
	228	3.66	0.57	232	3.18	0.87
	260	5.38	0.15	264	7.29	0.21
l-MTPA (d-d)	<u>162</u>	100	0.23	<u>166</u>	76.5	0.06
	163	24.0	4.51	167	88.4	0.34
	206	6.00	0.83	211	9.21	0.33
	228	3.82	0.63	232	3.14	0.93
	260	5.48	0.23	264	6.87	0.30
TMS	<u>116</u>	100	0.21	<u>120</u>	100	0.22
	236	6.22	0.85	241	3.53	0.11
t-BDMS	100	6.60	2.89	104	5.80	7.35
	158	100	0.04	162	100	0.52
	236	3.98	0.49	241	4.40	0.08
	278	2.10	0.13	286	2.00	0.26
TFA/	163	20.5	1.61	168	20.7	0.24
t-BDMS	<u>254</u>	100	0.04	<u>258</u>	100	0.23
	332	8.12	0.98	337	9.36	0.20
	389	12.1	0.11	394	14.0	0.03
PFP/	163	12.4	0.75	168	18.1	0.56
t-BDMS	<u>304</u>	100	0.03	<u>308</u>	100	0.50
	382	9.50	0.69	387	10.7	0.36
	439	14.9	2.75	444	13.6	0.20
HFB/	163	16.3	0.73	168	25.0	0.12
t-BDMS	<u>354</u>	100	0.03	<u>358</u>	100	0.17
	432	8.44	0.24	437	7.97	0.30
	489	10.6	0.12	494	8.32	0.45

[a–d] See the corresponding footnotes in Table I-1a.

Table I-6. Relative intensity and cross-contribution data[a] of ions[b] with potential for designating the analyte and the adapted internal standard — MDMA/MDMA-d₅

CD Group[c]	MDMA			MDMA-d₅		
	Ion (m/z)[d]	Rel. int.	Analog's cont.	Ion (m/z)[d]	Rel. int.	Analog's cont.
None	<u>58</u>	100	2.38	<u>62</u>	100	0.67
Acetyl	<u>58</u>	100	1.20	<u>62</u>	100	0.35
	162	89.3	2.37	164	46.1	2.24
	235	2.70	1.77	240	2.32	0.00
TCA	162	100	1.43	164	66.4	3.79
	<u>204</u>	82.0	0.14	<u>208</u>	94.7	3.60
TFA	110	29.7	0.37	113	30.0	0.08
	<u>154</u>	100	0.04	<u>158</u>	100	0.02
	289	14.0	0.04	294	15.1	0.00
PFP	<u>204</u>	100	0.02	<u>208</u>	100	0.02
	339	9.81	0.09	344	10.1	0.00
HFB	162	71.7	1.56	164	39.7	3.44
	210	37.8	0.60	213	36.5	0.02
	<u>254</u>	100	0.03	<u>258</u>	100	0.00
	389	7.35	0.05	394	7.32	0.00

Appendix Two — Ion Intensity Cross-Contribution Data

Table I-6. (Continued)

CD Group[c]	MDMA			MDMA-d$_5$		
	Ion $(m/z)^d$	Rel. int.	Analog's cont.	Ion $(m/z)^d$	Rel. int.	Analog's cont.
4-CB	<u>162</u>	100	1.91	<u>164</u>	84.6	2.97
	262	17.0	0.17	266	24.9	0.14
	280	19.6	0.14	284	29.0	0.05
	308	66.6	0.11	312	100	0.01
	443	2.04	0.11	448	3.18	0.00
PFB	<u>162</u>	43.2	2.64	<u>164</u>	23.6	1.27
	252	39.7	0.57	256	40.4	0.02
	253	4.62	2.36	257	4.62	0.24
	387	1.36	3.96	392	1.51	0.00
Propylformyl	58	38.2	0.68	62	41.5	0.57
	102	35.5	0.36	106	41.5	3.10
	<u>144</u>	100	0.02	<u>148</u>	100	0.92
	220	4.71	3.73	225	4.91	0.25
	279	7.82	0.32	284	6.78	0.32
l-TPC (*l-l*)	<u>58</u>	39.0	0.30	<u>62</u>	40.0	0.35
	251	12.7	0.08	255	13.0	0.07
	386	2.71	0.04	391	3.37	0.03
l-TPC (*d-d*)	<u>58</u>	30.8	0.33	<u>62</u>	38.6	0.43
	251	11.4	0.06	255	9.86	0.06
	386	3.42	0.05	391	2.44	0.08
l-MTPA (*l-l*)	<u>162</u>	53.4	1.95	<u>164</u>	31.4	2.05
	200	5.91	0.86	204	4.31	0.58
	274	16.7	0.38	278	18.6	0.05
l-MTPA (*d-d*)	<u>162</u>	52.4	2.07	<u>164</u>	30.7	2.10
	200	5.90	0.94	204	4.44	0.71
	274	16.7	0.43	278	17.7	0.05
TMS	<u>130</u>	100	0.11	<u>134</u>	100	0.40
	250	3.90	1.26	255	5.98	0.23
t-BDMS	172	100	0.01	176	100	0.30
	173	15.8	0.43	177	16.0	1.30

[a–d] See the corresponding footnotes in Table I-1a.

Table I-7a. Relative intensity and cross-contribution data[a] of ions[b] with potential for designating the analyte and the adapted internal standard — MDEA/MDEA-d$_5$

CD Group[c]	MDEA			MDEA-d$_5$		
	Ion $(m/z)^d$	Rel. int.	Analog's cont.	Ion $(m/z)^d$	Rel. int.	Analog's cont.
None	<u>72</u>	100	1.22	<u>77</u>	100	7.21
Acetyl	<u>72</u>	100	—	<u>77</u>	100	—
	114	28.4	0.54	119	29.5	0.90
TCA	<u>216</u>	100	0.10	<u>221</u>	100	2.04
	218	93.3	0.10	223	93.0	0.30
TFA	<u>168</u>	100	0.02	<u>173</u>	100	0.14
	303	8.26	0.02	308	9.29	0.00
PFP	<u>218</u>	100	0.01	<u>223</u>	100	0.05
	353	6.16	0.06	358	6.00	0.00
HFB	<u>268</u>	100	0.03	<u>273</u>	100	0.07
	403	4.92	0.14	408	4.54	1.33
4-CB	276	10.7	0.21	281	11.7	0.55
	294	13.0	0.17	299	15.1	0.12
	<u>322</u>	81.6	0.04	<u>327</u>	100	0.03
	323	10.8	0.18	328	13.1	0.08

Table I — Stimulants

Table I-7a. (Continued)

CD Group[c]	MDEA			MDEA-d$_5$		
	Ion (m/z)[d]	Rel. int.	Analog's cont.	Ion (m/z)[d]	Rel. int.	Analog's cont.
PFB	<u>266</u>	46.8	0.07	<u>271</u>	46.0	0.23
	267	6.59	1.54	272	5.63	1.55
Propylformyl	<u>158</u>	100	—	<u>163</u>	100	—
	116	32.0	0.36	121	31.9	1.80
	234	2.55	0.17	239	3.00	0.35
	293	3.42	0.04	298	3.45	0.07
l-TPC (l-l)	<u>72</u>	60.1	—	<u>77</u>	56.1	—
	265	14.9	0.05	270	18.4	0.05
	400	1.22	0.00	405	2.03	0.00
l-TPC (d-d)	<u>72</u>	54.0	—	<u>77</u>	50.1	—
	265	12.8	0.08	270	15.5	0.07
	400	1.57	0.00	405	2.57	0.05
l-MTPA (l-l)	214	6.84	0.34	219	6.59	0.34
	262	2.52	4.45	267	2.52	0.00
	<u>288</u>	27.5	0.18	<u>293</u>	26.7	0.07
	289	4.31	0.27	294	4.16	0.28
l-MTPA (d-d)	214	6.51	0.48	219	6.72	0.34
	<u>288</u>	27.7	0.23	<u>293</u>	26.6	0.12
	289	4.26	0.29	294	4.27	0.35
TMS	<u>144</u>	100	0.06	<u>149</u>	100	0.70
	264	5.92	0.26	269	7.09	0.13

[a–d] See the corresponding footnotes in Table I-1a.

Table I-7b. Relative intensity and cross-contribution data[a] of ions[b] with potential for designating the analyte and the adapted internal standard — MDEA/MDEA-d$_6$

CD Group[c]	MDEA			MDEA-d$_6$		
	Ion (m/z)[d]	Rel. int.	Analog's cont.	Ion (m/z)[d]	Rel. int.	Analog's cont.
None	<u>72</u>	100	1.00	<u>78</u>	100	1.67
Acetyl	<u>72</u>	100	0.72	<u>78</u>	100	2.00
	114	28.4	0.63	120	29.2	0.59
	162	57.1	1.64	165	75.1	0.16
TCA	162	75.7	1.24	165	78.6	0.34
	<u>216</u>	100	0.09	<u>222</u>	100	3.28
	218	93.3	0.11	224	92.6	0.06
	220	2.28	1.31	226	30.4	0.10
	351	2.64	0.04	357	3.12	3.89
TFA	140	36.2	0.25	144	26.4	0.39
	162	63.0	1.09	165	68.5	0.12
	<u>168</u>	100	0.10	<u>174</u>	100	0.19
	303	8.26	0.01	309	9.36	0.00
PFP	<u>162</u>	58.6	0.75	<u>165</u>	63.1	0.23
	190	41.1	0.29	194	29.1	0.05
	218	100	0.06	224	100	0.01
	353	6.16	0.00	359	5.73	0.00
HFB	162	55.5	1.26	165	59.8	0.17
	240	43.0	0.67	244	29.5	0.02
	<u>268</u>	100	0.04	<u>274</u>	100	0.01
	403	4.92	0.05	409	4.38	0.00

Table I-7b. (Continued)

CD Group[c]	MDEA			MDEA-d[6]		
	Ion (m/z)[d]	Rel. int.	Analog's cont.	Ion (m/z)[d]	Rel. int.	Analog's cont.
4-CB	162	100	0.95	165	100	0.27
	163	31.2	3.01	166	31.3	0.45
	276	10.7	0.09	281	8.28	0.61
	294	13.0	0.11	300	11.1	0.09
	322	81.6	0.04	328	82.9	0.01
	323	10.8	0.10	329	10.8	0.05
PFB	162	39.3	1.02	165	40.7	0.18
	266	46.8	0.43	272	50.3	0.07
	267	6.59	0.00	273	6.48	0.42
Propylformyl	72	40.7	0.15	78	41.5	4.59
	116	32.0	0.20	122	33.4	1.19
	158	100	0.01	164	100	0.92
	234	2.55	0.11	240	2.78	0.19
	293	3.42	0.02	299	3.37	0.06
l-TPC (l-l)	72	60.1	0.56	78	49.3	3.72
	162	65.8	0.94	165	59.3	1.23
	265	14.9	0.16	271	13.6	0.24
l-TPC (d-d)	72	54.0	0.62	78	41.6	4.16
	162	85.3	0.90	165	75.3	0.99
	265	12.8	0.11	271	11.7	0.26
l-MTPA (l-l)	162	44.5	0.68	165	45.1	0.26
	163	11.2	2.39	166	11.2	0.24
	214	6.48	0.49	217	6.56	0.10
	288	27.5	0.14	294	25.8	0.07
	289	4.31	0.21	295	4.18	0.00
l-MTPA (d-d)	162	41.8	1.06	165	42.7	0.31
	163	10.0	3.07	166	10.3	0.30
	214	6.51	0.48	217	6.58	0.18
	288	27.7	0.24	294	26.2	0.08
	289	4.26	0.35	295	4.18	0.00
TMS	144	100	0.07	150	100	0.17
	264	5.92	0.23	270	6.75	0.20

[a–d] See the corresponding footnotes in Table I-1a.

Table I-8. Relative intensity and cross-contribution data[a] of ions[b] with potential for designating the analyte and the adapted internal standard — MBDB/MBDB-d[5]

CD Group[c]	MBDB			MBDB-d[5]		
	Ion (m/z)[d]	Rel. int.	Analog's cont.	Ion (m/z)[d]	Rel. int.	Analog's cont.
None	72	100	4.43	76	100	0.66
Acetyl	72	100	0.70	76	100	0.61
	114	34.6	0.59	118	32.3	3.39
	249	1.27	0.98	254	1.24	0.71
TCA	176	100	1.83	178	61.5	1.72
	218	91.4	0.32	222	93.8	2.13
	351	3.74	0.33	356	4.25	2.89

Table I — Stimulants

Table I-8. (Continued)

CD Group[c]	MBDB			MBDB-d₅		
	Ion (m/z)[d]	Rel. int.	Analog's cont.	Ion (m/z)[d]	Rel. int.	Analog's cont.
TFA	110	18.2	0.37	113	18.1	0.45
	<u>168</u>	100	0.03	<u>172</u>	100	0.07
	303	11.6	0.02	308	12.1	0.00
PFP	160	28.3	0.43	163	28.2	0.65
	176	67.8	1.03	178	37.5	4.02
	190	5.25	1.87	194	3.29	0.66
	<u>218</u>	100	0.02	<u>222</u>	100	0.03
	353	7.75	0.04	358	8.10	0.10
HFB	176	61.9	0.95	178	34.1	3.21
	210	27.6	1.63	213	26.2	0.20
	<u>268</u>	100	0.02	<u>272</u>	100	0.03
	403	5.86	0.03	408	6.00	0.00
4-CB	<u>176</u>	100	1.97	<u>178</u>	71.2	2.15
	276	28.4	0.13	280	33.8	0.19
	294	24.5	0.32	298	30.9	0.03
	322	75.3	0.09	326	100	0.02
	457	2.35	0.30	462	3.43	0.00
PFB	176	46.0	2.56	178	25.4	3.57
	<u>266</u>	51.8	0.45	<u>270</u>	53.2	0.08
Propylformyl	72	42.1	0.30	76	39.2	1.49
	116	44.5	0.85	120	43.3	0.89
	<u>158</u>	100	0.00	<u>162</u>	100	0.31
	234	3.15	0.75	239	3.82	0.03
	293	4.26	0.02	298	5.36	0.00
l-TPC (l-l)	<u>72</u>	58.3	0.37	<u>76</u>	63.2	0.59
	265	16.3	0.00	269	16.3	0.02
	400	3.21	0.00	405	3.27	0.00
l-TPC (d-d)	<u>72</u>	52.2	0.43	<u>76</u>	52.2	0.63
	265	12.5	0.00	269	13.3	0.03
	400	2.54	0.00	405	3.04	0.10
l-MTPA (l-l)	<u>176</u>	55.8	2.36	<u>178</u>	30.4	1.47
	248	4.41	2.84	252	2.62	0.29
	288	20.2	1.05	292	20.4	0.07
l-MTPA (d-d)	<u>176</u>	55.3	2.51	<u>178</u>	31.2	1.43
	248	4.35	4.08	252	2.60	0.33
	288	20.8	1.25	292	21.1	0.07
TMS	<u>144</u>	100	0.08	<u>148</u>	100	0.65
	250	2.14	1.22	255	2.07	0.00
	264	4.53	0.29	269	4.42	0.10

[a–d] See the corresponding footnotes in Table I-1a.

Table I-9. Relative intensity and cross-contribution data[a] of ions[b] with potential for designating the analyte and the adapted internal standard — Selegiline/selegiline-d₈

CD Group[c]	Selegiline			Selegiline-d₈		
	Ion (m/z)[d]	Rel. int.	Analog's cont.	Ion (m/z)[d]	Rel. int.	Analog's cont.
None	<u>96</u>	100	0.49	<u>103</u>	100	1.03

[a–d] See the corresponding footnotes in Table I-1a.

Table I-10. Relative intensity and cross-contribution data[a] of ions[b] with potential for designating the analyte and the adapted internal standard — N-Desmethylselegiline/N-desmethylselegiline-d₁₁

CD Group[c]	N-Desmethylselegiline			N-Desmethylselegiline-d₁₁		
	Ion (m/z)[d]	Rel. int.	Analog's cont.	Ion (m/z)[d]	Rel. int.	Analog's cont.
None	<u>82</u>	100	3.61	<u>86</u>	100	0.14
	91	16.3	0.73	98	15.0	0.30
Acetyl	<u>82</u>	100	2.21	<u>86</u>	100	0.30
	91	16.9	1.60	98	18.7	2.69
	124	44.6	1.25	128	48.9	1.53
TCA	91	49.8	1.25	98	51.7	0.05
	<u>228</u>	96.0	0.79	<u>232</u>	95.5	3.55
TFA	91	24.2	0.50	98	24.8	0.32
	118	41.1	0.47	128	36.8	2.90
	<u>178</u>	100	0.16	<u>182</u>	100	0.04
PFP	91	29.3	1.43	98	26.5	0.20
	118	37.9	1.50	128	29.2	4.83
	<u>228</u>	100	1.15	<u>232</u>	100	0.00
HFB	91	27.3	1.39	98	32.6	0.11
	118	35.4	2.75	128	32.6	2.79
	<u>278</u>	100	1.20	<u>282</u>	100	1.04
4-CB	91	55.0	0.99	98	54.5	0.00
	118	43.9	1.85	128	31.9	3.01
	304	4.90	0.76	308	4.67	2.64
	<u>332</u>	100	0.19	<u>336</u>	100	0.06
TMS	91	11.4	1.64	98	12.8	1.40
	<u>154</u>	100	0.99	<u>158</u>	100	0.01
	230	3.02	1.99	241	3.53	0.79

[a–d] See the corresponding footnotes in Table I-1a.

Table I-11. Relative intensity and cross-contribution data[a] of ions[b] with potential for designating the analyte and the adapted internal standard — Fenfluramine/fenfluramine-d₁₀

CD Group[c]	Fenfluramine			Fenfluramine-d₁₀		
	Ion (m/z)[d]	Rel. int.	Analog's cont.	Ion (m/z)[d]	Rel. int.	Analog's cont.
None	<u>72</u>	100	0.58	<u>81</u>	100	0.73
	73	6.51	3.96	82	7.32	3.84
	230	2.73	1.66	239	1.10	3.14
Acetyl	<u>72</u>	100	0.13	<u>81</u>	100	0.08
	114	58.8	0.69	123	55.1	0.10
	216	3.35	0.36	223	2.53	0.00
	254	2.39	0.29	264	2.56	0.00
TCA	159	58.0	2.70	161	32.5	0.53
	<u>216</u>	100	0.25	<u>225</u>	100	0.01
	218	95.4	0.29	227	94.6	0.07
	220	30.7	1.29	229	30.6	0.84
TFA	159	13.5	2.68	161	4.91	2.73
	<u>168</u>	100	0.61	<u>177</u>	100	0.10
	186	4.60	2.61	191	2.27	1.98
	308	4.48	0.00	318	4.21	0.00
PFP	159	13.7	2.25	161	5.66	2.72
	<u>218</u>	100	0.11	<u>227</u>	100	1.54
	358	4.02	0.11	368	3.82	1.22

Table I — Stimulants

Table I-11. (Continued)

CD Group[c]	Fenfluramine			Fenfluramine-d[10]		
	Ion (m/z)[d]	Rel. int.	Analog's cont.	Ion (m/z)[d]	Rel. int.	Analog's cont.
HFB	159	15.4	2.94	161	6.66	1.19
	240	26.3	0.74	245	22.2	0.03
	<u>268</u>	100	0.13	<u>277</u>	100	0.00
	269	9.04	0.38	278	8.92	0.06
	408	3.39	0.00	418	3.32	0.00
4-CB	159	39.5	3.83	161	20.5	0.84
	220	8.98	2.45	224	9.44	3.26
	276	8.12	0.47	284	5.48	0.89
	294	9.55	0.57	303	7.79	0.23
	<u>322</u>	100	0.29	<u>331</u>	100	0.00

[a–d] See the corresponding footnotes in Table I-1a.

Table I-12. Relative intensity and cross-contribution data[a] of ions[b] with potential for designating the analyte and the adapted internal standard — Norcocaine/norcocaine-d[3]

CD Group[c]	Norcocaine			Norcocaine-d[3]		
	Ion (m/z)[d]	Rel. int.	Analog's cont.	Ion (m/z)[d]	Rel. int.	Analog's cont.
None	<u>168</u>	100	1.18	<u>171</u>	100	0.16
	289	11.0	0.85	292	10.6	0.26
TFA	100	32.5	1.44	103	34.1	1.19
	194	30.5	0.58	197	32.0	0.35
	<u>263</u>	38.1	0.09	<u>266</u>	39.7	0.18
	280	5.00	0.14	283	5.26	0.37
	316	3.47	0.07	319	3.61	0.34
	385	2.43	0.00	388	2.67	0.38
PFP	<u>100</u>	32.1	3.25	<u>103</u>	33.9	1.18
	194	30.8	1.22	197	31.8	0.23
	435	2.77	0.97	438	2.81	0.41
HFB	194	27.8	0.45	197	26.5	0.70
	334	9.10	2.25	337	7.38	0.60
	<u>363</u>	43.0	0.07	<u>366</u>	37.2	0.22
	380	4.28	0.53	383	3.91	0.26
	485	2.98	0.05	488	2.55	0.29
TMS	<u>240</u>	100	0.62	<u>243</u>	100	0.75
	256	7.56	0.67	259	7.98	1.89
	346	31.6	0.51	349	36.8	1.00
	361	15.7	0.28	364	19.0	0.99

[a–d] See the corresponding footnotes in Table I-1a.

Table I-13. Relative intensity and cross-contribution data[a] of ions[b] with potential for designating the analyte and the adapted internal standard — Cocaine/cocaine-d[3]

CD Group[c]	Cocaine			Cocaine-d[3]		
	Ion (m/z)[d]	Rel. int.	Analog's cont.	Ion (m/z)[d]	Rel. int.	Analog's cont.
None	82	91.1	4.15	85	94.3	0.42
	<u>182</u>	100	0.82	<u>185</u>	100	0.10
	198	11.9	0.99	201	9.98	0.73
	272	10.6	1.10	275	10.5	0.70
	303	26.4	0.65	306	26.0	0.36

[a–d] See the corresponding footnotes in Table I-1a.

Appendix Two — Ion Intensity Cross-Contribution Data

Table I-14a. Relative intensity and cross-contribution data[a] of ions[b] with potential for designating the analyte and the adapted internal standard — Cocaethylene/cocaethylene-d[3]

CD Group[c]	Cocaethylene			Cocaethylene-d[3]		
	Ion (m/z)[d]	Rel. int.	Analog's cont.	Ion (m/z)[d]	Rel. int.	Analog's cont.
None	82	100	3.20	85	100	0.48
	196	96.9	0.81	199	96.2	0.14
	212	12.6	1.01	215	10.1	0.72
	272	18.0	0.45	275	18.0	0.29
	317	27.8	0.56	320	28.4	0.37

[a–d] See the corresponding footnotes in Table I-1a.

Table I-14b. Relative intensity and cross-contribution data[a] of ions[b] with potential for designating the analyte and the adapted internal standard — Cocaethylene/cocaethylene-d[8]

CD Group[c]	Cocaethylene			Cocaethylene-d[8]		
	Ion (m/z)[d]	Rel. int.	Analog's cont.	Ion (m/z)[d]	Rel. int.	Analog's cont.
None	82	100	2.83	85	100	0.48
	196	96.6	0.07	204	94.5	0.07
	212	12.6	0.08	220	9.92	0.03
	272	18.0	0.09	275	14.4	0.34
	317	27.8	0.07	325	26.8	0.04

[a–d] See the corresponding footnotes in Table I-1a.

Table I-15. Relative intensity and cross-contribution data[a] of ions[b] with potential for designating the analyte and the adapted internal standard — Ecgonine methyl ester/ecgonine methyl ester-d[3]

CD Group[c]	Ecgonine methyl ester			Ecgonine methyl ester-d[3]		
	Ion (m/z)[d]	Rel. int.	Analog's cont.	Ion (m/z)[d]	Rel. int.	Analog's cont.
None	82	100	—	85	100	—
	96	68.1	4.40	99	69.8	0.86
	182	8.24	2.60	185	7.83	1.02
	199	15.6	2.26	202	15.2	0.12
TFA	182	100	4.78	185	100	0.12
	264	12.7	0.72	267	13.4	3.79
	295	25.3	0.47	298	24.0	0.18
PFP	182	100	4.60	185	100	0.09
	314	11.2	0.16	317	12.0	4.18
	345	20.3	0.19	348	21.8	0.17
HFB	182	100	—	185	100	—
	364	11.7	0.15	367	11.4	3.79
	395	19.4	0.21	398	18.7	0.18
TMS	82	100	4.40	85	100	1.57
	83	85.9	2.68	86	66.1	0.41
	96	75.5	1.81	99	73.9	1.56
	271	12.4	0.51	274	11.2	0.98
t-BDMS	82	100	2.54	85	100	1.71
	83	54.9	2.63	86	45.2	0.67
	96	41.2	2.25	99	42.5	2.57
	182	26.5	0.96	185	25.7	3.67
	256	22.1	0.52	259	20.1	0.73
	313	5.37	0.55	316	4.46	0.94

[a–d] See the corresponding footnotes in Table I-1a.

Table I — Stimulants

Table I-16a. Relative intensity and cross-contribution data[a] of ions[b] with potential for designating the analyte and the adapted internal standard — Benzoylecgonine/benzoylecgonine-d$_3$

CD Group[c]	Benzoylecgonine			Benzoylecgonine-d$_3$		
	Ion (m/z)[d]	Rel. int.	Analog's cont.	Ion (m/z)[d]	Rel. int.	Analog's cont.
Methyl	82	83.6	3.51	85	83.5	1.42
	83	33.4	3.93	86	22.3	0.72
	<u>182</u>	100	3.60	<u>185</u>	100	0.08
	198	12.4	0.31	201	9.82	0.59
	272	10.4	0.10	275	10.3	0.75
	303	26.2	0.12	306	26.4	0.29
Ethyl	82	79.7	2.81	85	89.4	0.52
	<u>196</u>	100	2.63	<u>199</u>	100	0.09
	212	13.2	0.33	215	10.8	0.45
	272	18.9	0.12	275	20.2	0.28
	317	29.7	0.13	320	32.5	0.33
Propyl	<u>82</u>	86.2	2.15	<u>85</u>	98.9	0.41
	210	100	0.44	213	100	0.12
	226	13.8	0.23	229	10.9	0.21
	272	21.4	0.18	275	21.1	0.22
	331	22.7	0.17	334	21.7	0.38
Butyl	82	100	2.32	85	92.1	0.49
	<u>224</u>	99.1	2.19	<u>227</u>	100	0.12
	240	14.6	0.13	243	12.1	0.54
	272	20.6	0.10	275	22.4	0.20
	345	25.5	0.11	348	29.8	0.30
Pentafluoro-1-propoxy	82	47.0	3.01	85	56.8	0.26
	272	13.0	1.75	275	12.7	0.69
	<u>300</u>	100	0.25	<u>303</u>	100	0.11
	316	13.0	0.19	319	9.60	0.17
	421	21.2	0.12	424	19.6	0.32
Hexafluoro-2-propoxy	164	11.5	1.21	167	11.8	3.35
	272	9.65	0.55	275	9.51	0.68
	<u>318</u>	100	—	<u>321</u>	100	—
	334	12.6	0.39	337	9.71	0.15
	439	20.5	0.10	442	20.5	0.20
TMS	<u>82</u>	100	1.95	<u>85</u>	100	0.52
	240	71.6	0.73	243	71.9	0.76
	256	11.6	0.45	259	9.36	0.89
	346	7.59	1.03	349	7.53	1.37
	361	21.7	0.18	364	21.5	1.36
t-BDMS	<u>82</u>	100	2.36	<u>85</u>	100	0.67
	83	35.5	4.03	86	25.3	0.80
	282	51.7	2.52	285	45.0	0.97
	298	9.15	0.25	301	6.21	1.43
	346	40.7	0.12	349	31.9	1.31
	403	28.6	0.15	406	20.2	1.41

[a–d] See the corresponding footnotes in Table I-1a.

Table I-16b. Relative intensity and cross-contribution data[a] of ions[b] with potential for designating the analyte and the adapted internal standard — Benzoylecgonine/benzoylecgonine-d$_8$

CD Group[c]	Benzoylecgonine			Benzoylecgonine-d$_8$		
	Ion (m/z)[d]	Rel. int.	Analog's cont.	Ion (m/z)[d]	Rel. int.	Analog's cont.
Methyl	<u>182</u>	100	3.86	<u>185</u>	100	0.15
	198	12.4	0.83	201	9.98	0.54
	272	10.4	0.09	280	10.1	0.10
	303	26.2	0.08	311	25.2	0.02

Table I-16b. (Continued)

CD Group[c]	| Benzoylecgonine			Benzoylecgonine-d$_8$		
	Ion (m/z)[d]	Rel. int.	Analog's cont.	Ion (m/z)[d]	Rel. int.	Analog's cont.
Ethyl	<u>196</u>	100	2.38	<u>199</u>	100	0.13
	212	13.2	0.57	215	10.6	0.44
	272	18.9	0.06	280	20.4	0.08
	317	29.7	0.07	325	33.8	0.04
Propyl	<u>82</u>	—	—	<u>85</u>	95.4	—
	210	100	0.25	213	100	0.14
	226	13.8	0.25	229	11.1	0.22
	272	21.4	0.11	280	20.5	0.06
	331	22.7	0.12	339	21.7	0.18
Butyl	105	37.5	1.69	110	32.7	4.86
	<u>224</u>	99.1	2.68	<u>227</u>	100	0.17
	240	14.6	0.21	243	12.2	0.62
	272	20.6	0.06	280	22.1	0.07
	345	25.5	0.06	353	29.5	0.01
PFPoxy	<u>300</u>	100	0.23	<u>303</u>	100	0.10
	316	13.0	0.17	319	9.75	0.16
	421	21.2	0.05	429	19.3	0.11
HFPoxy	164	11.5	0.89	167	13.3	4.73
	272	9.65	0.29	280	9.26	0.08
	334	12.6	0.33	337	10.3	0.16
	<u>439</u>	20.5	0.01	<u>447</u>	24.1	0.00
TMS	<u>82</u>	—	—	85	100	—
	122	10.3	4.43	125	8.73	1.81
	240	71.6	0.66	243	72.6	0.79
	256	11.6	0.40	259	9.54	1.00
	346	7.59	0.16	354	7.86	0.24
t-BDMS	<u>82</u>	—	—	85	100	—
	204	20.9	0.24	212	13.9	1.93
	282	51.7	2.63	285	50.9	0.78
	346	40.7	0.06	354	39.6	0.01
	403	28.6	0.05	411	27.5	0.00

[a–d] See the corresponding footnotes in Table I-1a.

Table I-17. Relative intensity and cross-contribution data[a] of ions[b] with potential for designating the analyte and the adapted internal standard — Ecgonine/ecgonine-d$_3$

CD Group[c]	| Ecgonine			Ecgonine-d$_3$		
	Ion (m/z)[d]	Rel. int.	Analog's cont.	Ion (m/z)[d]	Rel. int.	Analog's cont.
[TMS]$_2$	82	96.3	5.00	85	100	1.10
	<u>83</u>	100	3.00	<u>86</u>	76.6	0.28
	96	59.3	3.27	99	58.2	1.69
	212	9.21	3.19	215	8.66	3.84
	329	8.99	0.86	332	8.53	1.80
[t-BDMS]$_2$	<u>82</u>	100	1.92	<u>85</u>	100	2.16
	83	69.2	2.04	86	56.0	0.45
	96	29.2	3.59	99	29.8	4.71
	356	42.3	0.49	359	38.2	2.47
	357	12.8	1.44	360	11.7	1.59
	413	2.80	0.49	416	2.54	3.61
HFPoxy/TFA	264	11.6	0.81	267	11.5	1.53
	<u>318</u>	100	1.75	<u>321</u>	100	0.09
	431	18.4	0.89	434	17.2	0.19

Table I — Stimulants

Table I-17. (Continued)

CD Group[c]	Ecgonine			Ecgonine-d$_3$		
	Ion (m/z)[d]	Rel. int.	Analog's cont.	Ion (m/z)[d]	Rel. int.	Analog's cont.
PFPoxy/PFP	300	100	2.47	303	100	0.09
	314	12.8	1.03	317	13.0	2.35
	463	14.9	0.56	466	14.2	0.21
HFPoxy/HFB	318	100	2.87	321	100	0.10
	364	10.6	0.62	367	10.5	0.17
	531	11.6	0.67	534	11.1	0.25

[a–d] See the corresponding footnotes in Table I-1a.

Table I-18. Relative intensity and cross-contribution data[a] of ions[b] with potential for designating the analyte and the adapted internal standard — Anhydroecgonine methyl ester/anhydroecgonine methyl ester-d$_3$

CD Group[c]	Anhydroecgonine methyl ester			Anhydroecgonine methyl ester-d$_3$		
	Ion (m/z)[d]	Rel. int.	Analog's cont.	Ion (m/z)[d]	Rel. int.	Analog's cont.
None	152	100	1.99	155	100	0.15
	181	33.5	1.58	184	34.9	0.11

[a–d] See the corresponding footnotes in Table I-1a.

Table I-19. Relative intensity and cross-contribution data[a] of ions[b] with potential for designating the analyte and the adapted internal standard — Caffeine/caffeine-^{13}C$_3$

CD Group[c]	Caffeine			Caffeine-^{13}C$_3$		
	Ion (m/z)[d]	Rel. int.	Analog's cont.	Ion (m/z)[d]	Rel. int.	Analog's cont.
None	109	46.7	3.84	111	43.1	2.68
	194	100	2.93	197	100	0.23

[a–d] See the corresponding footnotes in Table I-1a.

Table I-20. Relative intensity and cross-contribution data[a] of ions[b] with potential for designating the analyte and the adapted internal standard — Methylphenidate/methylphenidate-d$_3$

CD Group[c]	Methylphenidate			Methylphenidate-d$_3$		
	Ion (m/z)[d]	Rel. int.	Analog's cont.	Ion (m/z)[d]	Rel. int.	Analog's cont.
None	150	1.85	—	153	2.01	—
TFA	150	7.94	0.69	153	7.61	3.83
PFP	150	5.66	0.41	153	5.71	1.76
HFB	150	3.74	1.98	153	3.69	3.32
4-CB	438	3.19	0.15	441	2.74	0.31
TMS	59	3.24	—	62	0.60	—

[a–d] See the corresponding footnotes in Table I-1a.

Table I-21. Relative intensity and cross-contribution data[a] of ions[b] with potential for designating the analyte and the adapted internal standard — Ritalinic acid/ritalinic acid-d_5

CD Group[c]	Ritalinic acid			Ritalinic acid-d_5		
	Ion (m/z)[d]	Rel. int.	Analog's cont.	Ion (m/z)[d]	Rel. int.	Analog's cont.
4-CB	424	1.09	0.46	429	1.27	1.97
	<u>452</u>	3.49	0.27	<u>457</u>	2.92	0.00
[TMS]$_2$	<u>118</u>	2.36	—	<u>123</u>	2.29	—
t-BDMS	91	12.4	0.42	96	13.6	0.12
	137	6.28	2.71	142	6.10	0.00
	165	4.70	0.64	170	4.03	0.00
	<u>193</u>	100	0.23	<u>198</u>	100	0.00
	194	16.6	0.18	199	16.6	0.00

[a–d] See the corresponding footnotes in Table I-1a.

Table I — Stimulants

Summary of Drugs, Isotopic Analogs, and Chemical Derivatization Groups Included in Table II (Opioids)

Compound	Isotopic analog	Chemical derivatization group[a]	Table #
Heroin	d_3, d_9	None	II-1
6-Acetylmorphine	d_3, d_6	None, Acetyl, TFA, propionyl, PFP, HFB, TMS, t-BDMS	II-2
Morphine	d_3, d_6	Ethyl, propyl, butyl, [acetyl]2, [TFA]2, propionyl, [propionyl]2, [PFP]2, [HFB]2, [TMS]2, t-BDMS, [t-BDMS]2, ethyl/acetyl, ethyl/TMS, propyl/TMS, propyl/t-BDMS, butyl/TMS, butyl/t-BDMS, acetyl/TMS, acetyl/t-BDMS, propionyl/TMS	II-3
Hydromorphone	d_3, d_6	Acetyl, [acetyl]2, [TFA]2, propionyl, PFP, [PFP]2, HFB, [HFB]2, TMS, [TMS]2, t-BDMS, [t-BDMS]2, MA/ethyl, MA/acetyl, MA/propionyl, MA/TMS, MA/t-BDMS, HA/[TMS]2	II-4
Oxymorphone	d_3	[acetyl]2, [acetyl]3, [TFA]2, propionyl, [propionyl]2, [propionyl]3, [PFP]2, [HFB]2, [TMS]2, [TMS]3, t-BDMS, MA/ethyl, MA/acetyl, MA/[acetyl]2, MA/propionyl, MA/[HFB]2, MA/[TMS]2, MA/[t-BDMS]2, MA/ethyl/propionyl, MA/ethyl/TMS, MA/ethyl/t-BDMS, MA/acetyl/TMS, MA/propionyl/TMS, HA/[TMS]3, HA/[ethyl]2/propionyl, HA/[ethyl]2/TMS	II-5
6-Acetylcodeine	d_3	None	II-6
Codeine	d_3, d_6, $^{13}C_1d_3$	None, acetyl, TFA, propionyl, PFP, HFB, TMS, t-BDMS	II-7
Hydrocodone	d_3, d_6	None, ethyl, acetyl, TMS, t-BDMS, MA, HA/TMS	II-8
Dihydrocodeine	d_3, d_6	None, acetyl, TFA, propionyl, PFP, HFB, TMS, t-BDMS	II-9
Oxycodone	d_3, d_6	None, acetyl, [acetyl]2, propionyl, TMS, [TMS]2, t-BDMS, [t-BDMS]2, MA, MA/propionyl, MA/TMS, HA/[propionyl]2, HA/[TMS]2, HA/ethyl/propionyl	II-10
Noroxycodone	d_3	None, [acetyl]2, [TFA]3, propionyl, [PFP]2, [HFB]2, [TMS]2, [TMS]3, MA/ethyl, MA/acetyl, MA/[TFA]2, MA/propionyl, MA/PFP, MA/[HFB]2, MA/[TMS]2, MA/t-BDMS, MA/ethyl/propionyl, MA/ethyl/TMS, MA/ethyl/t-BDMS, MA/acetyl/TMS, MA/propionyl/TMS, HA/[ethyl]2/TMS	II-11
Buprenorphine	d_4	Methyl, ethyl, acetyl, MBTFA, PFP, HFB, TMS, [TMS]2, t-BDMS	II-12
Norbuprenorphine	d_3	[Methyl]2, [ethyl]2, [acetyl]2, [MBTFA]2, [PFP]2, [HFB]2, [TMS]2, [TMS]3, t-BDMS	II-13
Fentanyl	d_5	None	II-14
Norfentanyl	d_5	None, acetyl, TCA, TFA, PFP, HFB, 4-CB, TMS, t-BDMS	II-15
Methadone	d_3, d_9	None	II-16
EDDP	d_3	None	II-17
Propoxyphene	d_5, d_7, d_{11}	None	II-18
Norpropoxyphene	d_5	None	II-19
Meperidine	d_4	None	II-20
Normeperidine	d_4	None, ethyl, propyl, butyl, acetyl, TCA, TFA, PFP, HFB, 4-CB, TMS, t-BDMS	II-21

[a] MA: methoxyimino; HA: hydroxylimino.

Tanle II — Opioids

Appendix Two — Table II
Cross-Contributions Between ions Designating the Drugs and Their Isotopically Labeled Analogs in Various Derivatization Forms — Opioids

Table II-1a. Relative intensity and cross-contribution data[a] of ions[b] with potential for designating the analyte and the adapted internal standard — Heroin/heroin-d$_3$

CD Group[c]	Heroin			Heroin-d$_3$		
	Ion (m/z)[d]	Rel. int.	Analog's cont.	Ion (m/z)[d]	Rel. int.	Analog's cont.
None	215	30.6	3.75	218	29.3	1.86
	268	57.8	2.62	271	56.2	1.41
	<u>327</u>	100	0.95	<u>330</u>	100	0.32
	328	21.1	3.89	331	21.4	0.16
	369	70.5	0.73	372	71.9	0.42

[a–d] See the corresponding footnotes in Table I-1a.

Table II-1b. Relative intensity and cross-contribution data[a] of ions[b] with potential for designating the analyte and the adapted internal standard — Heroin/heroin-d$_9$

CD Group[c]	Heroin			Heroin-d$_9$		
	Ion (m/z)[d]	Rel. int.	Analog's cont.	Ion (m/z)[d]	Rel. int.	Analog's cont.
None	268	58.0	1.82	272	64.0	0.19
	310	51.0	0.02	316	76.0	0.01
	<u>327</u>	100	0.12	<u>334</u>	100	0.01
	369	69.0	0.00	378	81.0	0.00

[a–d] See the corresponding footnotes in Table I-1a.

Table II-2a. Relative intensity and cross-contribution data[a] of ions[b] with potential for designating the analyte and the adapted internal standard — 6-Acetylmorphine/6-acetylmorphine-d$_3$

CD Group[c]	6-Acetylmorphine			6-Acetylmorphine-d$_3$		
	Ion (m/z)[d]	Rel. int.	Analog's cont.	Ion (m/z)[d]	Rel. int.	Analog's cont.
None	268	89.7	2.07	271	86.1	0.61
	<u>327</u>	100	1.04	<u>330</u>	100	0.24
	328	22.8	2.29	331	20.5	0.23
Acetyl	215	38.1	3.76	218	36.7	1.85
	268	69.6	2.61	271	68.8	1.10
	310	55.9	0.61	313	56.7	0.64
	<u>327</u>	100	0.93	<u>330</u>	100	0.34
	369	68.6	0.49	372	68.9	0.45
TFA	204	29.4	2.66	207	31.6	4.69
	311	29.0	1.77	314	29.6	0.47
	<u>364</u>	100	1.36	<u>367</u>	100	0.49
	380	10.4	1.57	383	10.6	0.64
	423	61.2	1.14	426	57.1	0.44
Propionyl	215	20.2	3.51	218	19.6	1.91
	<u>268</u>	49.7	2.48	<u>271</u>	47.6	1.65
	383	50.7	0.19	386	51.2	0.47
	384	12.5	2.50	387	12.7	0.18
PFP	361	30.1	3.55	364	29.9	0.36
	<u>414</u>	100	3.16	<u>417</u>	100	0.47
	473	57.1	2.87	476	57.2	0.21

Tanle II — Opioids

Table II-2a. (Continued)

CD Group[c]	6-Acetylmorphine			6-Acetylmorphine-d₃		
	Ion (m/z)[d]	Rel. int.	Analog's cont.	Ion (m/z)[d]	Rel. int.	Analog's cont.
HFB	411	30.9	1.79	414	29.8	0.45
	<u>464</u>	100	1.56	<u>467</u>	100	0.51
	480	10.6	1.75	483	10.5	0.69
	523	54.1	1.19	526	53.7	0.50
TMS	<u>399</u>	100	2.13	<u>402</u>	100	1.38
t-BDMS	<u>342</u>	100	0.75	<u>345</u>	100	1.14
	441	61.4	0.20	444	65.4	1.72
	442	20.6	2.42	445	21.6	0.77

[a–d] See the corresponding footnotes in Table I-1a.

Table II-2b. Relative intensity and cross-contribution data[a] of ions[b] with potential for designating the analyte and the adapted internal standard — 6-Acetylmorphine/6-acetylmorphine-d₆

CD Group[c]	6-Acetylmorphine			6-Acetylmorphine-d₆		
	Ion (m/z)[d]	Rel. int.	Analog's cont.	Ion (m/z)[d]	Rel. int.	Analog's cont.
None	<u>268</u>	89.7	1.87	<u>271</u>	100	2.06
	327	100	1.38	333	91.6	0.04
	328	22.8	1.15	334	20.7	0.00
Acetyl	215	38.1	4.01	218	38.4	1.80
	268	69.6	1.49	271	69.3	1.11
	310	55.9	0.34	313	58.1	0.67
	<u>327</u>	100	0.39	<u>333</u>	100	0.04
	369	68.6	0.30	375	68.0	0.04
TFA	311	29.0	0.87	314	28.3	1.33
	<u>364</u>	100	0.31	<u>367</u>	100	1.35
	380	10.4	0.42	383	6.78	2.68
	423	61.2	0.02	429	58.6	0.09
Propionyl	215	20.2	3.66	218	21.7	2.10
	<u>268</u>	49.7	1.72	<u>271</u>	51.0	1.78
	383	50.7	0.15	389	50.1	0.00
	384	12.5	0.19	390	12.5	0.00
PFP	361	30.1	3.40	364	29.3	2.63
	<u>414</u>	100	2.77	<u>417</u>	100	2.33
	430	10.4	4.05	433	7.08	4.04
	473	57.1	2.54	479	55.6	0.00
HFB	411	30.9	3.10	414	30.4	1.76
	<u>464</u>	100	2.50	<u>467</u>	100	1.86
	480	10.6	2.81	483	6.79	3.24
	523	54.1	2.18	529	51.4	0.00
TMS	<u>399</u>	100	4.96	<u>405</u>	100	0.05
t-BDMS	<u>342</u>	100	0.80	<u>346</u>	100	0.17
	384	42.5	2.95	390	48.3	0.00
	441	61.4	0.12	447	73.9	0.00
	442	20.6	0.14	448	25.0	0.00

[a–d] See the corresponding footnotes in Table I-1a.

Table II-3a. Relative intensity and cross-contribution data[a] of ions[b] with potential for designating the analyte and the adapted internal standard — Morphine/morphine-d$_3$

CD Group[c]	Morphine			Morphine-d$_3$		
	Ion (m/z)[d]	Rel. int.	Analog's cont.	Ion (m/z)[d]	Rel. int.	Analog's cont.
Ethyl	313	100	1.27	316	100	0.38
Propyl	327	100	0.47	330	100	0.30
Butyl	341	100	0.77	344	100	0.32
	342	23.3	0.09	345	23.5	0.13
[Acetyl]$_2$	215	36.8	3.49	218	38.5	1.81
	268	67.6	2.16	271	67.7	1.05
	310	55.1	0.42	313	57.0	0.53
	327	100	0.59	330	100	0.31
	369	66.6	0.32	372	67.8	0.44
[TFA]$_2$	311	6.24	1.43	314	6.54	0.42
	364	100	0.17	367	100	0.29
	380	6.32	0.35	383	6.51	0.39
	458	1.51	0.00	461	1.80	2.31
	477	37.4	0.13	480	37.9	0.41
Propionyl	268	95.0	2.43	271	89.1	0.45
	341	100	2.47	344	100	0.37
[Propionyl]$_2$	268	51.1	1.82	271	48.0	0.96
	324	39.7	0.28	327	39.7	2.55
	341	100	0.48	344	100	0.34
	342	22.0	4.13	345	22.5	0.19
	397	45.9	0.25	400	48.1	0.53
[PFP]$_2$	361	5.35	1.45	364	5.40	5.45
	414	100	0.31	417	100	0.33
	430	6.98	0.94	433	7.11	0.34
	558	3.77	0.00	561	3.95	1.71
	577	26.1	0.28	580	26.2	0.56
[HFB]$_2$	266	4.41	4.06	269	4.56	2.33
	411	5.36	2.41	414	5.43	0.71
	464	100	1.46	467	100	0.35
	480	7.42	1.43	483	8.15	0.43
	658	5.19	1.22	661	5.94	1.16
	677	18.0	1.38	680	18.8	0.61
[TMS]$_2$	236	57.8	—	239	54.3	—
	401	25.8	2.89	404	26.2	2.88
	429	100	0.48	432	100	3.50
t-BDMS	342	100	3.08	345	100	1.14
	399	53.8	0.65	402	52.1	1.58
[t-BDMS]$_2$	146	17.9	—	149	18.4	—
	456	53.2	2.08	459	46.7	4.31
	485	3.92	0.00	488	3.94	0.00
	513	6.51	0.00	516	6.24	0.00
Ethyl/acetyl	243	21.3	4.65	246	20.2	0.58
	296	56.8	0.81	299	53.7	1.98
	326	15.6	0.46	329	14.0	0.51
	355	100	0.37	358	100	0.36
Ethyl/TMS	146	37.6	1.50	149	37.0	2.11
	357	26.8	4.04	360	26.4	1.40
	385	100	0.18	388	100	1.65
	386	29.5	1.78	389	29.9	0.78
Propyl/TMS	206	60.2	—	209	51.9	—
	146	39.7	1.98	149	34.3	2.25
	196	46.1	3.92	199	39.5	4.60
	371	28.1	0.74	374	27.3	1.44
	399	100	0.16	402	100	1.64

Tanle II — Opioids

Table II-3a. (Continued)

CD Group[c]	Morphine Ion (m/z)[d]	Rel. int.	Analog's cont.	Morphine-d₃ Ion (m/z)[d]	Rel. int.	Analog's cont.
Propyl/t-BDMS	146	39.2	4.15	149	33.0	2.59
	_384_	64.5	0.00	_387_	71.2	0.00
	398	9.02	0.00	401	9.05	0.00
	441	10.7	0.00	444	11.9	0.00
Butyl/TMS	_220_	63.9	—	_223_	64.1	—
	146	41.5	2.98	149	43.1	2.87
	234	42.1	2.90	237	43.1	4.29
	385	29.8	3.03	388	28.9	2.37
	413	100	1.44	416	100	1.80
Butyl/t-BDMS	_398_	85.1	0.00	_401_	79.9	0.00
	427	8.36	0.00	430	7.52	0.00
	455	17.1	0.00	458	16.2	0.00
Acetyl/TMS	324	10.5	3.16	327	11.1	0.86
	_399_	100	0.51	_402_	100	3.17
Acetyl/t-BDMS	_342_	100	1.46	_345_	100	1.15
	441	70.6	0.94	444	73.4	1.58
	442	22.8	3.59	445	24.5	0.79
Propionyl/TMS	_357_	100	1.92	_360_	100	1.12
	413	78.0	0.36	416	82.1	1.43

[a–d] See the corresponding footnotes in Table I-1a.

Table II-3b. Relative intensity and cross-contribution data[a] of ions[b] with potential for designating the analyte and the adapted internal standard — Morphine/morphine-d₆

CD Group[c]	Morphine Ion (m/z)[d]	Rel. int.	Analog's cont.	Morphine-d₆ Ion (m/z)[d]	Rel. int.	Analog's cont.
Ethyl	284	24.5	4.87	290	22.8	0.12
	_313_	100	1.47	_319_	100	0.04
	314	20.9	2.04	320	21.8	0.23
Propyl	284	23.9	1.70	290	22.7	0.36
	_327_	100	0.60	_333_	100	0.02
	328	22.7	1.02	334	22.4	0.00
Butyl	_341_	100	0.21	_347_	100	0.05
	342	23.3	0.22	348	23.4	0.09
[Acetyl]₂	268	67.6	1.75	274	68.9	0.02
	284	9.18	0.95	290	8.64	0.00
	310	55.1	0.41	316	60.0	0.00
	_327_	100	0.35	_333_	100	0.00
	369	66.6	0.32	375	68.0	0.00
[TFA]₂	311	6.24	2.77	317	6.70	1.22
	_364_	100	0.12	_370_	100	0.00
	380	6.32	0.24	386	6.21	0.07
	477	37.4	0.09	483	34.9	0.00
Propionyl	268	95.0	1.87	274	91.4	0.00
	_341_	100	2.15	_347_	100	0.00
	342	21.3	2.60	348	2.00	0.00
[Propionyl]₂	268	51.1	2.69	274	48.8	0.03
	324	39.7	0.21	330	40.6	0.03
	_341_	100	0.32	_347_	100	0.00
	342	22.0	0.84	348	22.1	0.00
	397	45.9	0.19	403	46.8	0.02

Appendix Two — Ion Intensity Cross-Contribution Data

Table II-3b. (Continued)

CD Group[c]	Morphine			Morphine-d[6]		
	Ion (m/z)[d]	Rel. int.	Analog's cont.	Ion (m/z)[d]	Rel. int.	Analog's cont.
[PFP][2]	361	5.35	3.51	367	5.54	1.27
	_414_	100	0.84	_420_	100	0.00
	430	6.98	1.66	436	6.92	0.02
	577	26.1	0.78	583	26.0	0.00
[HFB][2]	266	4.41	4.39	272	3.79	1.13
	_464_	100	3.88	_470_	100	0.01
	480	7.42	3.64	486	8.14	0.00
	658	5.19	3.78	664	6.41	2.94
	677	18.0	3.74	683	20.5	0.00
[TMS][2]	_236_	57.8	—	_239_	60.7	—
	324	18.2	3.69	330	17.3	2.36
	401	25.8	2.44	404	30.6	2.85
	414	44.9	1.30	420	42.4	0.00
	429	100	1.02	435	100	0.00
t-BDMS	162	22.8	3.87	168	20.6	4.87
	_342_	100	3.03	_348_	100	0.07
	399	53.8	0.65	405	51.8	0.00
[t-BDMS][2]	_413_	61.8	—	_415_	63.8	—
	146	17.5	2.76	149	20.4	4.73
	456	53.2	1.27	462	54.0	0.00
	513	6.51	0.00	519	6.75	0.00
Ethyl/acetyl	266	15.9	0.00	272	12.9	0.00
	296	76.3	0.00	302	82.3	0.00
	312	9.75	0.00	318	10.6	0.00
	326	19.6	0.00	332	15.8	0.00
	_355_	100	3.50	_361_	100	0.00
	356	19.4	4.27	362	27.5	0.00
Ethyl/TMS	_146_	37.6	2.96	_149_	44.1	2.43
	357	26.8	1.13	360	28.6	1.53
	385	100	0.09	391	100	0.11
	386	29.5	0.30	392	30.4	0.14
Propyl/TMS	_206_	60.2	–	_209_	59.7	—
	146	39.7	2.74	149	38.3	2.55
	371	28.1	0.56	374	29.7	1.49
	399	100	0.09	405	100	0.04
Propyl/t-BDMS	_384_	64.5	0.00	_390_	59.1	0.00
	441	10.7	0.00	447	15.2	0.00
Butyl/TMS	_220_	63.9	—	_223_	71.0	—
	146	41.5	3.54	149	45.3	3.19
	385	29.8	2.52	388	32.6	2.57
	413	100	0.97	419	100	0.48
Butyl/t-BDMS	_398_	85.1	0.00	_404_	67.8	0.00
	455	17.1	0.00	461	6.41	0.00
Acetyl/TMS	287	47.2	2.48	293	37.7	4.69
	340	75.8	0.00	346	71.3	0.00
	_399_	98.3	0.00	_405_	79.7	0.00
Acetyl/t-BDMS	_342_	100	0.00	_348_	100	0.00
	343	24.7	0.00	349	28.0	0.00
	384	45.6	0.00	388	36.5	0.00
	441	50.2	0.00	447	48.3	0.00
Propionyl/TMS	_384_	64.5	—	_390_	59.1	—

[a–d] See the corresponding footnotes in Table I-1a.

Tanle II — Opioids

Table II-4a. Relative intensity and cross-contribution data[a] of ions[b] with potential for designating the analyte and the adapted internal standard — Hydromorphone/hydromrophone-d₃

CD Group[c]	Hydromorphone			Hydromorphone-d₃		
	Ion (m/z)[d]	Rel. int.	Analog's cont.	Ion (m/z)[d]	Rel. int.	Analog's cont.
Acetyl	256	10.3	4.52	259	10.2	3.47
	<u>285</u>	100	2.40	<u>288</u>	100	0.37
	327	26.3	1.84	330	26.7	0.41
[Acetyl]₂ (enol)	<u>237</u>	100	—	<u>330</u>	100	—
[TFA]₂ (enol)	258	36.1	3.05	261	37.2	1.21
	364	27.3	1.99	367	27.6	3.41
	380	49.3	0.36	383	50.4	0.29
	<u>477</u>	100	0.52	<u>480</u>	100	0.37
Propionyl	<u>285</u>	100	—	<u>288</u>	100	—
PFP	<u>431</u>	100	—	<u>434</u>	100	—
[PFP]₂ (enol)	308	41.8	1.74	311	39.3	1.33
	414	37.6	1.45	417	36.4	0.62
	430	63.9	0.28	433	63.8	0.33
	<u>577</u>	100	0.52	<u>580</u>	100	0.43
HFB	<u>481</u>	100	—	<u>484</u>	100	—
[HFB]₂ (enol)	358	61.5	3.57	361	60.4	2.51
	464	56.0	3.06	467	57.9	3.43
	<u>480</u>	100	1.34	<u>483</u>	95.3	1.62
	677	95.9	1.50	680	100	1.30
TMS	342	25.9	3.44	345	24.8	1.85
	<u>357</u>	100	0.78	<u>360</u>	100	1.64
[TMS]₂ (enol)	<u>234</u>	55.3	2.25	<u>237</u>	58.2	4.19
	429	100	0.60	432	100	3.77
	430	35.5	3.59	433	36.5	2.45
t-BDMS	<u>342</u>	28.1	—	<u>345</u>	26.8	—
	399	7.23	2.60	402	6.81	2.48
[t-BDMS]₂ (enol)	<u>456</u>	100	2.18	<u>459</u>	100	3.82
	513	8.27	3.46	516	8.39	4.61
Methoxyimino/ ethyl	<u>342</u>	100	0.62	<u>345</u>	100	0.31
	343	21.4	4.34	346	21.6	0.13
Methoxyimino/ acetyl	<u>314</u>	100	0.42	<u>314</u>	100	0.00
	356	44.9	0.00	359	49.7	0.00
Methoxyimino/ propionyl	<u>314</u>	100	1.18	<u>317</u>	100	0.29
	370	38.9	0.54	373	38.3	0.43
Methoxyimino/ TMS	355	82.7	1.48	358	80.6	4.86
	371	45.6	2.25	374	42.0	1.26
	<u>386</u>	100	0.33	<u>389</u>	100	1.34
	387	29.5	2.97	390	27.0	0.58
Methoxyimino/ t-BDMS	<u>371</u>	31.3	1.41	<u>374</u>	100	1.33
	428	11.2	0.33	431	33.5	1.67
Hydroxyimino/ [TMS]₂	<u>355</u>	79.3	0.55	<u>358</u>	64.5	1.62
	429	34.6	1.22	432	37.5	2.79
	444	64.8	0.00	447	56.5	2.91

[a–d] See the corresponding footnotes in Table I-1a.

Table II-4b. Relative intensity and cross-contribution data[a] of ions[b] with potential for designating the analyte and the adapted internal standard — Hydromorphone/hydromorphone-d6

CD Group[c]	Hydromorphone			Hydromorphone-d6		
	Ion (m/z)[d]	Rel. int.	Analog's cont.	Ion (m/z)[d]	Rel. int.	Analog's cont.
Acetyl	<u>285</u>	100	1.93	<u>291</u>	100	0.10
	327	26.3	1.65	333	25.5	0.04
[Acetyl]2 (enol)	326	31.4	4.18	332	30.1	0.72
	<u>327</u>	100	1.71	<u>333</u>	100	0.07
	369	44.5	2.45	375	44.9	0.38
[TFA]2 (enol)	258	36.1	2.57	264	32.7	2.27
	364	27.3	0.32	370	29.1	0.03
	380	49.3	0.15	386	55.1	0.00
	476	45.6	0.08	482	33.1	0.00
	<u>477</u>	100	0.09	<u>483</u>	100	0.00
Propionyl	<u>285</u>	100	—	<u>291</u>	100	—
PFP	<u>431</u>	100	—	<u>437</u>	100	—
[PFP]2 (enol)	308	41.8	1.58	314	37.3	0.23
	414	37.6	0.30	420	39.0	0.03
	430	63.9	0.05	436	68.9	0.02
	<u>577</u>	100	0.12	<u>583</u>	100	0.00
	578	24.1	1.00	584	25.3	0.00
HFB	<u>425</u>	75.9	—	<u>431</u>	72.8	—
[HFB]2 (enol)	464	56.0	1.35	470	57.4	0.07
	<u>480</u>	100	0.85	<u>486</u>	98.0	0.04
	677	95.9	0.61	683	100	0.00
	678	25.1	2.87	684	26.7	0.00
TMS	314	19.7	3.49	320	17.5	0.58
	342	25.7	0.87	348	25.9	0.34
	<u>357</u>	100	0.43	<u>363</u>	100	0.16
	358	27.8	1.18	364	27.9	0.08
[TMS]2 (enol)	184	24.8	3.42	188	28.0	1.86
	<u>234</u>	55.3	2.67	<u>240</u>	54.4	3.62
	324	17.1	3.95	330	18.0	3.08
	414	91.6	1.71	420	72.8	0.11
	429	100	0.12	435	100	0.14
	430	35.5	0.76	436	36.3	0.11
t-BDMS	<u>299</u>	100	—	<u>301</u>	100	—
	342	28.1	2.91	348	28.0	0.66
	399	7.23	1.35	405	6.86	0.59
[t-BDMS]2 (enol)	<u>456</u>	100	1.75	<u>462</u>	100	0.32
	457	38.4	4.51	463	38.1	0.33
	513	8.23	3.18	519	7.85	0.60
Methoxyimino/ ethyl	311	85.0	0.39	317	85.7	0.02
	<u>342</u>	100	0.12	<u>348</u>	100	0.00
	343	21.4	1.20	349	23.1	0.00
Methoxyimino/ acetyl	283	79.8	0.55	289	75.1	0.00
	<u>314</u>	100	0.31	<u>320</u>	100	0.00
	325	19.5	0.00	331	15.5	0.00
	356	44.9	0.00	362	42.3	0.00
Methoxyimino/ propionyl	283	58.7	0.56	289	58.4	0.04
	<u>314</u>	100	0.50	<u>320</u>	100	0.00
	315	20.8	2.32	321	20.9	0.21
	339	9.48	0.22	345	9.24	0.31
	370	38.9	0.12	376	37.3	0.01

Tanle II — Opioids

Table II-4b. (Continued)

CD Group[c]	Hydromorphone			Hydromorphone-d$_6$		
	Ion (m/z)[d]	Rel. int.	Analog's cont.	Ion (m/z)[d]	Rel. int.	Analog's cont.
Methoxyimino/	355	90.7	5.60	361	86.4	0.09
TMS	356	24.5	7.88	362	24.0	0.10
	371	40.3	0.83	377	41.3	0.14
	<u>386</u>	100	0.06	<u>392</u>	100	0.06
	387	29.7	1.37	393	41.3	0.04
Methoxyimino/	<u>371</u>	31.3	1.36	<u>377</u>	100	0.09
t-BDMS	397	5.45	0.24	403	15.0	0.44
	428	11.2	0.04	434	32.9	0.00
Hydroxyimino/	339	12.5	0.00	345	8.63	1.16
[TMS]$_2$	<u>355</u>	79.3	0.00	<u>361</u>	59.8	0.75
	356	19.7	3.47	362	15.2	0.83
	429	34.6	0.08	435	34.4	0.00
	444	64.8	0.00	450	48.6	0.00

[a–d] See the corresponding footnotes in Table I-1a.

Table II-5. Relative intensity and cross-contribution data[a] of ions[b] with potential for designating the analyte and the adapted internal standard — Oxymorphone/oxymorphone-d$_3$

CD Group[c]	Oxymorphone			Oxymorphone-d$_3$		
	Ion (m/z)[d]	Rel. int.	Analog's cont.	Ion (m/z)[d]	Rel. int.	Analog's cont.
[Acetyl]$_2$	300	37.6	2.10	303	37.6	0.50
	<u>343</u>	100	1.31	<u>346</u>	100	0.45
	344	21.4	2.82	347	21.1	0.41
	385	30.4	1.08	388	31.0	0.51
[Acetyl]$_3$ (enol)	342	23.9	1.78	345	23.1	0.87
	<u>385</u>	100	0.72	<u>388</u>	100	0.43
	386	23.5	1.61	389	23.4	0.32
	427	54.1	0.74	430	54.0	0.85
[TFA]$_2$	396	26.6	2.01	399	26.0	0.89
	<u>493</u>	100	1.72	<u>496</u>	100	0.50
Propionyl	<u>301</u>	100	—	<u>304</u>	100	—
[Propionyl]$_2$	<u>357</u>	100	—	<u>360</u>	100	—
[Propionyl]$_3$ (enol)	356	15.5	3.42	359	14.9	2.11
	396	6.88	2.66	399	7.26	2.48
	<u>413</u>	100	2.15	<u>416</u>	100	1.67
	414	25.2	3.38	417	25.0	1.37
	469	33.7	1.99	472	34.6	1.85
[PFP]$_2$	446	33.1	2.62	449	37.5	0.43
	<u>593</u>	100	2.64	<u>596</u>	100	0.52
[HFB]$_2$	308	21.0	2.88	311	19.5	1.67
	496	52.5	3.16	499	56.7	0.80
	<u>693</u>	100	3.38	<u>696</u>	100	0.64
[TMS]$_2$	430	20.6	1.82	433	20.6	2.80
	<u>445</u>	100	0.34	<u>448</u>	100	2.84
	446	36.2	1.86	449	35.8	1.74
[TMS]$_3$ (enol)	<u>502</u>	77.0	—	<u>505</u>	73.6	—
	518	40.4	2.45	521	39.1	3.07
t-BDMS	<u>358</u>	100	4.89	<u>361</u>	100	1.50

Table II-5. (Continued)

CD Group[c]	Oxymorphone			Oxymorphone-d₃		
	Ion (m/z)[d]	Rel. int.	Analog's cont.	Ion (m/z)[d]	Rel. int.	Analog's cont.
Methoxyimino/ ethyl	329	10.1	1.65	332	11.8	0.07
	<u>358</u>	100	0.00	<u>361</u>	100	0.00
	359	24.1	0.00	362	21.7	0.03
Methoxyimino/ acetyl	329	19.9	4.26	332	14.8	0.00
	341	6.75	0.00	344	6.15	0.00
	355	9.07	0.00	358	7.25	0.00
	<u>372</u>	100	0.03	<u>375</u>	100	0.00
Methoxyimino/ [acetyl]₂	329	17.8	0.00	332	19.6	0.33
	341	6.20	0.00	344	5.10	0.00
	355	6.89	0.00	358	6.92	3.28
	371	25.6	0.00	374	25.8	0.00
	<u>372</u>	100	0.00	<u>375</u>	100	0.40
	414	70.4	0.00	417	65.1	0.53
Methoxyimino/ propionyl	299	23.0	1.85	302	11.9	0.00
	355	7.97	0.00	358	4.61	0.00
	<u>386</u>	53.0	4.22	<u>389</u>	51.4	0.00
Methoxyimino/ [HFB]₂	412	16.9	0.00	415	15.3	0.00
	477	4.00	0.00	480	3.19	0.00
	509	5.47	0.00	512	4.22	0.00
	<u>525</u>	18.0	0.00	<u>528</u>	17.3	0.00
	722	16.9	0.00	725	15.4	0.00
Methoxyimino/ [TMS]₂	459	14.6	0.69	462	15.2	2.90
	<u>474</u>	100	0.17	<u>477</u>	100	3.25
	475	36.5	1.72	478	38.1	1.86
Methoximino/ [t-BDMS]₂	440	34.5	0.00	443	23.2	0.00
	<u>501</u>	57.0	0.00	<u>504</u>	71.2	0.00
	558	5.96	0.00	561	9.04	0.00
Methoxyimino/ ethyl/propionyl	244	25.6	4.32	247	22.1	0.75
	341	7.94	3.84	344	8.75	0.92
	357	35.1	1.32	360	33.0	0.33
	385	21.2	0.36	388	21.4	0.42
	<u>414</u>	100	0.00	<u>417</u>	100	0.58
	415	24.5	1.32	418	25.5	0.17
Methoxyimino/ ethyl/TMS	415	14.7	0.64	418	12.1	1.31
	<u>430</u>	100	0.06	<u>433</u>	100	1.31
Methoxyimino/ ethyl/t-BDMS	329	10.1	1.65	332	11.8	0.07
	<u>358</u>	100	0.00	<u>361</u>	100	0.00
	359	24.1	0.00	362	21.7	0.03
Methoxyimino/ acetyl/TMS	215	59.0	3.98	218	64.9	2.96
	402	62.5	0.79	405	63.3	1.38
	<u>444</u>	100	0.06	<u>447</u>	100	1.57
	445	30.9	2.86	448	31.5	0.75
Methoxyimino/ propionyl/TMS	215	42.8	4.96	218	41.6	1.87
	<u>402</u>	100	0.94	<u>405</u>	100	1.40
	403	31.8	1.88	406	25.8	0.70
	443	15.6	0.09	446	18.0	1.52
	458	91.9	0.06	461	90.8	1.64
Hydroxylimino/ [TMS]₃	459	4.44	0.00	462	4.92	4.52
	<u>533</u>	31.6	0.64	<u>536</u>	32.1	2.93
Hydroxylimino/ [ethyl]₂/propionyl	371	34.1	1.79	374	58.7	0.00
	399	23.4	0.00	402	27.6	0.00
	<u>428</u>	100	2.58	<u>431</u>	100	0.53
	429	30.0	0.00	432	27.3	0.00

Tanle II — Opioids

Table II-5. (Continued)

CD Group[c]	Oxymorphone			Oxymorphone-d$_3$		
	Ion (m/z)[d]	Rel. int.	Analog's cont.	Ion (m/z)[d]	Rel. int.	Analog's cont.
Hydroxylimino/	309	5.79	0.00	312	10.7	0.00
[ethyl]$_2$/TMS	399	7.97	0.00	402	10.0	0.00
	429	15.1	0.00	432	16.4	0.00
	<u>444</u>	100	0.00	<u>447</u>	100	0.00
	445	32.1	0.00	448	25.9	0.00

[a–d] See the corresponding footnotes in Table I-1a.

Table II-6. Relative intensity and cross-contribution data[a] of ions[b] with potential for designating the analyte and the adapted internal standard — 6-Acetylcodeine/6-acetylcodeine-d$_3$

CD Group[c]	6-Acetylcodeine			6-Acetylcodeine-d$_3$		
	Ion (m/z)[d]	Rel. int.	Analog's cont.	Ion (m/z)[d]	Rel. int.	Analog's cont.
None	229	59.7	3.75	232	43.4	0.75
	282	100	1.45	285	85.5	1.58
	<u>341</u>	89.4	1.50	<u>344</u>	100	0.39

[a–d] See the corresponding footnotes in Table I-1a.

Table II-7a. Relative intensity and cross-contribution data[a] of ions[b] with potential for designating the analyte and the adapted internal standard — Codeine/codeine-d$_3$

CD Group[c]	Codeine			Codeine-d$_3$		
	Ion (m/z)[d]	Rel. int.	Analog's cont.	Ion (m/z)[d]	Rel. int.	Analog's cont.
None	<u>299</u>	100	0.92	<u>302</u>	100	0.25
Acetyl	229	39.5	4.22	232	37.9	0.69
	282	76.5	0.98	285	75.9	3.10
	298	8.92	2.23	301	8.57	0.47
	<u>341</u>	100	0.54	<u>344</u>	100	0.32
	342	22.8	4.24	345	22.8	0.14
TFA	<u>282</u>	100	0.64	<u>285</u>	100	0.36
	395	83.3	0.49	398	81.2	0.33
Propionyl	229	30.2	3.46	232	33.2	0.68
	282	71.1	0.77	285	75.5	0.54
	<u>355</u>	100	0.57	<u>358</u>	100	0.38
	356	23.6	3.88	359	24.2	0.16
PFP	<u>282</u>	100	2.49	<u>285</u>	100	0.24
	445	56.7	2.31	448	57.4	0.37
HFB	<u>282</u>	100	1.96	<u>285</u>	100	0.25
	495	62.0	1.25	498	66.0	0.43
TMS	343	18.6	4.36	346	18.7	1.08
	<u>371</u>	100	0.83	<u>374</u>	100	1.17
t-BDMS	<u>356</u>	49.2	2.94	<u>359</u>	48.3	1.23
	413	3.21	2.70	416	2.95	2.65

[a–d] See the corresponding footnotes in Table I-1a.

Table II-7b. Relative intensity and cross-contribution data[a] of ions[b] with potential for designating the analyte and the adapted internal standard — Codeine/codeine-d_6

CD Group[c]	Codeine			Codeine-d_6		
	Ion (m/z)[d]	Rel. int.	Analog's cont.	Ion (m/z)[d]	Rel. int.	Analog's cont.
None	282	10.4	2.71	288	11.0	0.05
	<u>299</u>	100	1.01	<u>305</u>	100	0.01
	300	20.2	1.27	306	21.1	0.08
Acetyl	282	76.5	0.46	288	77.4	0.02
	298	8.92	0.62	304	8.59	0.06
	<u>341</u>	100	0.35	<u>347</u>	100	0.00
	342	22.8	0.39	348	23.1	0.00
TFA	266	10.4	1.95	269	9.70	3.36
	<u>282</u>	100	0.22	<u>288</u>	100	0.02
	338	5.73	1.03	341	5.45	1.93
	395	83.3	0.13	401	77.7	0.00
Propionyl	282	71.1	0.35	288	68.3	0.01
	298	12.8	0.40	304	8.90	0.04
	<u>355</u>	100	0.37	<u>361</u>	100	0.00
	356	23.6	0.50	362	23.7	0.00
PFP	<u>282</u>	100	3.37	<u>288</u>	100	0.01
	388	4.17	4.16	391	4.82	0.31
	445	56.7	3.20	451	62.3	0.00
HFB	<u>282</u>	100	3.14	<u>288</u>	100	0.02
	438	4.48	3.36	441	4.97	0.57
	495	62.0	2.24	501	58.2	0.05
TMS	<u>371</u>	100	0.40	<u>377</u>	100	0.07
t-BDMS	<u>313</u>	100	0.82	<u>316</u>	100	1.02
	314	25.4	2.00	317	25.2	0.66
	356	49.2	0.55	362	49.1	0.00
	357	13.5	2.33	363	13.6	0.00

[a–d] See the corresponding footnotes in Table I-1a.

Table II-7c. Relative intensity and cross-contribution data[a] of ions[b] with potential for designating the analyte and the adapted internal standard — Codeine/codeine-$^{13}C_1d_3$

CD Group[c]	Codeine			Codeine-$^{13}C_1d_3$		
	Ion (m/z)[d]	Rel. int.	Analog's cont.	Ion (m/z)[d]	Rel. int.	Analog's cont.
None	229	27.8	3.09	233	26.4	0.84
	282	11.9	1.29	286	11.1	1.52
	<u>299</u>	100	0.15	<u>303</u>	100	0.13
	300	19.5	1.29	304	19.8	0.11
Acetyl	229	42.4	2.67	233	43.7	0.67
	282	81.5	1.05	286	81.2	0.20
	298	9.06	2.38	302	9.31	0.06
	<u>341</u>	100	0.61	<u>345</u>	100	0.04
	342	22.5	0.89	346	21.8	0.02
TFA	<u>282</u>	100	0.50	<u>286</u>	100	0.13
	283	20.2	1.64	287	19.1	0.35
	395	50.7	0.16	399	54.0	0.13
	396	11.1	0.46	400	11.5	0.10
Propionyl	229	30.2	1.84	233	32.3	0.68
	282	71.1	0.48	286	73.9	0.13
	<u>355</u>	100	0.34	<u>359</u>	100	0.04
	356	23.6	0.63	360	22.7	0.07

Tanle II — Opioids

Table II-7c. (Continued)

CD Group[c]	Codeine			Codeine-[13]C[1]d[3]		
	Ion (m/z)[d]	Rel. int.	Analog's cont.	Ion (m/z)[d]	Rel. int.	Analog's cont.
PFP	282	100	0.49	286	100	0.08
	283	20.5	1.61	287	19.2	0.10
	445	39.0	0.13	449	35.3	0.04
	446	9.06	0.49	450	7.84	0.00
HFB	58	28.1	0.89	62	27.8	0.99
	282	100	1.45	286	100	0.04
	283	20.9	3.97	287	19.5	0.13
	495	35.6	0.45	499	32.3	0.07
	496	8.82	0.59	500	7.20	0.00
TMS	178	75.8	—	182	66.6	—
	343	21.6	1.38	347	21.1	0.16
	371	100	0.13	375	100	0.17
	372	29.1	0.94	376	28.0	0.10
t-BDMS	313	100	—	316	100	—
	356	45.9	2.55	360	51.5	0.43
	357	13.4	3.99	361	12.6	0.33

[a-d] See the corresponding footnotes in Table I-1a.

Table II-8a. Relative intensity and cross-contribution data[a] of ions[b] with potential for designating the analyte and the adapted internal standard — Hydrocodone/hydrocodone-d[3]

CD Group[c]	Hydrocodone			Hydrocodone-d[3]		
	Ion (m/z)[d]	Rel. int.	Analog's cont.	Ion (m/z)[d]	Rel. int.	Analog's cont.
None	299	100	2.39	302	100	0.27
Ethyl	282	21.0	2.28	285	19.4	2.98
	298	54.6	0.82	301	49.5	0.37
	327	100	0.44	330	100	0.17
Acetyl	298	77.0	0.11	301	79.3	0.00
	341	100	0.00	344	100	0.00
	342	16.8	0.00	345	24.3	0.00
TMS	234	60.5	1.12	237	66.9	2.79
	371	100	0.34	374	100	1.19
t-BDMS	276	18.2	0.00	279	16.8	0.00
	356	93.3	0.00	359	59.7	0.00
	357	27.1	0.00	360	14.7	0.00
	398	7.28	0.00	401	4.32	0.00
	413	7.16	0.00	416	5.44	0.00
Methoxyimino	297	80.7	1.91	300	70.9	4.75
	328	100	0.47	331	100	0.34
Hydroxylimino/	297	100	1.29	300	100	0.90
TMS	386	89.7	0.19	389	89.5	1.31

[a-d] See the corresponding footnotes in Table I-1a.

Table II-8b. Relative intensity and cross-contribution data[a] of ions[b] with potential for designating the analyte and the adapted internal standard — Hydrocodone/hydrocodone-d$_6$

CD Group[c]	Hydrocodone			Hydrocodone–d$_6$		
	Ion (m/z)[d]	Rel. int.	Analog's cont.	Ion (m/z)[d]	Rel. int.	Analog's cont.
None	242	48.3	4.30	245	32.3	2.97
	256	11.1	4.80	262	7.85	0.33
	284	13.3	2.70	287	10.2	1.86
	<u>299</u>	100	2.16	<u>305</u>	100	0.04
Ethyl	282	21.0	1.15	288	19.8	0.04
	298	54.6	0.79	304	49.3	0.04
	312	46.3	0.07	315	38.7	0.28
	326	21.3	0.00	332	25.4	0.00
	<u>327</u>	100	0.05	<u>337</u>	100	0.00
Acetyl	298	80.0	0.00	304	96.6	0.00
	326	22.0	0.00	329	8.95	0.00
	<u>341</u>	100	0.00	<u>347</u>	100	0.00
TMS	234	60.5	2.12	237	61.3	3.96
	313	23.1	4.46	317	23.1	1.61
	356	45.0	0.18	359	31.8	1.50
	370	36.6	0.00	376	37.7	0.04
	<u>371</u>	100	0.00	<u>377</u>	100	0.00
t-BDMS	<u>313</u>	100	0.00	<u>316</u>	86.1	0.00
	356	93.3	0.00	362	86.5	0.00
	357	27.1	0.00	363	22.5	0.00
	398	7.28	0.00	401	4.47	0.00
	413	7.16	0.00	419	6.68	0.00
Methoxyimino	297	80.7	0.08	303	77.4	0.00
	298	21.6	0.00	304	16.7	0.00
	<u>328</u>	100	0.00	<u>334</u>	100	0.00
	329	26.8	0.00	335	20.5	0.00
Hydroxylimino/ TMS	<u>297</u>	100	0.40	<u>303</u>	100	1.24
	298	22.4	2.35	304	22.7	1.61
	329	14.6	2.27	332	15.2	2.89
	371	18.4	4.67	377	13.8	0.00
	386	89.7	0.05	392	87.7	0.00

[a–d] See the corresponding footnotes in Table I-1a.

Table II-9a. Relative intensity and cross-contribution data[a] of ions[b] with potential for designating the analyte and the adapted internal standard — Dihydrocodeine/dihydrocodeine-d$_3$

CD Group[c]	Dihydrocodeine			Dihydrocodeine–d$_3$		
	Ion (m/z)[d]	Rel. int.	Analog's cont.	Ion (m/z)[d]	Rel. int.	Analog's cont.
None	<u>301</u>	100	—	<u>304</u>	100	—
Acetyl	300	30.9	1.17	303	31.1	0.67
	<u>343</u>	100	0.79	<u>346</u>	100	0.36
	344	22.0	4.57	347	30.9	0.12
TFA	284	39.2	1.53	287	34.2	0.73
	300	13.5	0.69	303	12.0	0.28
	<u>397</u>	100	0.41	<u>400</u>	100	0.36
	398	23.0	4.46	401	31.3	0.18
Propionyl	284	31.9	1.47	287	32.2	2.49
	<u>357</u>	100	0.69	<u>360</u>	100	0.38

Table II-9a. (Continued)

CD Group[c]	Dihydrocodeine			Dihydrocodeine–d₃		
	Ion (m/z)[d]	Rel. int.	Analog's cont.	Ion (m/z)[d]	Rel. int.	Analog's cont.
PFP	284	43.7	1.72	287	45.0	0.51
	300	14.0	1.27	303	14.4	0.52
	<u>447</u>	100	0.72	<u>450</u>	100	0.38
HFB	284	43.0	2.17	287	44.3	0.72
	300	14.5	1.41	303	14.8	0.84
	360	4.57	1.61	363	4.58	0.31
	<u>497</u>	10.0	1.01	<u>500</u>	100	0.40
TMS	146	22.7	3.78	149	20.3	2.54
	<u>373</u>	100	0.37	<u>376</u>	100	1.47
	374	29.2	3.29	377	38.1	0.60
t-BDMS	<u>358</u>	83.0	3.91	<u>361</u>	82.5	1.23

[a–d] See the corresponding footnotes in Table I-1a.

Table II-9b. Relative intensity and cross-contribution data[a] of ions[b] with potential for designating the analyte and the adapted internal standard — Dihydrocodeine/dihydrocodeine-d₆

CD Group[c]	Dihydrocodeine			Dihydrocodeine–d₆		
	Ion (m/z)[d]	Rel. int.	Analog's cont.	Ion (m/z)[d]	Rel. int.	Analog's cont.
None	284	14.0	0.55	290	14.5	0.11
	300	18.7	0.62	306	19.9	0.28
	<u>301</u>	100	0.47	<u>307</u>	100	0.05
	302	20.2	0.78	308	20.2	0.03
Acetyl	284	37.2	0.46	290	37.7	0.08
	300	34.1	0.43	306	33.1	0.00
	328	8.40	0.78	331	5.39	0.65
	<u>343</u>	100	0.40	<u>349</u>	100	0.00
	344	22.8	0.44	350	22.4	0.00
TFA	284	36.5	0.48	290	39.9	0.04
	300	12.7	0.63	306	13.6	0.19
	340	9.74	1.62	343	10.5	0.44
	382	9.15	0.80	385	5.59	0.88
	<u>397</u>	100	0.31	<u>403</u>	100	0.00
Propionyl	284	31.9	0.60	290	32.2	0.00
	300	38.8	0.47	306	32.5	0.00
	342	7.23	1.12	345	4.63	2.73
	<u>357</u>	100	0.39	<u>363</u>	100	0.00
	358	23.4	0.44	364	23.9	0.00
PFP	284	43.3	2.00	290	44.0	0.06
	300	14.2	2.55	306	13.6	0.11
	390	8.67	3.17	393	8.97	0.33
	432	8.05	2.60	435	4.96	1.16
	<u>447</u>	100	1.83	<u>453</u>	100	0.00
HFB	<u>497</u>	100	—	<u>503</u>	100	—
TMS	<u>373</u>	100	—	<u>379</u>	100	—
t-BDMS	<u>315</u>	100	0.80	<u>318</u>	100	1.07
	316	27.9	1.64	319	25.1	0.58
	358	81.8	0.72	364	79.2	0.00
	415	8.89	2.96	421	7.93	0.00

[a–d] See the corresponding footnotes in Table I-1a.

Table II-10a. Relative intensity and cross-contribution data[a] of ions[b] with potential for designating the analyte and the adapted internal standard — Oxycodone/oxycodone-d$_3$

CD Group[c]	Oxycodone			Oxycodone-d$_3$		
	Ion (m/z)[d]	Rel. int.	Analog's cont.	Ion (m/z)[d]	Rel. int.	Analog's cont.
None	315	100	4.67	318	100	0.52
Acetyl	314	55.4	1.13	317	53.7	0.38
	357	100	0.61	360	100	0.42
	358	22.8	2.94	361	23.2	0.34
[Acetyl]$_2$ (enol)	356	39.5	2.57	359	39.3	0.90
	399	100	0.77	402	100	0.57
Propionyl	298	22.5	4.25	301	19.9	1.03
	314	65.2	0.92	317	62.0	0.36
	315	14.9	2.75	318	14.5	0.18
	371	100	0.86	374	100	0.44
	372	23.8	3.20	375	23.2	0.20
TMS	387	100	0.41	390	100	1.33
	388	29.7	1.98	391	28.7	0.76
[TMS]$_2$ (enol)	459	100	1.13	462	100	3.11
t-BDMS	372	100	—	375	100	—
[t-BDMS]$_2$ (enol)	486	78.8	1.63	489	99.4	1.15
	487	63.7	2.23	490	42.2	0.00
	543	7.65	0.00	546	2.68	0.00
Methoxyimino	287	9.12	0.93	290	10.1	0.63
	313	12.9	0.39	316	12.4	2.80
	344	100	0.34	347	100	0.35
Methoxyimino/	295	19.4	3.58	298	15.1	0.00
propionyl	343	53.7	1.96	346	40.6	0.00
	400	100	1.79	403	100	0.00
Methoxyimino/	416	100	0.13	419	100	1.43
TMS	417	29.9	1.87	420	28.2	0.64
Hydroxylimino/	230	45.0	0.00	233	60.2	0.00
[propionyl]$_2$	295	50.0	0.00	298	48.2	0.00
	313	14.1	0.00	316	17.0	0.00
	328	30.3	0.00	331	32.1	0.00
	386	11.4	0.00	389	12.7	0.00
	442	21.5	0.00	445	20.8	0.00
Hydroxylimino/	385	11.0	2.03	388	9.67	3.75
[TMS]$_2$	474	82.6	0.06	477	90.6	3.11
	475	32.3	0.55	478	33.7	1.63
Hydroxylimino/	357	43.5	0.00	360	43.1	0.00
ethyl/propionyl	414	100	0.00	417	100	0.00
	415	20.2	0.00	418	24.0	0.00

[a–d] See the corresponding footnotes in Table I-1a.

Table II-10b. Relative intensity and cross-contribution data[a] of ions[b] with potential for designating the analyte and the adapted internal standard — Oxycodone/oxycodone-d$_6$

CD Group[c]	Oxycodone			Oxycodone-d$_6$		
	Ion (m/z)[d]	Rel. int.	Analog's cont.	Ion (m/z)[d]	Rel. int.	Analog's cont.
None	315	100	—	321	100	-
Acetyl	298	17.0	0.76	304	16.0	0.21
	314	55.4	0.29	320	53.4	0.02
	357	100	0.32	363	100	0.01
	358	22.8	0.49	364	21.6	0.00
[Acetyl]$_2$ (enol)	296	19.1	1.10	302	16.7	0.82
	340	10.4	1.26	346	10.1	0.65
	356	39.5	0.88	362	38.7	0.16
	399	100	0.38	405	100	0.05
	400	24.9	0.81	406	23.2	0.31
Propionyl	298	22.5	1.06	304	19.0	0.00
	314	65.2	0.96	320	57.5	0.00
	315	14.9	1.07	321	12.2	0.00
	371	100	0.93	377	100	0.00
	372	23.8	0.99	378	22.7	0.00
TMS	387	100	0.24	393	100	0.05
	388	29.7	0.48	394	27.6	0.26
[TMS]$_2$ (enol)	444	24.4	1.64	450	20.9	0.38
	459	100	1.48	465	100	0.13
	460	36.8	1.94	466	36.0	0.55
t-BDMS	372	100	—	378	100	—
[t-BDMS]$_2$ (enol)	486	78.8	0.00	492	85.7	0.00
	487	63.7	0.00	493	29.3	0.00
	543	7.65	0.00	549	7.05	0.00
Methoxyimino	287	9.12	2.47	290	10.4	0.60
	313	12.9	0.66	319	13.0	0.00
	344	100	3.57	350	100	0.12
	345	20.3	1.07	351	21.0	1.75
Methoxyimino/ propionyl	230	78.1	0.00	236	53.6	0.00
	295	19.4	0.00	301	22.3	0.00
	343	53.7	0.00	349	12.1	0.00
	400	100	0.00	406	100	0.00
	401	29.4	0.00	407	16.1	0.00
Methoxyimino/ TMS	326	6.00	4.25	332	6.17	3.23
	401	15.2	0.00	407	14.0	0.03
	416	100	0.00	422	100	0.07
	417	29.9	0.00	423	29.3	0.06
Hydroxylimino/ [propionyl]$_2$	139	100	0.00	142	100	0.00
	230	45.0	0.00	236	29.3	0.00
	295	50.0	0.00	301	71.5	0.00
	328	30.3	0.00	334	28.1	0.00
	442	21.5	0.00	448	36.9	0.00
Hydroxylimino/ [TMS]$_2$	385	11.0	0.39	391	7.91	0.35
	459	19.4	0.00	465	16.4	0.00
	474	82.6	0.00	480	65.8	0.00
	475	32.3	0.49	481	26.7	0.00
Hydroxylimino/ ethyl/propionyl	230	42.4	0.00	236	53.1	3.38
	357	10.7	0.00	363	54.7	0.00
	414	100	0.00	420	100	0.00
	415	20.2	0.00	421	32.1	0.00

[a–d] See the corresponding footnotes in Table I-1a.

Table II-11. Relative intensity and cross-contribution data[a] of ions[b] with potential for designating the analyte and the adapted internal standard — Noroxycodone/noroxycodone-d₃

CD Group[c]	Noroxycodone			Noroxycodone-d₃		
	Ion (m/z)[d]	Rel. int.	Analog's cont.	Ion (m/z)[d]	Rel. int.	Analog's cont.
None	<u>301</u>	100	—	<u>304</u>	100	—
[Acetyl]₂	<u>239</u>	70.3	—	<u>242</u>	55.7	—
[TFA]₃ (enol)	<u>336</u>	37.9	—	<u>339</u>	33.5	—
	362	33.1	3.28	365	29.3	1.39
	475	12.0	0.41	478	10.9	3.32
	589	39.3	0.32	592	33.6	0.70
Propionyl	301	17.8	2.84	304	17.8	0.52
	<u>357</u>	100	1.59	<u>360</u>	100	0.42
	358	22.5	2.79	361	22.1	0.31
[PFP]₂	<u>239</u>	100	1.12	<u>242</u>	100	2.31
	430	6.99	1.15	433	7.25	0.14
	593	20.1	0.87	596	20.2	0.21
[HFB]₂	<u>239</u>	100	3.54	<u>242</u>	100	2.38
	449	3.78	4.82	452	3.59	2.68
	480	6.00	4.42	483	6.56	0.91
	693	11.1	3.29	696	12.2	1.21
[TMS]₂	<u>445</u>	73.3	—	<u>448</u>	76.9	—
[TMS]₃ (enol)	<u>445</u>	35.8	1.56	<u>448</u>	32.1	2.67
	446	16.6	3.28	449	12.3	1.39
	517	16.4	0.64	520	15.9	4.41
	518	8.69	2.50	521	7.47	2.40
Methoxyimino/ ethyl	343	51.9	4.97	346	46.0	0.08
	<u>358</u>	100	1.50	<u>361</u>	100	0.08
Methoxyimino/ acetyl	<u>299</u>	100	1.71	<u>302</u>	100	0.32
	341	36.1	1.59	344	35.5	0.44
	354	6.58	2.02	357	8.42	3.58
	372	98.1	0.66	375	95.3	0.39
Methoxyimino/ [TFA]₂	<u>268</u>	20.9	0.00	<u>271</u>	23.3	0.00
	377	16.5	0.00	380	10.7	0.00
	409	4.91	0.00	412	4.96	0.00
	522	14.0	0.00	525	16.6	0.00
Methoxyimino/ propionyl	<u>299</u>	100	1.28	<u>302</u>	100	0.30
	355	20.0	1.04	358	20.9	0.68
	386	58.0	0.72	389	65.3	0.41
Methoxyimino/ PFP	<u>476</u>	24.4	0.00	<u>479</u>	25.5	0.00
Methoxyimino/ [HFB]₂	<u>477</u>	15.4	0.97	<u>480</u>	17.1	1.59
	509	5.53	0.44	512	5.65	2.35
	691	2.23	0.00	694	2.71	1.33
	722	8.51	0.69	725	10.1	0.59
Methoxyimino/ [TMS]₂	359	29.2	2.63	362	27.9	1.57
	<u>474</u>	100	0.36	<u>477</u>	100	3.15
	475	38.2	1.04	478	36.9	1.28
Methoxyimino/ t-BDMS	283	9.48	0.00	286	10.7	0.00
	357	6.27	0.00	360	14.6	0.00
	<u>387</u>	30.7	0.00	<u>390</u>	30.7	0.00
Methoxyimino/ ethyl/propionyl	244	36.0	3.92	247	31.7	0.00
	357	41.8	4.77	360	40.9	0.00
	<u>414</u>	100	0.42	<u>417</u>	100	1.06
Methoxyimino/ ethyl/TMS	359	16.2	4.84	362	15.5	1.95
	<u>430</u>	100	0.41	<u>433</u>	100	1.46

Tanle II — Opioids

Table II-11. (Continued)

CD Group[c]	Ion (m/z)[d]	Noroxycodone Rel. int.	Analog's cont.	Ion (m/z)[d]	Noroxycodone-d₃ Rel. int.	Analog's cont.
Methoxyimino/	342	100	1.02	345	100	0.53
ethyl/t-BDMS	343	22.1	1.26	346	22.0	0.00
Methoxyimino/	371	36.2	0.00	373	13.2	0.00
acetyl/TMS	413	44.8	0.00	416	32.0	0.00
	444	32.2	0.00	447	14.7	0.00
Methoxyimino/	269	73.7	—	272	73.3	—
propionyl/TMS	427	51.2	0.00	430	54.4	0.00
	458	55.4	0.00	461	53.1	0.00
Hydroxylimino/	243	30.6	0.00	246	23.2	0.00
[ethyl]₂/TMS	373	11.0	0.00	376	19.0	0.00
	429	23.6	0.00	432	15.4	0.00
	444	100	0.00	447	100	0.00
	445	39.8	0.00	448	26.9	0.00

[a–d] See the corresponding footnotes in Table I-1a.

Table II-12. Relative intensity and cross-contribution data[a] of ions[b] with potential for designating the analyte and the adapted internal standard — Buprenorphine/buprenorphine-d₄

CD Group[c]	Ion (m/z)[d]	Buprenorphine Rel. int.	Analog's cont.	Ion (m/z)[d]	Buprenorphine-d₄ Rel. int.	Analog's cont.
Methyl	366	8.99	2.88	370	10.8	0.10
	392	100	0.19	396	100	0.01
	424	32.4	1.89	428	37.8	0.01
	434	5.32	0.03	438	8.70	0.60
	448	9.02	0.89	452	8.15	0.02
	481	6.09	0.39	485	8.02	0.03
Ethyl	380	7.91	3.22	384	10.3	3.72
	394	15.3	2.21	398	15.6	3.09
	406	100	1.15	410	100	1.12
	438	37.2	2.08	442	34.8	1.22
	448	5.75	1.82	452	6.04	3.60
	495	6.58	1.26	499	7.06	3.08
Acetyl	394	12.1	2.37	398	13.1	0.64
	408	21.1	1.37	412	20.9	0.71
	420	100	0.40	424	100	0.33
	421	27.9	0.76	425	27.6	0.41
	452	58.5	2.61	456	59.8	0.10
TFA (MBTFA)	474	100	0.09	478	100	3.15
	506	43.3	0.36	510	38.6	1.20
	516	7.46	0.51	520	3.38	2.26
	548	3.84	0.10	552	4.18	2.38
PFP	498	9.85	1.51	602	3.52	4.89
	512	20.2	0.01	516	20.0	0.02
	524	100	0.29	528	100	0.78
	556	32.0	1.04	560	28.6	0.35
HFB	562	15.6	4.42	566	19.8	1.08
	574	100	2.65	578	100	0.91
	606	41.7	4.75	610	31.2	1.32
	630	3.87	0.00	634	1.11	1.91
	663	4.08	2.80	667	2.42	0.08

Table II-12. (Continued)

CD Group[c]	Buprenorphine			Buprenorphine-d$_4$		
	Ion (m/z)[d]	Rel. int.	Analog's cont.	Ion (m/z)[d]	Rel. int.	Analog's cont.
TMS (MSTFA)	424	9.86	2.24	428	10.2	3.23
	438	13.3	3.27	442	13.1	4.65
	<u>450</u>	100	0.48	<u>454</u>	100	1.26
	482	32.6	1.48	486	33.0	0.91
	492	15.6	0.18	496	16.3	1.48
	506	25.2	0.03	510	23.6	1.47
	524	8.46	4.78	528	7.29	1.11
	539	6.68	0.08	543	6.20	0.69
[TMS]$_2$	438	30.8	3.50	442	14.4	2.50
	506	41.5	1.70	510	26.3	0.95
	554	48.3	0.00	558	35.4	1.46
	<u>555</u>	52.1	2.25	<u>559</u>	14.6	0.00
	611	11.7	0.00	615	2.23	0.00
t-BDMS	<u>492</u>	100	0.33	<u>496</u>	100	0.47
	493	38.2	0.99	497	36.8	0.00
	506	34.5	0.60	510	34.3	1.33
	524	39.4	2.05	528	37.7	0.00
	581	5.09	0.00	585	4.93	0.00

[a–d] See the corresponding footnotes in Table I-1a.

Table II-13. Relative intensity and cross-contribution data[a] of ions[b] with potential for designating the analyte and the adapted internal standard — Norbuprenorphine/norbuprenorphine-d$_3$

CD Group[c]	Norbuprenorphine			Norbuprenorphine-d$_3$		
	Ion (m/z)[d]	Rel. int.	Analog's cont.	Ion (m/z)[d]	Rel. int.	Analog's cont.
[Methyl]$_2$	<u>384</u>	72.5	4.38	<u>387</u>	51.2	1.37
[Ethyl]$_2$	<u>412</u>	41.9	4.92	<u>415</u>	50.5	1.12
	469	8.62	1.59	472	0.44	1.66
[Acetyl]$_2$	422	10.6	2.66	425	9.53	1.84
	<u>440</u>	100	1.20	<u>443</u>	100	0.68
	441	27.6	1.79	444	26.6	0.23
	482	2.65	0.07	485	1.85	0.90
[TFA]$_2$ (MBTFA)	<u>548</u>	100	0.77	<u>551</u>	100	3.31
	530	5.87	0.92	533	15.4	4.68
[PFP]$_2$	<u>648</u>	54.1	2.63	<u>651</u>	91.3	0.24
	630	9.19	0.59	633	6.80	0.00
[HFB]$_2$	<u>748</u>	41.0	0.17	<u>751</u>	47.6	1.02
	730	3.72	1.49	733	4.02	2.17
[TMS]$_2$ (BSTFA)	<u>500</u>	37.8	4.34	<u>503</u>	27.8	3.97
	557	6.43	4.82	560	5.01	4.30
[TMS]$_3$	<u>524</u>	42.6	—	<u>527</u>	16.9	—
t-BDMS	452	41.8	1.71	455	41.3	2.36
	<u>470</u>	59.4	1.45	<u>473</u>	60.8	2.09
	527	3.10	0.00	530	2.79	0.00

[a–d] See the corresponding footnotes in Table I-1a.

Table II-14. Relative intensity and cross-contribution data[a] of ions[b] with potential for designating the analyte and the adapted internal standard — Fentanyl/fentanyl-d[5]

CD Group[c]	Fentanyl			Fentanyl-d[5]		
	Ion (m/z)[d]	Rel. int.	Analog's cont.	Ion (m/z)[d]	Rel. int.	Analog's cont.
None	146	54.0	0.81	151	54.0	0.20
	189	40.0	1.00	194	38.0	0.19
	<u>245</u>	100	0.19	<u>250</u>	100	0.13

[a–d] See the corresponding footnotes in Table I-1a.

Table II-15. Relative intensity and cross-contribution data[a] of ions[b] with potential for designating the analyte and the adapted internal standard — Norfentanyl/norfentanyl-d[5]

CD Group[c]	Norfentanyl			Norfentanyl-d[5]		
	Ion (m/z)[d]	Rel. int.	Analog's cont.	Ion (m/z)[d]	Rel. int.	Analog's cont.
None	<u>93</u>	64.8	4.08	<u>98</u>	48.5	0.71
	159	44.6	3.82	164	37.9	0.05
	175	55.0	4.65	180	50.1	0.02
	232	2.19	3.42	237	2.45	0.28
Acetyl	93	46.2	1.74	98	45.0	2.62
	<u>132</u>	73.5	1.20	<u>137</u>	64.2	0.07
	158	65.7	0.70	163	59.2	1.01
	175	14.6	1.60	180	13.0	0.20
	231	66.2	0.73	236	62.1	0.06
	274	8.33	2.18	279	8.42	0.06
TCA	93	15.1	3.39	98	15.4	0.55
	<u>132</u>	29.3	1.62	<u>137</u>	28.1	3.97
	175	9.06	4.51	180	8.33	2.33
	249	5.45	0.91	254	5.02	0.22
	285	4.70	2.26	290	4.12	2.13
	340	9.27	0.23	345	8.16	3.26
TFA	93	38.8	1.43	98	39.0	1.37
	104	10.6	2.56	109	9.05	0.89
	132	29.6	2.69	137	27.9	0.13
	<u>150</u>	100	1.65	<u>155</u>	100	0.11
	272	9.09	0.68	277	8.84	0.00
	328	6.32	0.51	333	6.69	0.00
PFP	93	27.0	1.21	98	26.7	1.02
	132	28.3	1.27	137	26.3	0.06
	<u>150</u>	100	1.18	<u>155</u>	100	0.08
	175	14.0	1.91	180	13.2	0.37
	322	7.10	0.42	327	7.66	0.85
	378	4.23	0.28	383	4.72	0.36
HFB	93	22.7	1.65	98	23.2	0.87
	132	28.5	1.46	137	26.9	0.09
	<u>150</u>	100	1.52	<u>155</u>	100	0.06
	175	13.0	1.68	180	12.4	0.30
	372	7.21	0.70	377	6.42	0.04
	428	3.63	0.48	433	3.25	0.05
4-CB	93	20.0	2.13	98	20.8	1.07
	132	28.1	2.08	137	26.7	0.25
	<u>150</u>	100	2.11	<u>155</u>	100	0.08
	175	11.9	1.93	180	11.3	0.51
	437	6.46	0.76	442	5.28	0.02

Table II-15. (Continued)

CD Group[c]	Norfentanyl			Norfentanyl-d[5]		
	Ion (m/z)[d]	Rel. int.	Analog's cont.	Ion (m/z)[d]	Rel. int.	Analog's cont.
TMS	206	22.5	1.46	211	19.8	0.18
	231	13.9	0.87	236	13.0	0.15
	<u>247</u>	49.9	0.52	<u>252</u>	46.3	0.08
	289	21.4	0.54	294	20.0	0.10
	304	9.92	3.54	309	8.83	0.27
t-BDMS	132	31.1	0.77	137	31.2	0.79
	<u>206</u>	100	0.42	<u>211</u>	100	0.02
	207	18.6	1.55	212	18.5	0.64
	231	4.32	0.42	236	4.69	0.23
	289	72.7	0.34	294	79.5	0.01
	290	17.7	0.38	295	19.6	0.02

[a–d] See the corresponding footnotes in Table I-1a.

Table II-16a. Relative intensity and cross-contribution data[a] of ions[b] with potential for designating the analyte and the adapted internal standard — Methadone/methadone-d[3]

CD Group[c]	Methadone			Methadone-d[3]		
	Ion (m/z)[d]	Rel. int.	Analog's cont.	Ion (m/z)[d]	Rel. int.	Analog's cont.
None	223	26.1	1.30	226	22.0	0.64
	<u>294</u>	5.68	0.28	<u>297</u>	27.1	0.47
	309	12.8	0.64	312	9.79	0.00

[a–d] See the corresponding footnotes in Table I-1a.

Table II-16b. Relative intensity and cross-contribution data[a] of ions[b] with potential for designating the analyte and the adapted internal standard — Methadone/methadone-d[9]

CD Group[c]	Methadone					Methadone-d[9]				
	Ion (m/z)[d]	Rel. int.	Analog's cont.			Ion (m/z)[d]	Rel. int.	Analog's cont.		
None	<u>72</u>	100	0.74	0.08[b]	0.48[c]	<u>78</u>	100	0.03	0.28[b]	0.05[c]
	223	2.00	55.8	5.70	44.7	226	4.00	0.04	0.37	0.03
	294	2.00	1.16	0.12	0.87	303	3.00	0.00	0.00	0.00
	309	0.15	1.62	0.16	3.60	318	0.30	0.00	0.00	0.00

[a–d] See the corresponding footnotes in Table I-1a.

Table II-17. Relative intensity and cross-contribution data[a] of ions[b] with potential for designating the analyte and the adapted internal standard — 2-ethylidine-1,5-dimethyl-3,3-diphenylpyrrolidine/2-ethylidine-1,5-dimethyl-3,3-diphenylpyrrolidine-d[3]

CD Group[c]	2-ethylidine-1,5-dimethyl-3,3-diphenylpyrrolidine			2-ethylidine-1,5-dimethyl-3,3-diphenylpyrrolidine-d[3]		
	Ion (m/z)[d]	Rel. int.	Analog's cont.	Ion (m/z)[d]	Rel. int.	Analog's cont.
None	276	100	1.18	279	95.5	2.11
	<u>277</u>	99.6	2.09	<u>280</u>	100	0.15

[a–d] See the corresponding footnotes in Table I-1a.

Tanle II — Opioids

Table II-18a. Relative intensity and cross-contribution data[a] of ions[b] with potential for designating the analyte and the adapted internal standard — Propoxyphene/propoxyphene-d$_5$

CD Group[c]	Propoxyphene			Propoxyphen-d$_5$		
	Ion (m/z)[d]	Rel. int.	Analog's cont.	Ion (m/z)[d]	Rel. int.	Analog's cont.
None	<u>208</u>	5.00	4.50	<u>213</u>	3.00	0.24
	250	1.00	0.36	255	1.00	0.25

[a-d] See the corresponding footnotes in Table I-1a.

Table II-18b. Relative intensity and cross-contribution data[a] of ions[b] with potential for designating the analyte and the adapted internal standard — Propoxyphene/propoxyphene-d$_7$

CD Group[c]	Propoxyphene			Propoxyphen-d$_7$		
	Ion (m/z)[d]	Rel. int.	Analog's cont.	Ion (m/z)[d]	Rel. int.	Analog's cont.
None	<u>91</u>	6.10	—	<u>98</u>	3.76	—
	193	4.00	3.53	200	1.91	3.06
	250	1.50	0.19	257	1.55	0.00

[a-d] See the corresponding footnotes in Table I-1a.

Table II-18c. Relative intensity and cross-contribution data[a] of ions[b] with potential for designating the analyte and the adapted internal standard — Propoxyphene/propoxyphene-d$_{11}$

CD Group[c]	Propoxyphene			Propoxyphen-d$_{11}$		
	Ion (m/z)[d]	Rel. int.	Analog's cont.	Ion (m/z)[d]	Rel. int.	Analog's cont.
None	<u>58</u>	100	0.21	<u>64</u>	100	0.14
	178	3.00	2.73	183	0.14	0.12
	250	1.00	0.00	261	1.22	1.04

[a-d] See the corresponding footnotes in Table I-1a.

Table II-19. Relative intensity and cross-contribution data[a] of ions[b] with potential for designating the analyte and the adapted internal standard — Norpropoxyphene/norpropoxyphene-d$_5$

CD Group[c]	Norpropoxyphene			Norpropoxyphen-d$_5$		
	Ion (m/z)[d]	Rel. int.	Analog's cont.	Ion (m/z)[d]	Rel. int.	Analog's cont.
None	178	6.00	3.19	183	5.00	0.94
	220	9.00	3.20	225	10.0	0.39
	<u>234</u>	65.0	0.86	<u>239</u>	71.0	0.06

[a-d] See the corresponding footnotes in Table I-1a.

Table II-20. Relative intensity and cross-contribution data[a] of ions[b] with potential for designating the analyte and the adapted internal standard — Meperidine/meperidine-d$_4$

CD Group[c]	Meperidine			Meperidine-d$_4$		
	Ion (m/z)[d]	Rel. int.	Analog's cont.	Ion (m/z)[d]	Rel. int.	Analog's cont.
None	<u>71</u>	100	—	<u>73</u>	100	—
	247	52.6	3.64	251	56.6	0.78

[a-d] See the corresponding footnotes in Table I-1a.

Appendix Two — Ion Intensity Cross-Contribution Data

Table II-21. Relative intensity and cross-contribution data[a] of ions[b] with potential for designating the analyte and the adapted internal standard — Normeperidine/normeperidine-d₄

CD Group[c]	Normeperidine Ion (m/z)[d]	Rel. int.	Analog's cont.	Normeperidine-d₄ Ion (m/z)[d]	Rel. int.	Analog's cont.
None	57	100	—	59	100	—
Ethyl	232	17.8	0.86	236	18.1	0.05
	246	100	0.17	250	100	0.06
	247	17.1	0.46	251	17.4	0.04
	260	14.6	0.18	264	15.2	0.33
	261	20.5	0.67	265	20.2	0.05
Propyl	202	4.66	2.12	206	4.85	0.04
	218	2.83	0.24	222	2.66	0.06
	246	100	0.08	250	100	0.02
	247	17.1	0.15	251	17.5	0.03
	275	2.94	2.24	279	3.02	0.06
Butyl	246	100	0.07	250	100	0.03
	247	18.4	0.31	251	18.1	0.11
	289	2.92	1.40	293	2.90	0.18
Acetyl	158	30.9	1.82	161	27.6	4.45
	187	100	0.70	191	100	0.34
	188	14.0	1.42	192	15.3	0.47
	202	28.0	0.63	206	32.3	0.17
	232	31.5	0.65	236	34.8	0.12
	275	30.2	0.61	279	34.2	0.08
TCA	342	100	—	346	100	—
	232	34.5	1.82	236	38.9	3.97
	344	66.3	0.38	348	66.4	0.37
TFA	143	72.7	2.56	146	55.2	0.73
	241	100	0.42	243	100	1.10
	255	33.5	0.52	259	32.4	0.18
	256	48.8	0.64	260	51.6	3.16
	329	38.5	0.04	333	38.6	0.02
PFP	143	66.6	2.72	146	54.0	2.34
	291	100	0.49	293	100	1.24
	305	30.7	0.53	309	30.4	0.18
	306	43.4	0.80	310	47.5	0.07
	379	36.5	0.03	383	38.5	0.02
HFB	341	100	—	343	100	—
	355	29.6	0.08	359	29.4	2.21
	356	42.6	0.10	360	46.9	4.23
	429	34.5	0.03	433	37.4	0.24
4-CB	143	44.2	4.21	146	40.2	1.80
	395	100	1.38	397	100	2.46
	410	29.8	1.45	414	33.5	0.04
	438	9.85	0.49	442	9.66	0.05
	483	31.6	0.49	487	28.5	0.05
TMS	276	68.9	0.13	280	64.1	0.21
	304	82.8	0.13	308	74.8	0.82
	305	82.0	0.13	309	75.1	0.25
t-BDMS	262	13.8	0.89	266	22.1	0.12
	274	7.82	0.37	278	8.26	0.28
	290	100	0.02	294	100	0.14
	291	23.4	0.04	295	23.9	0.02

[a–c] See the corresponding footnotes in Table II-1A.

Tanle II — Opioids

**Summary of Drugs, Isotopic Analogs, and Chemical Derivatization Groups Included in
Table III (Hallucinogens)**

Compound	Isotopic analog	Chemical derivatization group	Table #
Cannabinol	d_3	Methyl, ethyl, propyl, butyl, propionyl	III-1
Tetrahydrocannabinol	d_3	Methyl, ethyl, propyl, butyl, TFA, propionyl, PFP, HFB, TMS, t-BDMS	III-2
THC-OH	d_3	[Methyl]$_2$, [ethyl]$_2$, [propyl]$_2$, [butyl]$_2$, [TFA]$_2$, propionyl, [PFP]$_2$, [HFB]$_2$, [TMS]$_2$, [t-BDMS]$_2$	III-3
THC-COOH	d_3, d_9	[Methyl]$_2$, [ethyl]$_2$, [propyl]$_2$, [butyl]$_2$, propionyl, [TMS]$_2$, [t-BDMS]$_2$, methyl/TFA, PFPoxy/PFP, HFPoxy/HFB	III-4
Ketamine	d_4	None, acetyl, TFA, HFB, PFB, TMS	III-5
Norketamine	d_4	None, acetyl, TCA, TFA, PFP, HFB, 4-CB, PFB, TMS, TFA/t-BDMS, PFP/t-BDMS, HFB/t-BDMS	III-6
Phencyclidine	d_5	Acetyl, TFA, HFB, PFB, TMS	III-7
LSD	d_3	None, TMS	III-8
Mescaline	d_9	Acetyl, TCA, TFA, PFP, HFB, 4-CB, [TMS]$_2$, t-BDMS, TFA/TMS, TFA/t-BDMS, PFP/TMS, PFP/t-BDMS, HFB/TMS, HFB/t-BDMS	III-9
Psilocin	d_{10}	None, acetyl, [acetyl]$_2$, [TMS]$_2$, t-BDMS, [t-BDMS]$_2$	III-10

Table III — Hallucinogens

Appendix Two — Table III
Cross-Contributions Between Ions Designating the Drugs and Their Isotopically Labeled Analogs in Various Derivatization Forms — Hallucinogens

Table III — Hallucinogens

Table III-1. Relative intensity and cross-contribution data[a] of ions[b] with potential for designating the analyte and the adapted internal standard — Cannabinol/cannabinol-d$_3$

CD Group[c]	Cannabinol			Cannabinol-d$_3$		
	Ion (m/z)[d]	Rel. int.	Analog's cont.	Ion (m/z)[d]	Rel. int.	Analog's cont.
Methyl	<u>309</u>	100	0.33	<u>312</u>	100	0.52
	310	23.6	4.70	313	24.0	0.30
	324	12.9	0.40	327	13.0	13.3
Ethyl	<u>323</u>	100	0.30	<u>326</u>	100	0.75
Propyl	<u>337</u>	100	0.61	<u>340</u>	100	2.24
Butyl	<u>351</u>	100	0.51	<u>354</u>	100	0.61
	366	12.0	0.65	369	12.0	2.10
Propionyl	<u>295</u>	100	2.24	<u>298</u>	100	0.21
	351	93.1	2.09	354	90.2	0.45
	352	23.4	10.3	355	23.2	0.18

[a–d] See the corresponding footnotes in Table I-1a.

Table III-2. Relative intensity and cross-contribution data[a] of ions[b] with potential for designating the analyte and the adapted internal standard — Tetrahydrocannabinol/tetrahydrocannabinol-d$_3$

CD Group[c]	Tetrahydrocannabinol			Tetrahydrocannabinol-d$_3$		
	Ion (m/z)[d]	Rel. int.	Analog's cont.	Ion (m/z)[d]	Rel. int.	Analog's cont.
Methyl	245	44.2	1.59	248	42.3	1.07
	285	33.0	2.56	288	32.4	0.49
	<u>313</u>	100	0.33	<u>316</u>	100	0.26
	328	74.1	0.09	331	74.8	0.28
Ethyl	259	40.2	3.14	262	39.5	1.14
	313	40.6	0.73	316	40.2	0.52
	<u>327</u>	100	0.00	<u>330</u>	100	0.29
	328	24.2	2.41	331	24.3	0.16
	342	88.7	0.00	345	91.8	0.35
Propyl	313	72.5	1.68	316	69.7	0.33
	<u>341</u>	100	0.77	<u>344</u>	100	0.39
	356	95.0	0.69	359	95.4	0.39
Butyl	313	61.7	4.98	316	64.1	0.82
	327	20.4	3.72	330	20.6	0.51
	<u>355</u>	100	1.49	<u>358</u>	100	0.42
	370	99.3	0.31	373	99.6	0.40
TFA	<u>297</u>	74.8	—	<u>300</u>	64.9	—
	313	33.4	1.13	316	31.3	1.13
	327	53.5	1.23	330	51.3	0.49
	367	83.4	0.73	370	80.8	0.33
	395	83.7	0.52	398	80.7	0.31
	410	100	0.09	413	100	0.39
Propionyl	313	50.2	1.48	316	49.2	0.93
	314	21.7	2.69	317	21.0	0.22
	<u>341</u>	100	—	<u>344</u>	100	—
	370	9.26	0.93	373	8.66	0.54
PFP	<u>377</u>	100	0.63	<u>380</u>	100	1.39
	378	20.1	4.53	381	20.2	1.01
	417	45.9	0.80	420	44.3	0.51
	445	10.6	1.03	448	9.78	0.48
	460	59.0	0.08	463	54.8	0.48

Table III-2. (Continued)

CD Group[c]	Tetrahydrocannabinol			Tetrahydrocannabinol-d[3]		
	Ion (m/z)[d]	Rel. int.	Analog's cont.	Ion (m/z)[d]	Rel. int.	Analog's cont.
HFB	<u>297</u>	100	1.39	<u>300</u>	100	0.47
	313	34.1	1.92	316	33.8	0.69
	427	54.0	1.44	430	53.5	0.07
	467	80.2	1.42	470	80.9	0.46
	495	73.7	1.77	498	73.5	0.42
	510	98.1	0.99	513	97.8	0.49
TMS	303	47.5	1.12	306	47.6	1.85
	343	30.6	3.44	346	29.0	1.39
	<u>371</u>	100	0.44	<u>374</u>	100	1.26
	386	91.7	0.09	389	92.7	1.24
	387	29.1	2.33	390	29.5	0.52
t-BDMS	345	18.4	2.86	348	18.1	2.15
	<u>371</u>	100	2.88	<u>374</u>	100	1.74
	413	30.4	0.87	416	30.5	1.61
	428	53.3	0.43	431	54.7	1.66

[a–d] See the corresponding footnotes in Table I-1a.

Table III-3. Relative intensity and cross-contribution data[a] of ions[b] with potential for designating the analyte and the adapted internal standard — THC-OH/THC-OH-d[3]

CD Group[c]	THC-OH			THC-OH-d[3]		
	Ion (m/z)[d]	Rel. int.	Analog's cont.	Ion (m/z)[d]	Rel. int.	Analog's cont.
[Methyl][2]	231	85.9	2.03	234	83.7	0.42
	<u>299</u>	100	1.96	<u>302</u>	100	0.17
	314	81.2	1.87	317	81.0	0.18
[Ethyl][2]	<u>377</u>	100	—	<u>340</u>	100	—
[Propyl][2]	<u>351</u>	100	—	<u>354</u>	100	—
[Butyl][2]	<u>365</u>	100	—	<u>368</u>	100	—
[TFA][2]	313	9.38	1.90	316	9.15	1.35
	340	8.02	3.72	343	7.81	1.87
	365	34.4	1.24	368	33.3	1.32
	<u>408</u>	100	0.07	<u>411</u>	100	1.06
	409	45.4	0.89	412	45.2	0.26
	522	7.86	0.00	525	7.84	0.55
Propionyl	<u>312</u>	100	1.47	<u>315</u>	100	0.51
	368	48.8	0.90	371	46.0	1.30
	369	18.4	5.34	372	17.6	0.54
[PFP][2]	363	9.41	1.09	366	9.04	1.26
	415	33.5	1.71	418	31.6	1.28
	<u>458</u>	100	0.11	<u>461</u>	100	1.19
	459	47.8	0.69	462	46.7	0.34
	622	5.45	0.00	625	5.36	0.77
[HFB][2]	413	9.50	2.44	416	8.78	2.62
	465	30.6	2.83	468	29.7	1.53
	<u>508</u>	100	1.11	<u>511</u>	100	1.27
	509	47.6	1.71	512	47.4	0.33
	722	3.34	1.23	725	3.68	0.81
[TMS][2]	<u>371</u>	100	0.58	<u>374</u>	100	1.40
	459	3.77	2.68	462	3.86	3.74
	474	5.03	0.62	477	5.27	3.91
[t-BDMS][2]	<u>413</u>	100	1.26	<u>416</u>	100	1.55

[aa–d] See the corresponding footnotes in Table I-1a.

Table III — Hallucinogens

Table III-4a. Relative intensity and cross-contribution data[a] of ions[b] with potential for designating the analyte and the adapted internal standard — THC-COOH/THC-COOH-d₃

CD Group[c]	Ion (m/z)[d]	THC-COOH Rel. int.	Analog's cont.	Ion (m/z)[d]	THC-COOH-d₃ Rel. int.	Analog's cont.
[Methyl]₂	<u>313</u>	100	—	<u>316</u>	100	—
	341	8.54	2.19	344	7.89	1.14
	357	64.4	0.64	360	62.1	0.40
	372	33.9	0.33	375	32.5	0.44
[Ethyl]₂	<u>327</u>	100	0.59	<u>330</u>	100	1.66
	371	32.7	0.94	374	33.8	0.42
	385	40.4	0.56	388	40.6	0.47
	400	25.9	0.28	403	26.0	0.61
[Propyl]₂	<u>341</u>	100	0.66	<u>344</u>	100	0.98
	385	38.9	0.61	388	38.9	0.49
	413	41.0	0.00	416	40.6	0.96
[Butyl]₂	<u>355</u>	100	2.91	<u>358</u>	100	0.58
	399	41.8	2.12	402	42.0	0.70
	441	41.3	2.74	444	41.8	0.62
	456	22.0	0.00	459	22.2	0.70
Propionyl	<u>314</u>	87.9	1.82	<u>317</u>	84.5	0.37
	337	2.60	3.12	340	2.96	1.06
	370	6.04	1.94	373	5.78	0.62
[TMS]₂	<u>371</u>	100	0.97	<u>374</u>	100	1.58
	473	36.3	0.91	476	37.2	3.31
	488	22.3	0.80	491	23.3	3.35
[t-BDMS]₂	<u>413</u>	95.5	1.55	<u>416</u>	96.5	2.67
	515	100	2.00	518	100	3.73
	516	43.3	7.80	519	42.4	2.00
	557	29.8	1.47	560	29.8	4.71
Methyl/TFA	379	20.5	1.70	382	21.1	2.42
	395	49.5	0.79	398	50.7	2.33
	411	9.54	4.33	414	9.35	0.42
	<u>439</u>	100	0.53	<u>442</u>	100	0.46
	454	46.3	0.12	457	46.5	0.48
PFPoxy/PFP	445	69.8	1.44	448	68.0	1.24
	<u>459</u>	100	0.95	<u>462</u>	100	0.59
	489	32.5	1.14	492	32.6	0.63
	579	23.3	1.71	582	22.8	0.58
	607	82.5	0.95	610	82.5	0.55
	622	71.6	0.19	625	72.2	0.62
HFPoxy/HFB	<u>477</u>	100	—	<u>480</u>	100	—
	495	28.0	4.17	498	27.8	1.96
	523	27.3	1.84	526	27.6	0.54
	539	39.0	0.83	542	39.2	0.63
	690	38.6	0.82	693	38.2	0.00

[a-d] See the corresponding footnotes in Table I-1a.

Table III-4b. Relative intensity and cross-contribution data[a] of ions[b] with potential for designating the analyte and the adapted internal standard — THC-COOH/THC-COOH-d₉

CD Group[c]	Ion (m/z)[d]	THC-COOH Rel. int.	Analog's cont.	Ion (m/z)[d]	THC-COOH-d₉ Rel. int.	Analog's cont.
[Methyl]₂	<u>313</u>	100	0.24	<u>322</u>	100	0.02
	314	24.1	1.79	323	22.2	0.73
	341	8.54	0.87	350	7.38	0.00
	357	64.4	0.00	363	46.2	0.00
	372	33.9	0.00	381	32.3	0.00

Table III-4b. (Continued)

CD Group[c]	THC-COOH			THC-COOH-d$_9$		
	Ion $(m/z)^d$	Rel. int.	Analog's cont.	Ion $(m/z)^d$	Rel. int.	Analog's cont.
[Ethyl]$_2$	<u>327</u>	100	0.08	<u>336</u>	100	0.01
	328	24.7	0.41	337	22.9	0.33
	355	8.32	0.28	364	7.82	0.00
	371	32.7	0.00	380	34.3	0.00
	385	40.4	0.02	391	39.5	0.00
	400	25.9	0.00	409	26.1	0.00
[Propyl]$_2$	<u>341</u>	100	0.00	<u>350</u>	100	0.01
	342	25.9	0.00	351	24.1	0.20
	369	8.97	0.00	378	11.3	0.00
	385	38.9	0.00	394	37.3	0.00
	413	41.0	0.00	419	40.7	0.00
	428	23.8	0.00	437	23.8	0.00
[Butyl]$_2$	<u>355</u>	100	2.91	<u>364</u>	100	0.58
	399	41.8	2.12	408	41.8	0.70
	441	41.3	2.74	447	40.6	0.62
	456	22.0	0.00	465	22.9	0.00
Propionyl	<u>258</u>	100	2.08	<u>264</u>	100	0.05
	259	18.2	4.78	265	18.0	1.19
	271	16.6	1.36	277	12.3	0.48
	299	18.1	0.80	305	16.0	0.19
	314	87.9	0.62	323	88.9	0.13
[TMS]$_2$	<u>371</u>	100	4.58	<u>380</u>	100	0.05
	473	36.3	2.70	479	27.8	0.09
	488	22.3	4.05	497	24.0	0.00
[t-BDMS]$_2$	<u>413</u>	95.5	1.47	<u>422</u>	94.6	0.13
	515	100	1.24	524	100	0.00
	516	43.3	1.96	525	40.7	0.00
	557	29.8	1.17	563	22.1	0.00
	572	36.1	1.18	581	36.8	0.00
Methyl/TFA	341	31.4	4.92	350	31.7	0.79
	379	20.5	0.25	385	14.1	2.56
	395	49.5	0.15	404	53.8	0.46
	411	9.54	3.64	414	9.44	0.41
	<u>439</u>	100	0.04	<u>445</u>	100	0.00
	454	46.3	0.00	463	50.0	0.00
PFPoxy/PFP	445	69.8	0.81	454	68.5	0.48
	<u>459</u>	100	0.66	<u>468</u>	100	0.27
	473	38.7	0.51	482	36.0	0.98
	489	32.5	2.25	498	31.3	4.27
	579	23.3	0.36	582	21.1	0.40
	607	82.5	0.06	613	80.8	0.00
	622	71.6	0.00	631	70.1	0.00
HFPoxy/HFB	<u>477</u>	100	0.79	<u>486</u>	100	0.08
	495	28.0	0.27	504	27.1	5.36
	523	27.3	0.29	532	25.4	0.15
	539	39.0	0.23	548	38.6	0.96
	675	23.7	0.50	681	24.9	0.00
	690	38.6	0.04	699	37.3	0.00

[a-d] See the corresponding footnotes in Table I-1a.

Table III — Hallucinogens

Table III-5. Relative intensity and cross-contribution data[a] of ions[b] with potential for designating the analyte and the adapted internal standard — Ketamine/ketamine-d_4

CD Group[c]	Ketamine			Ketamine-d_4		
	Ion (m/z)[d]	Rel. int.	Analog's cont.	Ion (m/z)[d]	Rel. int.	Analog's cont.
None	138	13.0	1.28	142	15.0	3.61
	<u>180</u>	100	0.75	<u>184</u>	100	0.66
	209	24.0	1.19	213	32.0	0.18
Acetyl	180	83.2	0.45	184	71.3	1.64
	<u>208</u>	100	0.36	<u>212</u>	100	0.60
	216	100	0.46	220	91.9	0.73
	251	9.43	1.04	255	9.04	0.59
TFA	236	40.3	0.21	240	44.9	3.62
	262	43.4	2.62	266	47.7	1.32
	<u>270</u>	74.2	0.93	<u>274</u>	85.8	1.25
	298	21.0	0.50	302	22.8	0.04
HFB	<u>152</u>	59.4	—	<u>156</u>	67.0	—
	236	46.3	0.18	240	53.8	3.11
	328	7.79	0.75	332	10.7	0.49
	362	41.5	0.14	366	47.9	1.06
	364	12.5	0.44	368	14.5	0.84
	370	44.9	0.32	374	48.9	1.62
	398	5.80	0.38	402	6.02	0.53
PFB	152	57.0	2.38	156	56.2	1.29
	326	5.91	2.38	330	5.54	2.40
	360	80.6	0.77	364	78.6	0.64
	362	26.4	1.64	366	26.3	0.21
	<u>368</u>	88.5	0.92	<u>372</u>	85.8	0.22
	369	19.2	1.42	373	18.2	1.31
	396	8.93	1.14	400	8.74	0.33
	403	9.76	3.13	407	9.05	1.32
	431	2.81	3.16	435	2.88	0.89
TMS	<u>152</u>	41.9	2.32	<u>156</u>	41.1	0.40
	294	8.63	3.42	298	8.21	3.67

[a–d] See the corresponding footnotes in Table I-1a.

Table III-6. Relative intensity and cross-contribution data[a] of ions[b] with potential for designating the analyte and the adapted internal standard — Norketamine/Norketamine-d_4

CD Group[c]	Norketamine			Norketamine-d_4		
	Ion (m/z)[d]	Rel. int.	Analog's cont.	Ion (m/z)[d]	Rel. int.	Analog's cont.
None	102	10.0	3.75	106	11.0	0.76
	<u>166</u>	100	0.85	<u>170</u>	100	0.38
	195	25.0	0.83	199	33.0	0.17
Acetyl	102	46.9	0.72	106	47.1	2.44
	194	15.6	0.77	198	13.4	2.59
	<u>202</u>	100	0.06	<u>206</u>	100	1.42
	203	14.2	0.35	207	14.7	5.01
	230	78.3	0.23	234	77.3	0.23
	231	12.1	0.13	235	12.8	0.06
TCA	102	28.0	2.02	106	22.8	0.32
	<u>306</u>	96.0	0.25	<u>310</u>	95.1	4.73
TFA	<u>284</u>	100	—	<u>288</u>	100	—
	102	76.6	0.85	106	72.7	1.39
	214	67.7	0.13	218	60.4	0.79
	239	47.0	0.43	243	38.3	2.48
	256	56.2	0.31	260	55.8	4.69
	275	37.1	0.15	279	37.0	1.89

Table III-6. (Continued)

CD Group[c]	Norketamine			Norketamine-d$_4$		
	Ion (m/z)[d]	Rel. int.	Analog's cont.	Ion (m/z)[d]	Rel. int.	Analog's cont.
PFP	264	55.3	1.03	268	54.1	0.10
	289	45.2	1.86	293	42.7	0.23
	290	50.5	8.83	294	5.00	0.84
	306	77.1	0.93	310	78.6	0.43
	325	44.9	0.98	329	45.2	0.48
	<u>334</u>	100	0.88	<u>338</u>	100	0.02
HFB	102	58.9	2.06	106	57.1	0.52
	314	69.1	0.39	318	69.8	0.09
	339	43.8	1.09	343	41.0	0.35
	356	100	0.29	360	100	0.26
	375	41.8	0.33	379	40.8	0.54
	377	13.1	2.85	381	12.8	0.10
	<u>384</u>	84.8	0.21	<u>388</u>	90.8	0.03
4-CB	<u>410</u>	100	0.49	<u>414</u>	100	0.16
	411	21.2	0.59	415	21.9	0.31
	429	20.2	0.53	433	22.0	1.28
	438	57.1	0.43	442	57.3	0.04
PFB	102	16.9	2.12	106	16.2	1.16
	312	12.3	1.31	316	11.3	1.33
	346	13.5	3.11	350	12.5	1.06
	<u>354</u>	98.5	0.49	<u>358</u>	95.1	0.06
	355	19.7	1.35	359	18.6	0.35
	382	66.5	0.43	386	62.5	0.02
TMS	210	31.4	2.10	214	32.5	2.24
	224	29.7	2.26	229	33.6	1.30
	<u>238</u>	100	0.64	<u>242</u>	100	1.46
	239	17.9	1.79	243	25.5	0.72
	267	11.9	0.63	271	15.4	1.83
	280	11.3	0.38	284	10.1	1.36
TFA/t-BDMS	336	25.8	0.71	340	26.4	3.69
	<u>376</u>	100	0.29	<u>380</u>	100	2.18
	378	38.3	0.46	382	37.8	0.17
PFP/t-BDMS	263	62.6	1.58	267	55.0	3.83
	296	51.0	2.25	300	51.2	10.7
	309	35.8	5.85	313	28.2	2.74
	324	35.4	2.89	328	34.3	4.44
	358	20.6	1.75	362	19.8	4.41
	<u>426</u>	100	0.04	<u>430</u>	100	2.41
	428	37.9	0.41	432	38.4	0.19
HFB/t-BDMS	263	44.4	0.75	267	57.2	2.52
	359	27.7	2.83	363	25.9	3.78
	374	35.3	0.62	378	40.4	4.61
	<u>476</u>	100	0.03	<u>480</u>	100	2.16
	478	38.8	0.51	482	38.5	0.12

[a-d] See the corresponding footnotes in Table I-1a.

Table III — Hallucinogens

Table III-7. Relative intensity and cross-contribution data[a] of ions[b] with potential for designating the analyte and the adapted internal standard — Phencyclidine/phencyclidine-d$_5$

	Phencyclidine			Phencyclidine-d$_5$		
CD Group[c]	Ion (m/z)[d]	Rel. int.	Analog's cont.	Ion (m/z)[d]	Rel. int.	Analog's cont.
None	91	30.0	0.53	96	28.0	1.01
	186	22.0	0.16	190	15.0	0.00
	<u>200</u>	100	0.03	<u>205</u>	100	0.02
	242	35.0	0.02	246	28.0	0.10

[a–d] See the corresponding footnotes in Table I-1a.

Table III-8. Relative intensity and cross-contribution data[a] of ions[b] with potential for designating the analyte and the adapted internal standard — LSD/LSD-d$_3$

	LSD			LSD-d$_3$		
CD Group[c]	Ion (m/z)[d]	Rel. int.	Analog's cont.	Ion (m/z)[d]	Rel. int.	Analog's cont.
None	<u>221</u>	100	1.48	<u>224</u>	93.0	4.11
	323	99.3	0.62	326	100	0.23
TMS	<u>293</u>	81.1	—	<u>296</u>	78.2	—
	395	93.8	0.69	398	90.0	1.56
	396	29.0	4.04	399	27.8	1.11

[a–d] See the corresponding footnotes in Table I-1a.

Table III-9. Relative intensity and cross-contribution data[a] of ions[b] with potential for designating the analyte and the adapted internal standard — Mescaline/mescaline-d$_9$

	Mescaline			Mescaline-d$_9$		
CD Group[c]	Ion (m/z)[d]	Rel. int.	Analog's cont.	Ion (m/z)[d]	Rel. int.	Analog's cont.
Acetyl	179	47.1	1.17	185	46.8	0.07
	181	47.9	1.29	190	48.0	0.15
	<u>194</u>	100	1.13	<u>203</u>	100	0.01
	195	13.0	1.32	204	10.5	0.14
	253	16.3	1.14	262	16.2	0.03
TCA	179	34.5	0.52	185	32.5	0.34
	181	100	0.46	190	100	0.29
	<u>194</u>	97.4	0.23	<u>203</u>	98.8	0.05
	195	20.8	1.04	204	19.1	1.05
	355	17.6	0.11	364	22.5	0.00
	357	17.1	0.05	366	21.9	0.02
TFA	179	22.2	0.36	185	22.0	0.06
	<u>181</u>	100	0.47	<u>190</u>	100	0.06
	194	35.3	0.38	203	34.4	0.06
	307	34.8	0.25	316	37.9	0.00
PFP	179	25.4	0.20	185	25.6	0.12
	<u>181</u>	100	0.33	<u>190</u>	100	0.16
	194	38.6	0.21	203	38.1	0.02
	357	38.8	0.11	366	40.2	0.00
HFB	179	27.3	0.35	185	27.1	0.08
	<u>181</u>	100	0.28	<u>190</u>	100	0.08
	194	42.5	0.21	203	41.3	0.03
	407	39.2	0.07	416	39.5	0.00

Table III-9. (Continued)

CD Group[c]	Mescaline			Mescaline-d$_9$		
	Ion (m/z)[d]	Rel. int.	Analog's cont.	Ion (m/z)[d]	Rel. int.	Analog's cont.
4-CB	179	40.2	0.70	185	36.2	0.15
	<u>181</u>	100	0.92	<u>190</u>	100	0.12
	194	94.8	0.75	203	92.3	0.04
	416	4.00	0.74	425	4.40	0.35
	461	27.8	0.37	470	31.4	0.00
[TMS]$_2$	266	1.06	1.96	275	1.05	0.08
	<u>340</u>	9.19	0.04	<u>349</u>	11.0	0.00
	354	0.24	1.65	363	0.29	0.60
t-BDMS	181	8.06	4.93	190	8.65	2.69
	<u>268</u>	37.1	0.50	<u>277</u>	38.0	0.05
	269	7.46	1.06	278	6.66	0.42
	310	3.03	0.33	319	3.04	0.08
TFA/TMS	<u>181</u>	100	0.49	<u>190</u>	100	0.14
	182	12.4	1.38	191	9.76	0.66
	194	4.45	2.64	203	4.47	0.50
	379	24.1	0.25	388	26.0	0.01
	380	5.76	0.34	389	5.49	0.06
TFA/t-BDMS	<u>181</u>	100	0.26	<u>190</u>	100	0.27
	182	13.1	2.46	191	10.0	1.61
	195	24.0	0.83	204	24.6	0.70
	220	13.5	0.46	229	14.9	0.39
	421	22.4	0.01	430	27.6	0.02
PFP/TMS	<u>181</u>	100	0.43	<u>190</u>	100	0.27
	182	12.9	0.95	191	10.5	0.96
	195	8.14	0.88	204	8.71	1.23
	429	30.3	0.15	438	30.2	0.02
	430	7.57	0.15	439	6.72	0.03
PFP/t-BDMS	<u>181</u>	100	0.25	<u>190</u>	100	0.32
	195	33.6	0.34	204	33.1	0.46
	414	4.69	0.04	423	3.42	0.05
	471	26.3	0.02	480	19.6	0.01
HFB/TMS	<u>181</u>	100	0.33	<u>190</u>	100	0.08
	182	11.8	0.98	191	9.42	2.92
	195	8.00	2.81	204	7.98	1.17
	479	24.3	0.17	488	21.2	0.00
	480	6.20	0.17	489	4.80	0.10
HFB/t-BDMS	<u>181</u>	100	0.34	<u>190</u>	100	0.25
	195	28.6	0.32	204	28.6	0.76
	521	22.4	0.04	530	17.0	0.23

[a-d] See the corresponding footnotes in Table I-1a.

Table III-10. Relative intensity and cross-contribution data[a] of ions[b] with potential for designating the analyte and the adapted internal standard — Psilocin/psilocin-d$_{10}$

CD Group[c]	Psilocin			Psilocin-d$_{10}$		
	Ion (m/z)[d]	Rel. int.	Analog's cont.	Ion (m/z)[d]	Rel. int.	Analog's cont.
None	<u>58</u>	100	1.45	<u>66</u>	100	0.25
	59	3.93	2.30	67	4.17	2.55
	204	22.4	1.21	214	21.6	0.09
	205	3.12	1.54	215	3.30	0.24

Table III — Hallucinogens

Table III-10. (Continued)

CD Group[c]	Psilocin Ion $(m/z)^d$	Rel. int.	Analog's cont.	Psilocin-d$_{10}$ Ion $(m/z)^d$	Rel. int.	Analog's cont.
Acetyl	58	100	0.26	66	100	0.10
	59	4.02	0.05	67	3.87	1.25
	146	4.87	3.13	148	2.47	2.00
	160	2.36	3.64	165	1.08	0.42
	246	3.74	0.21	256	3.11	0.02
[Acetyl]$_2$	58	100	0.22	66	100	0.08
	59	3.76	0.96	67	3.78	1.73
	146	3.67	3.52	148	2.11	2.34
	160	2.09	3.92	165	0.91	0.85
	288	0.58	2.70	298	0.46	1.47
[TMS]$_2$	58	56.7	0.38	66	64.3	0.03
	290	100	—	292	100	—
	333	4.97	0.11	343	6.22	0.50
	348	28.2	0.04	358	32.3	0.01
	349	8.84	0.05	359	10.1	0.05
t-BDMS	58	100	0.52	66	100	0.06
	303	1.28	0.36	313	1.43	4.48
	318	16.2	0.18	328	15.8	0.15
	319	4.23	0.23	329	4.23	0.33
[t-BDMS]$_2$	58	100	0.60	66	100	0.01
	432	11.1	0.16	442	11.7	0.00
	433	4.27	0.00	443	4.42	0.87

$^{a-d}$ See the corresponding footnotes in Table I-1a.

**Summary of Drugs, Isotopic Analogs, and Chemical Derivatization Groups Included in
Table IV (Depressant/Hypnotics)**

Compound	Isotopic analog	Chemical derivatization group	Table #
Pentobarbital	d_5	None, [methyl]$_2$, [ethyl]$_2$, [propyl]$_2$, [butyl]$_2$, [TMS]$_2$, [t-BDMS]$_2$	IV-1
Phenobarbital	d_5, d_5 (ring)	[Methyl]$_2$, [ethyl]$_2$, [propyl]$_2$, [butyl]$_2$, [TMS]$_2$, [t-BDMS]$_2$	IV-2
Butalbital	d_5, $^{13}C_4$	None, [methyl]$_2$, [ethyl]$_2$, [propyl]$_2$, [butyl]$_2$, [TMS]$_2$, [t-BDMS]$_2$	IV-3
Sceobarbital	d_5, $^{13}C_4$	None, [methyl]$_2$, [ethyl]$_2$, [propyl]$_2$, [butyl]$_2$, [TMS]$_2$, [t-BDMS]$_2$	IV-4
Methohexital	d_5	None, methyl, ethyl, propyl, butyl, TMS, t-BDMS	IV-5
γ-Hydroxybutyric acid	d_6	[TMS]$_2$, [t-BDMS]$_2$	IV-6
γ-Butyrolactone	d_6	None	IV-7

Table IV — Depressants/Hypnotics

Appendix Two — Table IV
Cross-Contributions Between ions Designating the Drugs and Their Isotopically Labeled Analogs in Various Derivatization Forms — Depressants/Hypnotics

Table IV — Depressants/Hypnotics

Table IV-1. Relative intensity and cross-contribution data[a] of ions[b] with potential for designating the analyte and the adapted internal standard — Pentobarbital/pentobarbital-d$_5$

CD Group[c]	Pentobarbital			Pentobarbital-d$_5$		
	Ion (m/z)[d]	Rel. int.	Analog's cont.	Ion (m/z)[d]	Rel. int.	Analog's cont.
None	<u>156</u>	100	2.83	<u>161</u>	100	1.41
	157	21.6	3.35	162	23.3	1.60
[Methyl]$_2$	<u>184</u>	82.5	4.73	<u>189</u>	100	0.00
[Ethyl]$_2$	184	5.63	1.59	189	6.29	0.01
	<u>197</u>	100	1.62	<u>199</u>	76.5	1.35
	212	93.0	3.39	217	100	0.00
	213	12.5	4.18	218	13.2	0.00
[Propyl]$_2$	97	24.6	3.74	102	23.9	0.02
	156	51.9	0.63	161	53.5	0.01
	181	15.0	1.86	186	16.9	0.14
	198	83.5	0.65	203	83.9	0.01
	225	23.4	3.41	227	18.9	1.75
	<u>240</u>	100	4.56	<u>245</u>	100	0.00
[Butyl]$_2$	97	26.2	2.66	102	25.4	0.45
	156	29.4	4.10	161	27.9	0.52
	195	70.1	0.52	200	67.0	0.07
	<u>251</u>	100	0.22	<u>256</u>	100	0.27
[TMS]$_2$	<u>169</u>	100	3.25	<u>171</u>	90.4	0.94
	184	82.5	2.97	189	100	0.00
[t-BDMS]$_2$	<u>327</u>	74.4	—	<u>332</u>	73.5	—

[a–d] See the corresponding footnotes in Table I-1a.

Table IV-2a. Relative intensity and cross-contribution data[a] of ions[b] with potential for designating the analyte and the adapted internal standard — Phenobarbital/phenobarbital-d$_5$

CD Group[c]	Phenobarbital			Phenobarbital-d$_5$		
	Ion (m/z)[d]	Rel. int.	Analog's cont.	Ion (m/z)[d]	Rel. int.	Analog's cont.
[Methyl]$_2$	<u>232</u>	100	—	<u>233</u>	100	—
	260	2.67	0.09	265	4.28	0.00
[Ethyl]$_2$	146	49.9	4.20	151	58.0	0.03
	<u>260</u>	100	—	<u>261</u>	100	—
	288	1.60	0.07	293	2.22	0.05
[Propyl]$_2$	<u>146</u>	87.1	2.49	<u>151</u>	100	0.05
	189	12.5	1.73	194	14.6	0.41
	275	13.0	1.54	280	16.8	0.55
[Butyl]$_2$	<u>146</u>	27.8	3.13	<u>151</u>	26.5	0.08
	189	9.10	3.73	194	7.38	0.39
	289	43.0	0.77	294	37.2	0.02
	344	2.98	0.11	349	2.55	0.06
[TMS]$_2$	<u>146</u>	100	—	<u>151</u>	100	—
[t-BDMS]$_2$	<u>403</u>	100	—	<u>408</u>	100	—

[a–d] See the corresponding footnotes in Table I-1a.

Appendix Two — Ion Intensity Cross-Contribution Data

Table IV-2b. Relative intensity and cross-contribution data[a] of ions[b] with potential for designating the analyte and the adapted internal standard — Phenobarbital/phenobarbital-d_5(ring)

CD Group[c]	Phenobarbital			Phenobarbital-d_5(ring)		
	Ion (m/z)[d]	Rel. int.	Analog's cont.	Ion (m/z)[d]	Rel. int.	Analog's cont.
[Methyl]$_2$	117	23.6	1.24	122	17.4	0.12
	146	17.8	0.52	151	16.4	0.02
	175	18.6	0.35	180	18.4	0.02
	188	9.82	0.26	193	9.97	0.02
	<u>232</u>	100	0.04	<u>237</u>	100	0.01
	260	2.67	0.05	265	2.38	0.04
[Ethyl]$_2$	103	11.3	0.92	108	10.8	0.22
	117	31.8	1.79	122	23.9	0.09
	146	49.9	1.85	151	47.8	0.03
	202	9.28	0.50	207	8.84	0.09
	232	17.4	0.47	237	16.1	0.06
	<u>260</u>	100	0.04	<u>265</u>	100	0.00
[Propyl]$_2$	117	42.5	1.09	122	35.5	0.08
	<u>146</u>	87.1	0.64	<u>151</u>	91.3	0.06
	204	23.3	0.35	209	26.8	1.00
	246	40.3	0.04	251	39.5	0.73
	275	13.0	1.29	280	12.8	0.62
	288	100	0.02	293	100	0.03
[Butyl]$_2$	117	52.3	1.07	122	41.1	0.27
	<u>146</u>	100	0.69	<u>151</u>	100	0.10
	260	17.0	0.62	265	17.0	0.53
	289	73.4	0.24	294	71.5	0.04
	299	29.0	0.11	304	28.5	0.50
	316	5.97	0.06	321	66.7	0.06
[TMS]$_2$	<u>146</u>	100	—	<u>151</u>	100	—
[t-BDMS]$_2$	<u>403</u>	100	—	<u>408</u>	100	—

[a-d] See the corresponding footnotes in Table I-1a.

Table IV-3a. Relative intensity and cross-contribution data[a] of ions[b] with potential for designating the analyte and the adapted internal standard — Butalbital/butalbital-d_5

CD Group[c]	Butalbital			Butalbital-d_5		
	Ion (m/z)[d]	Rel. int.	Analog's cont.	Ion (m/z)[d]	Rel. int.	Analog's cont.
None	<u>168</u>	100	—	<u>173</u>	100	—
[Methyl]$_2$	181	26.8	4.46	184	17.9	0.22
	<u>196</u>	100	1.09	<u>201</u>	100	0.00
	237	2.42	1.54	242	2.76	0.00
[Ethyl]$_2$	209	15.1	3.86	212	15.1	0.77
	223	59.7	4.16	228	63.3	0.01
	<u>224</u>	100	0.65	<u>229</u>	100	0.00
	265	3.81	0.53	270	3.86	0.00
[Propyl]$_2$	210	58.4	1.11	215	64.8	0.03
	251	81.2	1.84	256	63.7	0.02
	<u>252</u>	100	0.61	<u>257</u>	100	0.00
	293	7.08	0.48	298	8.05	0.00
[Butyl]$_2$	224	22.5	4.77	229	21.9	0.46
	<u>263</u>	100	0.16	<u>268</u>	100	0.02
	279	64.0	1.41	284	40.1	0.06
	280	47.6	1.28	285	40.2	0.01
	293	24.0	3.61	298	25.0	0.08
[TMS]$_2$	<u>353</u>	100	—	<u>358</u>	100	—

Table IV — Depressants/Hypnotics

Table IV-3a. (Continued)

CD Group[c]	Butalbital			Butalbital-d₅		
	Ion (m/z)[d]	Rel. int.	Analog's cont.	Ion (m/z)[d]	Rel. int.	Analog's cont.
[t-BDMS]₂	<u>395</u>	100	—	<u>400</u>	100	—

a–d See the corresponding footnotes in Table I-1a.

Table IV-3b. Relative intensity and cross-contribution data[a] of ions[b] with potential for designating the analyte and the adapted internal standard — Butalbital/butalbital-$^{13}C_4$

CD Group[c]	Butalbital			Butalbital-$^{13}C_4$		
	Ion (m/z)[d]	Rel. int.	Analog's cont.	Ion (m/z)[d]	Rel. int.	Analog's cont.
None	<u>168</u>	100	—	<u>172</u>	100	—
[Methyl]₂	138	17.5	2.01	141	20.1	2.79
	169	14.0	2.15	173	14.3	0.07
	181	27.8	0.39	185	28.6	0.02
	195	71.2	0.35	199	76.5	0.11
	<u>196</u>	100	0.16	<u>200</u>	100	0.01
	209	15.9	0.50	213	16.1	0.35
[Ethyl]₂	196	15.7	1.48	200	16.1	0.10
	209	15.1	1.37	213	15.1	0.14
	223	59.7	1.29	227	63.3	0.17
	<u>224</u>	100	0.88	<u>228</u>	100	0.01
	237	12.9	1.50	241	13.2	0.57
	265	3.81	1.40	269	3.86	0.00
[Propyl]₂	210	60.1	0.82	214	61.0	0.25
	251	86.4	0.77	255	90.7	0.15
	<u>252</u>	100	0.55	<u>256</u>	100	0.01
	265	22.1	1.91	269	22.4	0.79
	293	6.27	1.01	297	6.43	0.04
[Butyl]₂	224	21.7	2.67	228	21.7	1.54
	<u>263</u>	100	0.46	<u>267</u>	100	0.19
	279	69.1	1.09	283	70.5	0.27
	280	48.2	0.74	284	46.0	0.04
	293	24.4	0.72	297	24.4	0.97
[TMS]₂	269	13.5	4.78	273	13.4	1.51
	297	23.5	2.50	301	23.3	0.60
	312	43.9	2.22	316	42.9	0.67
	325	30.6	2.06	329	30.3	3.13
	<u>353</u>	100	2.12	<u>357</u>	100	0.65
	354	29.5	2.18	358	26.1	0.44
[t-BDMS]₂	297	12.5	1.29	301	11.9	0.37
	<u>395</u>	100	1.23	<u>399</u>	100	0.52
	396	39.4	1.56	400	34.1	0.38
	397	15.6	3.64	401	26.6	0.62
	437	11.4	0.80	441	11.0	0.61

a–d See the corresponding footnotes in Table I-1a.

Table IV-4a. Relative intensity and cross-contribution data[a] of ions[b] with potential for designating the analyte and the adapted internal standard — Secobarbital/secobarbital-d$_5$

CD Group[c]	Secobarbital			Secobarbital-d$_5$		
	Ion (m/z)[d]	Rel. int.	Analog's cont.	Ion (m/z)[d]	Rel. int.	Analog's cont.
None	<u>168</u>	100	3.78	<u>173</u>	100	1.93
[Methyl]$_2$	138	18.0	4.77	143	13.2	3.05
	<u>196</u>	100	1.55	<u>201</u>	100	0.00
[Ethyl]$_2$	196	15.3	4.54	201	16.7	0.41
	<u>224</u>	100	1.63	<u>229</u>	100	0.36
[Propyl]$_2$	210	50.1	1.40	215	54.9	0.07
	<u>252</u>	100	2.37	<u>257</u>	100	0.03
	322	5.35	0.26	327	6.87	0.55
[Butyl]$_2$	224	26.1	4.77	229	29.1	0.09
	263	84.4	0.40	268	100	2.14
	<u>279</u>	100	—	<u>284</u>	63.2	—
	350	7.39	0.14	355	9.29	0.65
[TMS]$_2$	<u>297</u>	100	—	<u>302</u>	100	—
	312	54.0	4.64	317	60.4	0.06
	339	50.3	4.17	344	62.5	0.36
	367	61.6	3.78	372	78.0	0.29
[t-BDMS]$_2$	<u>339</u>	49.5	—	<u>344</u>	47.8	—

[a–d] See the corresponding footnotes in Table I-1a.

Table IV-4b. Relative intensity and cross-contribution data[a] of ions[b] with potential for designating the analyte and the adapted internal standard — Secobarbital/secobarbital-$^{13}C_4$

CD Group[c]	Secobarbital			Secobarbital-$^{13}C_4$		
	Ion (m/z)[d]	Rel. int.	Analog's cont.	Ion (m/z)[d]	Rel. int.	Analog's cont.
None	<u>168</u>	100	—	<u>172</u>	100	—
[Methyl]$_2$	138	19.1	3.10	141	20.9	2.26
	181	42.1	1.79	185	44.1	0.17
	195	74.8	0.50	199	81.1	0.12
	<u>196</u>	100	0.55	<u>200</u>	100	0.01
[Ethyl]$_2$	196	14.1	2.12	200	14.1	0.11
	209	21.0	0.59	213	21.8	0.23
	223	67.4	0.41	227	72.1	0.16
	<u>224</u>	100	0.48	<u>228</u>	100	0.01
[Propyl]$_2$	210	48.5	1.06	214	48.9	0.20
	237	11.5	1.85	241	12.0	0.79
	251	94.0	0.48	255	98.9	0.15
	<u>252</u>	100	0.58	<u>256</u>	100	0.01
[Butyl]$_2$	168	23.1	4.01	172	22.2	0.17
	224	24.1	2.74	228	23.5	0.34
	<u>279</u>	100	0.56	<u>283</u>	100	0.68
	280	84.5	0.65	284	79.4	0.15
[TMS]$_2$	<u>297</u>	100	1.48	<u>301</u>	100	0.68
	311	30.4	1.47	315	32.4	3.67
	312	58.8	1.25	316	58.1	0.47
	339	53.9	1.39	343	53.7	2.32
	367	64.5	1.12	371	64.7	0.78

Table IV — Depressants/Hypnotics

Table IV-4b. (Continued)

CD Group[c]	Secobarbital			Secobarbital-$^{13}C_3$		
	Ion (m/z)[d]	Rel. int.	Analog's cont.	Ion (m/z)[d]	Rel. int.	Analog's cont.
[t-BDMS]$_2$	281	9.33	2.00	285	9.13	0.86
	<u>339</u>	53.4	1.07	<u>343</u>	52.3	0.45
	381	4.82	0.97	385	4.65	0.98
	409	100	1.06	413	100	0.53
	410	36.1	1.26	414	32.5	0.53
	451	8.81	0.85	455	8.54	4.85

$^{a-d}$ See the corresponding footnotes in Table I-1a.

Table IV-5. Relative intensity and cross-contribution data^a of ions^b with potential for designating the analyte and the adapted internal standard — Methohexital/methohexital-d$_5$

CD Group[c]	Methohexital			Methohexital-d$_5$		
	Ion (m/z)[d]	Rel. int.	Analog's cont.	Ion (m/z)[d]	Rel. int.	Analog's cont.
None	233	51.8	1.20	238	31.7	0.08
	<u>247</u>	52.4	0.43	<u>252</u>	41.1	0.09
	261	15.0	0.15	266	9.01	0.00
Methyl	247	42.1	3.48	252	21.3	0.31
	<u>261</u>	46.0	0.55	<u>266</u>	34.4	0.30
	275	15.6	0.04	280	8.84	0.51
Ethyl	<u>209</u>	38.5	—	<u>214</u>	20.9	—
	275	39.4	0.86	280	27.8	0.02
	289	14.8	0.00	294	7.51	0.10
Propyl	<u>223</u>	40.7	—	<u>228</u>	20.8	—
	289	39.2	0.86	294	29.2	0.02
	303	14.4	0.18	308	7.55	0.29
Butyl	<u>237</u>	39.0	—	<u>242</u>	19.6	—
	303	39.9	2.02	308	30.7	0.17
	318	15.3	0.50	323	13.6	0.35
TMS	<u>239</u>	100	2.33	<u>244</u>	100	0.11
	305	13.6	3.91	310	10.5	0.10
	319	38.7	0.17	324	39.1	0.01
	333	7.56	0.20	338	4.59	0.23
t-BDMS	<u>239</u>	100	0.62	<u>244</u>	100	0.04
	240	18.5	2.26	245	18.1	0.39
	319	52.2	0.51	324	51.6	0.04

$^{a-d}$ See the corresponding footnotes in Table I-1a.

Table IV-6. Relative intensity and cross-contribution data[a] of ions[b] with potential for designating the analyte and the adapted internal standard — γ-Hydroxybutyric acid/γ-hydroxybutyric acid-d$_6$

CD Group[c]	γ-Hydroxybutyric acid			γ-Hydroxybutyric acid-d$_6$		
	Ion (m/z)[d]	Rel. int.	Analog's cont.	Ion (m/z)[d]	Rel. int.	Analog's cont.
[TMS]$_2$	<u>233</u>	27.9	1.49	<u>239</u>	29.9	0.75
	234	5.55	1.60	240	6.12	0.79
	235	2.52	2.71	241	2.84	0.84
[t-BDMS]$_2$	<u>275</u>	76.6	0.06	<u>281</u>	100	0.01
	276	18.7	0.08	282	24.4	0.01
	277	7.78	0.37	283	10.1	0.03
	317	3.00	0.09	323	4.39	0.27

[a-d] See the corresponding footnotes in Table I-1a.

Table IV-7. Relative intensity and cross-contribution data[a] of ions[b] with potential for designating the analyte and the adapted internal standard — γ-Butyrolactone/γ-butyrolactone-d$_6$

CD Group[c]	γ-Butyrolactone			γ-Butyrolactone-d$_6$		
	Ion (m/z)[d]	Rel. int.	Analog's cont.	Ion (m/z)[d]	Rel. int.	Analog's cont.
None	<u>42</u>	100	—	48	100	—
	56	36.2	0.73	60	32.4	0.00
	86	85.7	0.16	92	65.6	0.00

[a-d] See the corresponding footnotes in Table I-1a.

Table IV — Depressants/Hypnotics

**Summary of Drugs, Isotopic Analogs, and Chemical Derivatization Groups Included in
Table V (Antianxiety Agents)**

Compound	Isotopic analog	Chemical derivatization group	Table #
Oxazepam	d$_5$	None, [Methyl]$_2$, [ethyl]$_2$, [propyl]$_2$, [butyl]$_2$, [TMS]$_2$, [*t*-BDMS]$_2$	V-1
Diazepam	d$_3$, d$_5$	None	V-2
Nordiazepam	d$_5$	None, methyl, ethyl, propyl, butyl, TMS, *t*-BDMS	V-3
Nitrazepam	d$_5$	Methyl, ethyl, propyl, butyl, TMS, *t*-BDMS	V-4
Temazepam	d$_5$	None, methyl, ethyl, propyl, butyl, acetyl, TMS, *t*-BDMS	V-5
Clonazepam	d$_4$	Methyl, ethyl, propyl, butyl, TMS, *t*-BDMS	V-6
7-Aminoclonazepam	d$_4$	[Methyl]$_3$, [ethyl]$_2$, [ethyl]$_3$, propyl, [propyl]$_2$, butyl, [butyl]$_2$, PFP, HFB, [TMS]$_2$, *t*-BDMS, [*t*-BDMS]$_2$ TFA/[TMS]$_2$, TFA/[*t*-BDMS]$_2$, [TFA]$_2$/*t*-BDMS, PFP/TMS, PFP/[TMS]$_2$, PFP/[*t*-BDMS]$_2$, HFB/[*t*-BDMS]$_2$	V-7
Prazepam	d$_5$	None	V-8
Lorazepam	d$_4$	[Methyl]$_2$, [ethyl]$_2$, [propyl]$_2$, [butyl]$_2$, HFB, [TMS]$_2$, [*t*-BDMS]$_2$	V-9
Flunitrazepam	d$_3$, d$_7$	None	V-10
7-Aminoflunitrazepam	d$_3$, d$_7$	None, [methyl]$_2$, ethyl, [ethyl]$_2$, propyl, butyl, acetyl, TFA, PFP, HFB, TMS, TFA/TMS, TFA/*t*-BDMS, PFP/TMS, PFP/*t*-BDMS, HFB/TMS, HFB/*t*-BDMS	V-11
N-Desalkylflurazepam	d$_4$	None, methyl, [methyl]$_2$, ethyl, propyl, butyl, acetyl, TMS, *t*-BDMS	V-12
N-Desmethylflunitrazepam	d$_4$	[Methyl]$_2$, ethyl, propyl, butyl, acetyl, TMS, *t*-BDMS	V-13
2-Hydroxyethylflurazepam	d$_4$	None, butyl, TMS, *t*-BDMS	V-14
Estazolam	d$_5$	None	V-15
Alprazolam	d$_5$	None	V-16
α-Hydroxyalprazolam	d$_5$	TMS, *t*-BDMS	V-17
α-Hydroxytriazolam	d$_4$	TMS, *t*-BDMS	V-18
Mianserin	d$_3$	None	V-19
Methaqualone	d$_7$	None	V-20
Haloperidol	d$_4$	TMS	V-21

Table V — Antianxiety Agents

Appendix Two — Table V
Cross-Contributions Between Ions Designating the Drugs and Their Isotopically Labeled Analogs
in Various Derivatization Forms — Antianxiety Agents

Table V — Antianxiety Agents

Table V-1. Relative intensity and cross-contribution data^a of ions^b with potential for designating the analyte and the adapted internal standard — Oxazepam/oxazepam-d$_5$

CD Group^c	Ion (m/z)	Oxazepam Rel. int.	Analog's cont.	Ion (m/z)	Oxazepam-d$_5$ Rel. int.	Analog's cont.
None	205	69.1	4.24	210	68.6	0.21
	241	71.4	4.36	246	81.2	0.22
	269	100	3.04	274	100	0.30
	270	74.0	4.29	275	93.0	0.20
[Methyl]$_2$	255	48.7	0.47	260	43.2	1.19
	256	32.6	0.83	261	31.6	0.16
	271	100	0.51	276	100	0.04
	273	32.4	4.67	278	35.1	0.34
	314	24.5	0.37	319	24.8	0.06
[Ethyl]$_2$	257	34.7	1.20	262	32.5	0.56
	270	19.1	1.49	275	19.3	0.46
	285	100	0.56	290	100	0.29
	287	34.1	3.11	292	33.3	0.00
	342	11.8	0.40	347	12.8	1.33
[Propyl]$_2$	241	18.4	2.06	246	18.7	0.34
	257	47.4	1.19	262	46.0	0.28
	285	20.7	1.09	290	20.8	0.21
	299	100	0.31	304	100	0.07
	370	9.48	1.05	375	8.26	0.15
[Butyl]$_2$	241	16.7	1.49	246	15.9	0.95
	257	45.1	0.89	262	42.6	1.87
	299	21.4	0.80	304	20.5	0.11
	313	100	0.20	318	100	0.07
	315	34.3	0.54	320	33.7	0.00
	398	5.83	1.29	403	5.37	1.59
[TMS]$_2$	313	26.0	1.54	318	42.9	0.35
	340	12.4	0.84	345	19.2	2.03
	401	14.9	3.75	406	31.8	0.40
	415	15.1	0.22	420	22.8	0.57
	429	100	—	433	100	—
	430	54.7	0.15	435	86.9	0.40
[t-BDMS]$_2$	313	19.6	3.02	318	20.7	1.73
	457	100	1.44	462	100	1.02
	458	36.8	1.52	463	37.7	0.55
	459	45.6	1.57	464	45.6	0.09

$^{a–d}$ See the corresponding footnotes in Table I-1a.

Table V-2a. Relative intensity and cross-contribution data^a of ions^b with potential for designating the analyte and the adapted internal standard — Diazepam/diazepam-d$_3$

CD Group^c	Ion (m/z)	Diazepam Rel. int.	Analog's cont.	Ion (m/z)	Diazepam-d$_3$ Rel. int.	Analog's cont.
None	256	100	—	259	100	—
	257	44.8	3.11	260	44.5	2.83

$^{a–d}$ See the corresponding footnotes in Table I-1a.

Table V — Antianxiety Agents

Table V-2b. Relative intensity and cross-contribution data[a] of ions[b] with potential for designating the analyte and the adapted internal standard — Diazepam/diazepam-d5

CD Group[c]	Ion (*m/z*)	Diazepam Rel. int.	Analog's cont.	Ion (*m/z*)	Diazepam-d5 Rel. int.	Analog's cont.
None	<u>256</u>	100	0.36	<u>261</u>	100	0.12
	258	36.3	1.69	263	36.1	0.02
	283	91.3	0.09	287	86.4	4.39
	285	40.7	1.25	289	79.3	0.04

[a-d] See the corresponding footnotes in Table I-1a.

Table V-3. Relative intensity and cross-contribution data[a] of ions[b] with potential for designating the analyte and the adapted internal standard — Nordiazepam/nordiazepam-d5

CD Group[c]	Ion (*m/z*)	Nordiazepam Rel. int.	Analog's cont.	Ion (*m/z*)	Nordiazepam-d5 Rel. int.	Analog's cont.
None	241	88.0	4.06	246	77.2	1.91
	<u>242</u>	100	3.74	<u>247</u>	100	0.16
	270	68.8	3.23	275	73.0	0.05
Methyl	255	44.1	1.18	260	41.5	3.42
	<u>256</u>	100	0.41	<u>261</u>	100	0.12
	257	46.4	2.69	262	44.6	0.03
	283	89.8	0.11	287	84.2	4.76
	284	69.4	0.13	289	78.1	0.04
Ethyl	<u>270</u>	100	0.16	<u>275</u>	100	0.14
	271	55.6	0.96	276	40.1	0.04
	297	98.3	0.08	301	89.8	3.73
	298	62.7	0.10	303	70.5	0.04
Propyl	<u>269</u>	90.3	0.16	<u>273</u>	82.9	3.14
	270	59.4	0.35	275	70.2	0.11
	284	81.9	0.15	289	84.2	0.14
	311	100	0.09	315	100	3.01
	312	58.7	0.11	317	71.7	0.03
Butyl	255	18.3	2.27	260	18.8	0.63
	<u>269</u>	100	0.83	<u>273</u>	98.0	2.91
	270	57.1	0.50	275	76.0	0.06
	298	60.8	0.37	303	64.8	0.19
	325	96.3	0.08	329	100	2.57
	326	43.6	0.09	331	58.2	0.04
TMS	91	7.72	4.54	96	8.76	0.73
	227	5.55	3.85	232	5.73	2.45
	327	20.2	0.67	332	21.6	0.26
	<u>341</u>	100	—	<u>345</u>	100	—
	342	57.8	0.56	346	29.6	3.43
	343	46.9	1.58	347	71.5	0.19
t-BDMS	313	3.52	2.95	318	4.15	0.42
	<u>327</u>	100	2.15	<u>332</u>	100	0.37
	328	27.5	2.37	333	28.3	0.19
	329	40.1	4.45	334	40.4	0.06
	369	3.07	1.95	374	3.22	0.40

[a-d] See the corresponding footnotes in Table I-1a.

465

Table V-4. Relative intensity and cross-contribution data[a] of ions[b] with potential for designating the analyte and the adapted internal standard — Nitrazepam/nitrazepam-d$_5$

CD Group[c]	Nitrazepam Ion (m/z)	Rel. int.	Analog's cont.	Nitrazepam-d$_5$ Ion (m/z)	Rel. int.	Analog's cont.
Methyl	220	53.0	0.00	225	41.3	4.55
	_267_	100	1.57	_272_	100	0.00
	294	86.6	0.88	298	81.4	0.11
	295	60.5	1.30	300	45.1	0.00
Ethyl	234	39.7	0.67	239	22.5	2.09
	281	65.6	0.66	286	66.5	0.02
	282	59.9	3.75	287	62.8	0.00
	_308_	100	0.41	_312_	100	0.08
	309	44.5	0.41	314	30.1	0.00
Propyl	295	54.0	0.47	300	56.6	0.02
	296	49.1	4.44	301	50.1	0.00
	_322_	100	0.21	_326_	100	0.10
Butyl	_280_	74.6	0.38	_284_	86.0	0.10
	309	43.7	0.47	314	45.3	0.03
	336	100	0.20	340	100	0.07
	337	35.4	0.25	341	24.5	0.09
TMS	306	34.2	4.13	310	28.6	1.87
	_352_	100	4.24	_356_	100	0.54
	353	61.7	4.25	357	28.2	0.24
t-BDMS	292	16.6	1.23	297	15.7	1.44
	_338_	100	0.98	_343_	100	0.02
	339	26.3	1.93	344	26.4	0.02
	380	0.78	1.58	385	2.89	0.09
	394	6.59	0.95	398	5.94	0.75

[a-d] See the corresponding footnotes in Table I-1a.

Table V-5. Relative intensity and cross-contribution data[a] of ions[b] with potential for designating the analyte and the adapted internal standard — Temazepam/temazepam-d$_5$

CD Group[c]	Temazepam Ion (m/z)	Rel. int.	Analog's cont.	Temazepam-d$_5$ Ion (m/z)	Rel. int.	Analog's cont.
None	_228_	100	—	_232_	78.7	—
	257	60.8	2.54	262	67.2	0.05
	300	89.4	0.37	305	100	0.11
Methyl	255	48.0	0.97	260	42.5	1.47
	256	33.2	1.71	261	33.9	0.11
	_271_	100	0.95	_276_	100	0.03
	314	29.5	0.59	319	26.3	0.00
Ethyl	255	30.0	0.76	260	30.0	3.25
	256	23.2	0.62	261	24.9	0.78
	257	24.7	1.37	262	24.4	0.44
	_271_	100	0.35	_276_	100	0.28
	273	33.7	4.00	278	34.0	0.19
	328	12.9	1.08	333	12.8	0.00
Propyl	255	33.1	4.54	260	32.3	3.20
	257	31.3	3.04	262	31.2	0.07
	_271_	100	2.12	_276_	100	0.04
Butyl	255	33.3	0.40	260	32.2	3.99
	257	37.6	1.44	262	37.3	0.07
	_271_	100	0.23	_276_	100	0.06
	300	12.3	0.70	305	12.6	0.98

Table V — Antianxiety Agents

Table V-5. (Continued)

CD Group[c]	Temazepam Ion (m/z)	Rel. int.	Analog's cont.	Temazepam-d₅ Ion (m/z)	Rel. int.	Analog's cont.
Acetyl	228	7.42	4.35	233	5.82	1.29
	256	22.7	1.10	261	24.8	0.41
	257	27.3	3.76	262	27.4	0.05
	<u>271</u>	100	0.66	<u>276</u>	100	0.04
	300	25.7	0.57	305	25.2	0.07
TMS	283	28.5	0.86	288	28.1	0.18
	<u>343</u>	100	0.58	<u>348</u>	100	0.31
	345	38.6	2.99	350	39.1	0.01
	357	21.3	0.50	362	22.2	0.32
	372	19.8	0.84	377	20.2	0.33
t-BDMS	255	28.6	2.14	260	27.7	1.22
	256	24.3	3.44	261	23.9	0.42
	283	48.6	0.60	288	46.8	0.17
	<u>357</u>	100	0.42	<u>362</u>	100	0.42
	359	38.7	0.59	364	39.8	0.02
	385	10.7	0.41	390	11.0	0.55

[a–d] See the corresponding footnotes in Table I-1a.

Table V-6. Relative intensity and cross-contribution data[a] of ions[b] with potential for designating the analyte and the adapted internal standard — Clonazepam/clonazepam-d₄

CD Group[c]	Clonazepam Ion (m/z)	Rel. int.	Analog's cont.	Clonazepam-d₄ Ion (m/z)	Rel. int.	Analog's cont.
Methyl	248	97.1	1.20	252	83.5	1.23
	<u>294</u>	100	0.75	<u>298</u>	100	1.37
	302	95.9	1.05	306	76.5	0.77
	329	91.3	0.74	333	83.1	0.70
Ethyl	234	50.1	3.37	238	45.1	3.30
	262	35.7	3.70	266	30.0	1.66
	280	62.8	0.76	284	56.4	2.56
	308	100	0.60	312	100	0.93
	316	60.5	2.34	320	46.5	0.48
	<u>342</u>	98.4	—	<u>345</u>	62.4	—
Propyl	234	58.3	2.38	238	60.6	4.79
	<u>315</u>	100	2.03	<u>319</u>	100	0.67
	357	50.6	0.63	361	49.3	0.76
Butyl	<u>280</u>	100	0.89	<u>284</u>	100	1.59
	315	83.3	0.61	319	73.7	0.96
	336	80.7	0.43	340	71.7	1.84
TMS	306	58.3	3.68	310	61.3	1.03
	352	78.1	3.69	356	87.9	3.23
	372	39.7	3.85	376	40.8	2.23
	<u>387</u>	82.0	3.57	<u>391</u>	86.2	2.33
t-BDMS	326	12.3	1.97	330	12.0	4.38
	<u>372</u>	100	1.35	<u>376</u>	100	2.37
	373	26.0	1.52	377	26.3	1.38
	374	39.3	1.58	378	38.8	0.16
	414	2.39	1.25	418	2.40	2.72

[a–d] See the corresponding footnotes in Table I-1a.

Table V-7. Relative intensity and cross-contribution data^a of ions^b with potential for designating the analyte and the adapted internal standard — 7-Aminoclonazepam/7-aminoclonazepam-d$_4$

CD Group^c	7-Aminoclonazepam			7-Aminoclonazepam-d$_4$		
	Ion (*m/z*)	Rel. int.	Analog's cont.	Ion (*m/z*)	Rel. int.	Analog's cont.
[Methyl]$_3$	284	14.0	1.74	288	14.2	4.79
	298	55.6	0.93	302	55.3	2.56
	299	23.6	0.72	303	24.1	0.53
	<u>327</u>	100	0.31	<u>331</u>	100	0.67
	328	24.7	0.96	332	21.1	0.23
[Ethyl]$_2$	306	19.5	0.70	310	22.6	3.89
	313	33.3	0.82	317	32.8	0.67
	<u>341</u>	100	0.45	<u>345</u>	100	0.73
	342	28.7	1.33	346	22.9	0.24
[Ethyl]$_3$	<u>354</u>	100	0.16	<u>358</u>	100	0.81
	356	35.6	1.90	360	34.3	0.09
	369	64.0	0.12	373	67.4	0.56
Propyl	250	39.7	2.62	254	40.4	3.91
	256	26.8	4.06	260	24.7	2.15
	285	28.0	1.40	289	28.3	0.77
	299	33.8	0.99	303	31.3	0.70
	<u>327</u>	100	0.98	<u>331</u>	100	0.73
	328	28.9	1.79	322	21.4	3.11
[Propyl]$_2$	298	27.7	2.00	302	26.3	1.04
	340	59.4	0.92	344	57.7	2.59
	341	27.4	2.03	345	26.9	0.64
	<u>369</u>	100	0.33	<u>373</u>	100	0.96
	370	28.9	0.79	374	24.8	0.32
Butyl	250	4.30	2.69	254	49.0	3.75
	256	32.2	3.85	260	33.2	2.47
	285	40.2	1.11	289	42.8	0.99
	306	22.1	1.65	310	25.6	1.21
	<u>341</u>	100	0.78	<u>345</u>	100	0.84
	342	28.7	1.51	346	23.2	0.36
[Butyl]$_2$	298	17.5	1.22	302	18.6	2.04
	312	17.6	2.81	316	18.0	1.32
	354	45.8	0.35	358	45.2	0.97
	369	11.4	1.48	373	11.6	1.05
	<u>397</u>	100	0.22	<u>401</u>	100	1.27
	398	29.1	0.59	402	27.0	0.38
PFP	368	23.5	0.26	372	23.2	0.69
	396	89.4	1.73	400	100	0.75
	<u>402</u>	99.6	3.99	<u>406</u>	97.7	4.81
	403	69.2	1.88	407	72.2	0.83
	431	100	1.52	435	99.8	0.71
HFB	418	23.2	3.06	422	23.8	0.00
	446	91.0	2.03	450	98.1	0.63
	<u>452</u>	97.5	—	<u>456</u>	99.4	—
	453	75.9	2.59	457	73.6	1.12
	481	100	2.10	485	100	0.72
[TMS]$_2$	314	30.5	1.30	318	29.1	3.08
	394	98.3	0.53	398	100	1.72
	395	34.1	2.22	399	34.7	3.12
	414	37.8	0.34	418	36.8	4.09
	<u>429</u>	100	0.49	<u>433</u>	98.8	4.41
	430	44.3	1.44	434	35.6	2.57

Table V — Antianxiety Agents

Table V-7. (Continued)

CD Group[c]	Ion (*m/z*)	7-Aminoclonazepam Rel. int.	Analog's cont.	Ion (*m/z*)	7-Aminoclonazepam-d₄ Rel. int.	Analog's cont.
t-BDMS	242	9.65	2.90	246	9.67	2.83
	328	8.99	4.06	332	9.26	2.04
	<u>342</u>	100	0.59	<u>346</u>	100	2.42
	343	27.3	1.42	347	27.9	1.35
	344	39.5	1.24	348	39.5	0.41
	399	17.7	0.56	403	18.1	2.91
[*t*-BDMS]₂	<u>456</u>	92.2	0.51	<u>460</u>	88.4	4.85
	457	35.5	0.83	461	34.1	2.83
	458	42.3	1.37	462	40.3	0.57
TFA/[TMS]₂	410	20.1	3.20	414	26.5	2.59
	491	19.8	1.77	495	29.2	1.25
	<u>525</u>	59.0	1.40	<u>529</u>	84.0	2.67
TFA/[*t*-BDMS]₂	368	15.1	4.70	372	12.4	2.37
	438	5.21	3.04	442	6.48	4.12
	<u>552</u>	76.9	—	<u>556</u>	82.8	—
	553	30.4	1.23	557	34.7	3.05
	554	35.5	1.53	558	39.0	0.54
[TFA]₂/*t*-BDMS	<u>590</u>	2.12	—	<u>594</u>	1.88	—
PFP/TMS	388	25.4	0.81	392	23.0	2.75
	<u>468</u>	100	0.23	<u>472</u>	100	0.29
	469	29.5	0.81	473	29.0	1.89
	503	69.0	0.37	507	64.4	1.34
PFP/[TMS]₂	460	40.1	2.79	464	35.2	3.14
	540	100	0.79	544	100	0.31
	541	37.2	1.39	545	36.7	1.53
	<u>575</u>	96.7	0.33	<u>579</u>	91.9	4.95
	576	44.9	1.13	580	35.3	2.85
PFP/[*t*-BDMS]₂	440	11.3	0.96	444	10.9	2.51
	<u>602</u>	100	—	<u>606</u>	100	—
	603	45.4	0.32	607	44.0	3.56
	604	50.4	0.68	608	48.8	0.64
HFB/[*t*-BDMS]₂	490	5.87	1.11	494	6.17	2.55
	<u>652</u>	35.1	—	<u>656</u>	30.7	—
	653	14.6	0.27	657	12.9	3.05
	654	16.4	0.57	658	14.5	0.75

[a–d] See the corresponding footnotes in Table I-1a.

Table V-8. Relative intensity and cross-contribution data[a] of ions[b] with potential for designating the analyte and the adapted internal standard — Prazepam/prazepam-d₅

CD Group[c]	Ion (*m/z*)	Prazepam Rel. int.	Analog's cont.	Ion (*m/z*)	Prazepam-d₅ Rel. int.	Analog's cont.
None	91	78.6	4.09	96	80.7	1.62
	241	41.1	3.62	246	45.0	1.75
	<u>269</u>	100	2.60	<u>273</u>	100	0.63
	295	81.2	2.22	300	82.4	0.56
	296	54.0	2.22	301	61.3	0.07
	324	38.7	1.74	329	48.5	0.08

[a–d] See the corresponding footnotes in Table I-1a.

Table V-9. Relative intensity and cross-contribution data[a] of ions[b] with potential for designating the analyte and the adapted internal standard — Lorazepam/lorazepam-d₄

CD Group[c]	Lorazepam Ion (m/z)	Rel. int.	Analog's cont.	Lorazepam-d₄ Ion (m/z)	Rel. int.	Analog's cont.
[Methyl]₂	255	9.21	4.27	259	8.83	1.89
	<u>305</u>	100	—	<u>309</u>	100	—
	307	65.3	1.17	311	65.1	0.27
	348	12.1	0.95	352	12.3	11.2
[Ethyl]₂	293	24.3	1.69	297	23.3	1.33
	<u>319</u>	100	—	<u>323</u>	100	—
	321	67.1	0.32	325	65.3	0.33
	341	8.17	0.88	345	7.68	1.08
[Propyl]₂	293	31.4	4.00	297	32.1	3.00
	<u>333</u>	100	—	<u>337</u>	100	—
	335	66.0	0.87	339	66.1	0.37
[Butyl]₂	<u>347</u>	100	—	<u>351</u>	100	—
	349	68.0	1.05	353	68.2	0.15
[HFB]₂	<u>407</u>	100	0.25	<u>411</u>	100	0.75
	409	36.1	1.40	413	33.4	0.15
[TMS]₂	<u>429</u>	100	0.52	<u>433</u>	100	4.78
	430	34.9	0.57	434	35.3	3.06
	431	44.6	0.86	435	43.9	3.47
[t-BDMS]₂	<u>491</u>	75.4	—	<u>495</u>	85.2	—
	493	58.7	0.57	497	65.3	2.34
	515	32.4	0.55	519	36.7	0.86

[a–d] See the corresponding footnotes in Table I-1a.

Table V-10a. Relative intensity and cross-contribution data[a] of ions[b] with potential for designating the analyte and the adapted internal standard — Flunitrazepam/flunitrazepam-d₃

CD Group[c]	Flunitrazepam Ion (m/z)	Rel. int.	Analog's cont.	Flunitrazepam-d₃ Ion (m/z)	Rel. int.	Analog's cont.
None	238	44.4	3.85	241	49.1	3.20
	285	91.8	2.16	288	97.6	1.91
	286	93.6	4.23	289	95.0	0.17
	<u>312</u>	100	0.95	<u>315</u>	100	1.48
	313	70.7	0.90	316	70.0	0.17

[a–d] See the corresponding footnotes in Table I-1a.

Table V-10b. Relative intensity and cross-contribution data[a] of ions[b] with potential for designating the analyte and the adapted internal standard — Flunitrazepam/flunitrazepam-d₇

CD Group[c]	Flunitrazepam Ion (m/z)	Rel. int.	Analog's cont.	Flunitrazepam-d₇ Ion (m/z)	Rel. int.	Analog's cont.
None	238	43.6	2.58	245	39.4	0.59
	266	56.0	0.18	272	44.7	0.94
	<u>285</u>	95.8	0.43	<u>292</u>	100	0.16
	312	100	0.77	318	99.4	0.00

[a–d] See the corresponding footnotes in Table I-1a.

Table V — Antianxiety Agents

Table V-11a. Relative intensity and cross-contribution data*a* of ions*b* with potential for designating the analyte and the adapted internal standard — 7-Aminoflunitrazepam/7-aminoflunitrazepam-d$_3$

CD Group*c*	7-Aminoflunitrazepam			7-Aminoflunitrazepam-d$_3$		
	Ion (*m/z*)	Rel. int.	Analog's cont.	Ion (*m/z*)	Rel. int.	Analog's cont.
None	283	100	—	286	100	—
[Methyl]$_2$	282	57.5	0.98	285	55.1	0.10
	283	39.2	4.29	286	37.8	0.01
	311	100	0.43	314	100	0.21
	312	20.3	1.94	315	20.5	0.07
Ethyl	283	41.7	0.42	286	42.5	0.40
	311	100	0.65	314	100	0.16
	312	20.7	3.57	315	20.7	0.00
[Ethyl]$_2$	310	7.16	2.39	313	5.32	0.04
	324	100	0.27	327	100	1.74
	339	60.0	2.59	342	54.4	0.13
Propyl	268	14.9	4.48	271	13.1	0.82
	296	100	0.75	299	100	0.46
	325	81.4	1.13	328	82.7	0.24
	326	19.0	3.68	329	18.7	0.06
Butyl	268	14.7	3.16	271	15.9	0.31
	296	100	—	299	100	—
	310	27.7	0.87	313	28.6	1.47
	339	95.5	0.99	342	97.7	0.47
Acetyl	297	67.4	0.64	300	64.5	0.29
	306	24.1	0.35	309	24.4	2.03
	324	53.1	0.44	327	53.2	4.61
	325	100	0.34	328	100	0.24
TFA	351	100	—	354	100	—
PFP	401	100	3.48	404	100	0.36
	410	34.5	2.42	413	35.1	3.52
	428	80.6	2.51	431	81.8	2.90
	429	91.8	2.94	432	93.7	0.25
HFB	451	100	—	454	100	—
TMS	327	47.3	2.92	330	49.3	1.31
	355	100	0.44	358	100	0.89
TFA/TMS	280	22.9	1.20	283	26.3	2.73
	423	81.9	0.53	426	84.1	1.22
	424	24.0	0.95	427	24.9	0.57
	432	17.9	0.56	435	18.7	2.50
	451	100	0.50	454	100	1.17
TFA/*t*-BDMS	386	20.7	4.19	389	20.3	1.76
	436	100	4.17	439	100	1.35
	493	39.7	4.12	496	41.6	1.61
PFP/TMS	352	15.0	1.12	355	15.3	2.59
	473	85.2	0.25	476	84.7	1.34
	474	25.6	0.71	477	25.2	0.55
	482	21.4	0.27	485	21.6	1.51
	501	100	0.26	504	100	1.06
PFP/*t*-BDMS	486	100	—	489	100	—
HFB/TMS	280	17.4	2.41	283	16.6	2.50
	402	15.4	1.08	405	14.7	1.44
	523	81.5	0.59	526	79.0	0.98
	524	25.0	1.30	527	24.2	0.42
	532	21.2	0.35	535	21.3	1.76
	551	100	0.27	554	100	1.31
HFB/*t*-BDMS	536	100	—	539	100	—

a–d See the corresponding footnotes in Table I-1a.

Appendix Two — Ion Intensity Cross-Contribution Data

Table V-11b. Relative intensity and cross-contribution data[a] of ions[b] with potential for designating the analyte and the adapted internal standard — 7-Aminoflunitrazepam/7-aminoflunitrazepam-d₇

CD Group[c]	7-Aminoflunitrazepam			7-Aminoflunitrazepam-d₇		
	Ion (*m/z*)	Rel. int.	Analog's cont.	Ion (*m/z*)	Rel. int.	Analog's cont.
None	255	62.9	0.60	262	68.1	0.51
	<u>283</u>	100	0.24	<u>290</u>	100	0.00
[Methyl]₂	266	14.8	1.18	273	12.7	0.10
	282	57.2	2.74	289	57.0	0.00
	310	23.5	0.33	316	19.8	0.03
	<u>311</u>	100	0.27	<u>318</u>	100	0.00
	312	20.5	0.36	319	20.6	0.01
Ethyl	268	18.5	1.51	275	9.13	0.47
	282	55.9	1.20	289	56.5	0.12
	283	43.0	4.79	290	43.5	0.76
	296	10.8	0.60	303	12.0	0.04
	310	29.9	0.25	316	26.4	0.09
	<u>311</u>	100	0.68	<u>318</u>	100	0.04
[Ethyl]₂	266	19.4	0.99	273	17.6	0.08
	<u>324</u>	100	0.19	<u>331</u>	100	0.00
	325	23.1	0.63	332	22.2	0.06
	339	60.2	0.19	346	59.0	0.01
Propyl	268	14.1	1.29	275	13.3	0.93
	<u>296</u>	100	0.42	<u>303</u>	100	0.05
	297	31.5	1.44	304	31.8	1.27
	325	80.7	0.32	332	83.9	0.00
	326	17.7	0.27	333	17.4	0.00
Butyl	268	13.4	2.13	275	14.2	1.46
	<u>296</u>	100	0.30	<u>303</u>	100	0.04
	297	18.9	4.80	304	20.4	0.25
	310	28.1	0.82	317	29.6	0.00
	339	92.5	0.04	346	99.1	0.00
	340	21.7	0.11	347	22.6	0.00
Acetyl	255	22.9	2.05	262	23.3	1.65
	296	27.6	3.72	303	30.4	0.06
	297	67.4	3.32	304	72.6	0.56
	324	53.1	0.06	330	50.1	0.02
	<u>325</u>	100	0.25	<u>332</u>	100	0.01
TFA	350	31.2	1.55	357	30.9	0.08
	<u>351</u>	100	1.41	<u>358</u>	100	0.45
	378	66.5	1.39	384	55.7	0.02
	379	86.5	1.41	386	72.3	0.03
PFP	400	37.5	2.40	407	36.3	0.47
	<u>401</u>	100	2.24	<u>408</u>	100	1.26
	428	75.5	2.30	434	64.6	1.06
	429	86.0	2.09	436	70.1	0.80
HFB	450	33.7	2.75	457	33.3	0.13
	<u>451</u>	100	2.61	<u>458</u>	100	0.28
	478	74.9	2.56	484	63.8	0.01
	479	85.2	2.69	486	68.9	0.10
TMS	326	43.0	1.96	333	44.7	0.07
	<u>355</u>	100	0.66	<u>362</u>	100	0.01
	356	27.3	0.71	363	27.3	0.02
TFA/TMS	423	81.9	0.47	430	91.5	0.37
	424	24.0	0.60	431	26.8	3.75
	450	43.5	0.57	456	39.9	0.06
	<u>451</u>	100	0.60	<u>458</u>	100	0.00

Table V — Antianxiety Agents

Table V-11b. (Continued)

CD Group[c]	7-Aminoflunitrazepam			7-Aminoflunitrazepam-d[7]		
	Ion $(m/z)^d$	Rel. int.	Analog's cont.	Ion $(m/z)^d$	Rel. int.	Analog's cont.
TFA/t-BDMS	<u>436</u>	100	0.19	<u>443</u>	100	0.01
	437	28.3	0.24	444	28.0	0.24
	493	37.5	0.18	500	40.6	0.00
PFP/TMS	352	15.0	1.30	359	16.4	0.28
	<u>473</u>	85.2	0.21	<u>480</u>	92.8	0.53
	474	25.6	0.29	481	27.8	4.99
	500	45.4	0.17	506	42.3	0.04
	501	100	0.20	508	100	0.00
	502	30.3	0.26	509	30.2	0.00
PFP/t-BDMS	<u>486</u>	100	—	<u>493</u>	100	—
HFB/TMS	402	15.4	1.36	409	17.7	0.12
	523	81.5	0.55	530	89.0	0.40
	524	25.0	2.11	531	27.9	3.60
	550	43.9	0.41	556	40.5	0.06
	<u>551</u>	100	0.40	<u>558</u>	100	0.00
HFB/t-BDMS	<u>296</u>	56.3	4.73	<u>299</u>	62.9	1.95
	536	100	2.95	543	100	0.22
	537	30.2	3.05	544	29.8	0.43
	593	29.0	2.81	600	27.3	0.00

$^{a-d}$ See the corresponding footnotes in Table I-1a.

Table V-12. Relative intensity and cross-contribution data[a] of ions[b] with potential for designating the analyte and the adapted internal standard — N-Desalkylflurazepam/N-desalkylflurazepam-d[4]

CD Group[c]	N-Desalkylflurazepam			N-Desalkylflurazepam-d[4]		
	Ion $(m/z)^d$	Rel. int.	Analog's cont.	Ion $(m/z)^d$	Rel. int.	Analog's cont.
None	<u>259</u>	100	—	<u>263</u>	100	—
Methyl	274	100	3.43	278	100	0.39
	275	44.9	3.10	279	44.1	0.08
	283	34.3	1.85	287	43.2	0.63
	<u>301</u>	88.0	—	<u>304</u>	75.0	—
	302	82.9	2.17	306	91.0	0.09
[Methyl]₂	239	12.5	1.43	243	9.39	2.49
	<u>275</u>	100	—	<u>279</u>	100	—
	297	28.1	0.37	301	37.4	0.00
	316	31.8	1.02	320	33.7	0.51
Ethyl	259	26.1	4.08	263	25.3	1.11
	288	100	0.45	292	100	1.28
	289	51.2	4.36	293	38.8	0.30
	297	36.9	0.44	301	46.5	2.13
	<u>315</u>	99.3	—	<u>318</u>	76.4	—
	316	80.2	0.17	320	80.6	0.41
Propyl	259	34.9	3.96	263	32.3	1.79
	<u>288</u>	100	0.53	<u>292</u>	100	0.30
	302	81.8	0.18	306	76.8	0.68
	311	38.5	0.38	315	49.2	0.63
	330	71.1	0.24	334	74.8	0.20
Butyl	<u>287</u>	100	—	<u>290</u>	86.3	—
	288	82.8	1.50	292	100	0.37
	316	45.4	0.69	320	48.4	1.68

Table V-12. (Continued)

CD Group[c]	N-Desalkylflurazepam			N-Desalkylflurazepam-d$_4$		
	Ion (m/z)[d]	Rel. int.	Analog's cont.	Ion (m/z)[d]	Rel. int.	Analog's cont.
Acetyl	260	59.4	0.99	264	56.7	0.80
	269	39.5	0.27	273	48.4	3.20
	<u>288</u>	100	0.26	<u>292</u>	100	0.46
	302	26.8	0.06	306	25.6	0.70
TMS	<u>360</u>	90.7	0.65	<u>364</u>	100	2.07
t-BDMS	<u>345</u>	100	1.08	<u>349</u>	100	0.24
	346	27.4	2.91	350	28.6	1.30
	347	40.7	2.25	351	41.5	0.12
	402	7.58	1.07	406	819	2.52

[a–d] See the corresponding footnotes in Table I-1a.

Table V-13. Relative intensity and cross-contribution data[a] of ions[b] with potential for designating the analyte and the adapted internal standard — N-Desmethylflunitrazepam/N-desmethylflunitrazepam-d$_4$

CD Group[c]	N-Desmethylflunitrazepam			N-Desmethylflunitrazepam-d$_4$		
	Ion (m/z)[d]	Rel. int.	Analog's cont.	Ion (m/z)[d]	Rel. int.	Analog's cont.
[Methyl]$_2$	238	28.9	3.40	242	26.6	1.07
	285	33.7	3.19	289	35.7	0.46
	<u>286</u>	100	0.67	<u>290</u>	100	0.09
	326	33.0	0.45	329	25.5	0.54
Ethyl	299	53.0	2.03	303	62.7	1.75
	300	54.5	2.46	304	63.1	0.02
	308	32.3	0.55	312	50.7	2.14
	<u>326</u>	100	0.52	<u>329</u>	100	0.99
	327	51.8	0.68	331	40.6	0.02
Propyl	298	74.3	4.47	301	72.5	2.55
	<u>299</u>	95.7	2.04	<u>303</u>	99.0	0.02
	313	48.8	2.33	317	56.9	0.17
	322	34.0	0.56	326	52.4	1.80
	340	100	0.45	343	100	1.20
	341	57.5	0.97	345	44.4	0.02
Butyl	<u>298</u>	100	1.74	<u>301</u>	100	1.78
	327	37.0	3.14	331	43.2	0.29
	354	93.7	0.09	357	92.6	1.00
Acetyl	213	17.3	2.86	216	12.0	0.66
	<u>260</u>	100	0.60	<u>262</u>	100	1.53
	302	26.7	0.45	306	30.8	0.10
TMS	324	29.7	3.45	327	26.4	2.92
	<u>371</u>	100	2.14	<u>375</u>	100	0.11
t-BDMS	310	12.5	2.43	314	11.9	0.77
	<u>356</u>	100	1.41	<u>360</u>	100	0.13
	357	26.5	1.65	361	26.4	0.06
	413	2.76	1.21	417	2.57	0.25

[a–d] See the corresponding footnotes in Table I-1a.

Table V — Antianxiety Agents

Table V-14. Relative intensity and cross-contribution data[a] of ions[b] with potential for designating the analyte and the adapted internal standard — 2-Hydroxyethylflurazepam/2-hydroxyethylflurazepam-d$_4$

CD Group[c]	2-Hydroxyethylflurazepam			2-Hydroxyethylflurazepam-d$_4$		
	Ion (m/z)[d]	Rel. int.	Analog's cont.	Ion (m/z)[d]	Rel. int.	Analog's cont.
None	<u>288</u>	100	2.85	292	100	0.64
	313	15.3	1.97	317	19.2	1.35
Butyl	183	16.9	4.24	187	11.3	1.48
	260	31.0	3.90	264	30.3	0.90
	<u>288</u>	100	1.11	<u>292</u>	100	0.67
TMS	260	26.6	4.74	264	25.5	1.11
	<u>288</u>	100	2.28	<u>292</u>	100	0.55
	360	16.0	1.09	364	15.5	3.42
t-BDMS	345	8.87	1.30	349	8.83	2.73
	<u>389</u>	100	0.71	<u>393</u>	100	2.39
	390	26.7	0.93	394	27.9	1.33
	391	38.7	1.76	395	39.1	0.10
	431	2.63	1.85	435	2.58	2.96

[aa–d] See the corresponding footnotes in Table I-1a.

Table V-15. Relative intensity and cross-contribution data[a] of ions[b] with potential for designating the analyte and the adapted internal standard — Estazolam/estazolam-d$_5$

CD Group[c]	Estazolam			Estazolam-d$_5$		
	Ion (m/z)[d]	Rel. int.	Analog's cont.	Ion (m/z)[d]	Rel. int.	Analog's cont.
None	205	66.5	4.90	210	66.1	0.52
	239	45.4	1.19	244	27.4	0.37
	<u>259</u>	100	0.78	<u>264</u>	98.9	0.77
	394	62.7	0.39	299	100	0.20

[a–d] See the corresponding footnotes in Table I-1a.

Table V-16. Relative intensity and cross-contribution data[a] of ions[b] with potential for designating the analyte and the adapted internal standard — Alprazolam/alprazolam-d$_5$

CD Group[c]	Alprazolam			Alprazolam-d$_5$		
	Ion (m/z)[d]	Rel. int.	Analog's cont.	Ion (m/z)[d]	Rel. int.	Analog's cont.
None	204	76.6	4.60	209	71.4	0.16
	273	58.5	4.52	278	30.4	2.78
	<u>279</u>	100	—	<u>284</u>	100	—
	308	70.7	0.50	313	74.8	0.07

[a–d] See the corresponding footnotes in Table I-1a.

Table V-17. Relative intensity and cross-contribution data[a] of ions[b] with potential for designating the analyte and the adapted internal standard — α-Hydroxyalprazolam/α-hydroxyalprazolam-d$_5$

CD Group[c]	α-Hydroxyalprazolam			α-Hydroxyalprazolam-d$_5$		
	Ion (m/z)[d]	Rel. int.	Analog's cont.	Ion (m/z)[d]	Rel. int.	Analog's cont.
TMS	364	3.59	4.84	369	3.90	3.92
	<u>381</u>	100	4.52	<u>386</u>	100	0.45
	383	38.1	4.91	388	39.3	0.10
	396	33.4	4.41	401	33.7	0.49
	398	12.6	4.46	403	12.8	0.09
t-BDMS	<u>381</u>	100	2.54	<u>386</u>	100	0.40
	382	27.9	2.60	387	28.2	0.16
	383	39.0	2.80	388	38.7	0.06
	423	2.89	2.31	428	2.79	0.51

[a–d] See the corresponding footnotes in Table I-1a.

Table V-18. Relative intensity and cross-contribution data[a] of ions[b] with potential for designating the analyte and the adapted internal standard — α-Hydroxytriazolam/α-hydroxytriazolam-d$_4$

CD Group[c]	α-Hydroxytriazolam			α-Hydroxytriazolam-d$_4$		
	Ion (m/z)[d]	Rel. int.	Analog's cont.	Ion (m/z)[d]	Rel. int.	Analog's cont.
TMS	<u>415</u>	100	—	<u>419</u>	100	—
t-BDMS	380	6.82	2.38	384	6.46	3.23
	<u>415</u>	100	—	<u>419</u>	100	—
	417	70.1	2.34	421	69.7	1.14

[a–d] See the corresponding footnotes in Table I-1a.

Table V-19. Relative intensity and cross-contribution data[a] of ions[b] with potential for designating the analyte and the adapted internal standard — Mianserin/mianserin-d$_3$

CD Group[c]	Mianserin			Mianserin-d$_3$		
	Ion (m/z)[d]	Rel. int.	Analog's cont.	Ion (m/z)[d]	Rel. int.	Analog's cont.
None	204	76.6	4.60	209	71.4	0.16
	273	58.5	4.52	278	30.4	2.78
	<u>264</u>	46.0	0.88	<u>267</u>	46.5	0.31

[a–d] See the corresponding footnotes in Table I-1a.

Table V — Antianxiety Agents

Table V-20. Relative intensity and cross-contribution data[a] of ions[b] with potential for designating the analyte and the adapted internal standard — Methaqualone/methaqualone-d$_7$

CD Group[c]	Methaqualone			Methaqualone-d$_7$		
	Ion (m/z)[d]	Rel. int.	Analog's cont.	Ion (m/z)[d]	Rel. int.	Analog's cont.
None	<u>235</u>	100	—	242	100	—

[a-d] See the corresponding footnotes in Table I-1a.

Table V-21. Relative intensity and cross-contribution data[a] of ions[b] with potential for designating the analyte and the adapted internal standard — Haloperidol/haloperidol-d$_4$

CD Group[c]	Haloperidol			Haloperidol-d$_4$		
	Ion (m/z)[d]	Rel. int.	Analog's cont.	Ion (m/z)[d]	Rel. int.	Analog's cont.
TMS	<u>123</u>	5.05	—	<u>127</u>	4.44	—

[a-d] See the corresponding footnotes in Table I-1a.

**Summary of Drug, Isotopic Analogs, and Chemical Derivatization Groups Included in
Table VI (Antidepressants)**

Compound	Isotopic analog	Chemical derivatization group	Table #
Imipramine	d₃	None	VI-1
Desipramine	d₃	None, acetyl, TCA, TFA, PFP, 4-CB, TMS, *t*-BDMS	VI-2
Trimipramine	d₃	None	VI-3
Clomipramine	d₃	None	VI-4
Nortriptyline	d₃	None, acetyl, TCA, TFA, PFP, HFB, 4-CB, TMS, *t*-BDMS	VI-5
Protriptyline	d₃	None, acetyl, TCA, TFA, PFP, HFB, 4-CB, TMS, *t*-BDMS	VI-6
Doxepin	d₃	None	VI-7
Dothiepin	d₃	None	VI-8
Amitriptyline	d₃	None	VI-9
Maprotiline	d₃	None, acetyl, TCA, TFA, PFP, HFB, 4-CB, TMS, *t*-BDMS	VI-10

Table VI — Antidepressants

Appendix Two — Table VI
Cross-Contributions Between Ions Designating the Drugs and Their Isotopically Labeled Analogs
in Various Derivatization Forms — Antidepressants

Table VI — Antidepressants

Table VI-1. Relative intensity and cross-contribution data[a] of ions[b] with potential for designating the analyte and the adapted internal standard — Imipramine/imipramine-d$_3$

CD Group[c]	Imipramie Ion (m/z)[d]	Rel. int.	Analog's cont.	Imipramine-d$_3$ Ion (m/z)[d]	Rel. int.	Analog's cont.
None	58	89.7	1.75	61	73.3	0.85
	85	47.4	1.55	88	37.7	1.40
	265	32.0	3.43	268	0.40	0.26
	280	20.6	0.79	283	21.9	0.16

[a–d] See the corresponding footnotes in Table I-1a.

Table VI-2. Relative intensity and cross-contribution data[a] of ions[b] with potential for designating the analyte and the adapted internal standard — Desipramine/desipramine-d$_3$

CD Group[c]	Desipramine Ion (m/z)[d]	Rel. int.	Analog's cont.	Desipramine-d$_3$ Ion (m/z)[d]	Rel. int.	Analog's cont.
None	44	46.4	—	47	30.0	—
	266	31.7	1.32	269	29.4	0.12
Acetyl	114	26.3	1.61	117	26.1	4.81
	308	36.4	0.99	311	34.2	0.22
TCA	303	1.42	0.93	306	1.43	1.69
	410	13.0	—	413	12.3	—
TFA	362	31.8	1.49	365	25.6	0.21
PFP	412	29.3	0.11	415	32.7	0.24
	413	7.06	4.00	416	8.03	0.08
4-CB	276	1.01	1.79	279	0.87	1.06
	294	1.42	1.25	297	1.20	0.50
	322	7.67	0.67	325	6.05	0.40
	501	0.47	4.70	504	0.37	2.76
	516	20.6	0.56	519	16.3	0.63
	517	5.96	3.99	520	4.77	0.22
TMS	116	44.8	—	119	41.6	—
	143	43.6	3.36	146	41.0	0.59
	338	13.5	2.03	341	10.7	1.44
t-BDMS	102	100	1.31	105	100	0.46
	380	7.84	1.06	383	6.94	1.93

[a–d] See the corresponding footnotes in Table I-1a.

Table VI-3. Relative intensity and cross-contribution data[a] of ions[b] with potential for designating the analyte and the adapted internal standard — Trimipramine/trimipramine-d$_3$

CD Group[c]	Trimipramine Ion (m/z)[d]	Rel. int.	Analog's cont.	Trimipramine-d$_3$ Ion (m/z)[d]	Rel. int.	Analog's cont.
None	58	100	1.86	61	17.5	0.01
	84	16.9	3.16	87	17.1	2.08
	294	16.3	1.61	297	21.9	0.18

[a–d] See the corresponding footnotes in Table I-1a.

Table VI-4. Relative intensity and cross-contribution data[a] of ions[b] with potential for designating the analyte and the adapted internal standard — Clomipramine/clomipramine-d₃

	Clomipramine			Clomipramine-d₃		
CD Group[c]	Ion (m/z)[d]	Rel. int.	Analog's cont.	Ion (m/z)[d]	Rel. int.	Analog's cont.
None	<u>58</u>	100	2.53	<u>61</u>	100	0.12
	85	47.0	2.86	88	48.9	1.37

[a–d] See the corresponding footnotes in Table I-1a.

Table VI-5. Relative intensity and cross-contribution data[a] of ions[b] with potential for designating the analyte and the adapted internal standard — Nortriptyline/nortriptyline-d₃

	Nortriptyline			Nortriptyline-d₃		
CD Group[c]	Ion (m/z)[d]	Rel. int.	Analog's cont.	Ion (m/z)[d]	Rel. int.	Analog's cont.
None	<u>44</u>	100	2.26	<u>47</u>	100	0.02
Acetyl	<u>44</u>	30.6	1.11	<u>47</u>	42.3	0.00
	305	7.33	0.45	308	6.92	0.71
	306	1.89	1.99	309	1.74	0.55
TCA	301	0.80	4.92	304	0.75	2.00
	372	0.24	1.40	375	0.23	4.67
	<u>407</u>	1.23	—	<u>410</u>	1.10	—
TFA	320	0.09	1.52	323	0.08	4.44
	<u>359</u>	2.47	0.50	<u>362</u>	2.18	0.19
PFP	<u>409</u>	1.63	0.70	<u>412</u>	1.49	0.25
HFB	<u>240</u>	10.4	0.32	<u>243</u>	10.6	0.60
	459	1.02	0.80	462	1.22	0.34
4-CB	<u>294</u>	3.45	—	<u>297</u>	3.25	—
	468	0.61	0.61	471	0.43	0.59
	513	0.50	1.00	516	0.34	0.79
TMS	<u>116</u>	100	0.45	<u>119</u>	100	0.27
	117	12.1	4.33	120	11.9	0.11
	320	2.68	0.39	323	3.17	1.27
	334	0.16	3.02	337	0.20	2.77
t-BDMS	102	5.15	2.95	105	5.22	2.94
	<u>158</u>	100	0.04	<u>161</u>	100	0.49
	159	16.4	1.17	162	16.4	0.30
	320	4.08	1.06	323	3.46	1.21
	362	1.81	0.13	365	1.48	1.61
	376	0.29	1.14	379	0.25	1.48

[a–d] See the corresponding footnotes in Table I-1a.

Table VI-6. Relative intensity and cross-contribution data[a] of ions[b] with potential for designating the analyte and the adapted internal standard — Protriptyline/protriptyline-d₃

	Protriptyline			Protriptyline-d₃		
CD Group[c]	Ion (m/z)[d]	Rel. int.	Analog's cont.	Ion (m/z)[d]	Rel. int.	Analog's cont.
None	44	40.6	1.87	47	57.3	0.03
	<u>70</u>	61.7	2.30	<u>73</u>	67.4	0.05
Acetyl	<u>114</u>	5.81	1.79	<u>117</u>	5.91	2.03
	305	10.6	0.57	308	8.92	0.38
	306	2.55	0.76	309	2.18	0.83
TCA	299	0.37	1.03	302	0.38	4.78
	<u>407</u>	2.25	—	<u>410</u>	2.10	—

Table VI — Antidepressants

Table VI-6. (Continued)

CD Group[c]	Protriptyline			Protriptyline-d[3]		
	Ion (m/z)[d]	Rel. int.	Analog's cont.	Ion (m/z)[d]	Rel. int.	Analog's cont.
TFA	345	0.06	3.64	348	0.06	0.00
	<u>359</u>	4.71	0.15	<u>362</u>	5.14	0.24
	360	1.10	0.47	363	1.22	0.07
PFP	336	0.05	1.58	369	0.05	0.74
	395	0.08	3.18	398	0.07	0.00
	<u>409</u>	5.00	0.14	<u>412</u>	3.81	0.25
	410	1.23	0.58	413	0.92	0.07
HFB	<u>240</u>	2.68	0.30	<u>243</u>	2.64	0.21
	268	0.71	0.60	271	0.70	1.21
	445	0.07	1.97	448	0.07	0.00
	459	5.29	0.06	462	5.61	0.35
	460	1.39	0.34	463	1.46	0.11
4-CB	322	1.17	1.92	325	0.96	0.87
	468	0.80	0.78	471	9.56	0.56
	<u>513</u>	4.21	0.74	<u>516</u>	2.88	0.49
	514	1.24	0.91	517	0.85	0.27
TMS	<u>116</u>	100	2.33	<u>119</u>	100	0.32
	142	51.3	1.98	145	52.7	0.39
	320	3.95	2.86	323	4.37	1.36
	335	2.46	2.18	338	2.87	2.16
t-BDMS	158	28.0	1.15	161	28.1	0.81
	184	41.3	1.06	187	42.4	1.75
	<u>320</u>	82.2	1.03	<u>323</u>	81.6	1.25
	321	24.2	1.52	324	24.4	0.47
	377	2.38	1.15	380	2.36	1.79

[a–d] See the corresponding footnotes in Table I-1a.

Table VI-7. Relative intensity and cross-contribution data[a] of ions[b] with potential for designating the analyte and the adapted internal standard — Doxepin/doxepin-d[3]

CD Group[c]	Doxepin			Doxepin-d[3]		
	Ion (m/z)[d]	Rel. int.	Analog's cont.	Ion (m/z)[d]	Rel. int.	Analog's cont.
None	<u>58</u>	100	1.86	<u>61</u>	17.5	0.01
	84	16.9	3.16	87	17.1	2.08
	294	16.3	1.61	297	21.9	0.18

[a–d] See the corresponding footnotes in Table I-1a.

Table VI-8. Relative intensity and cross-contribution data[a] of ions[b] with potential for designating the analyte and the adapted internal standard — Dothiepin/dothiepin-d[3]

CD Group[c]	Dothiepin			Dothiepin-d[3]		
	Ion (m/z)[d]	Rel. int.	Analog's cont.	Ion (m/z)[d]	Rel. int.	Analog's cont.
None	<u>58</u>	100	1.84	<u>61</u>	100	0.04

[a–d] See the corresponding footnotes in Table I-1a.

Table VI-9. Relative intensity and cross-contribution data[a] of ions[b] with potential for designating the analyte and the adapted internal standard — Amitriptyline/amitriptyline-d₃

CD Group[c]	Amitriptyline			Amitriptyline-d₃		
	Ion (m/z)[d]	Rel. int.	Analog's cont.	Ion (m/z)[d]	Rel. int.	Analog's cont.
None	<u>58</u>	100	0.38	<u>61</u>	100	0.02

[a–d] See the corresponding footnotes in Table I-1a.

Table VI-10. Relative intensity and cross-contribution data[a] of ions[b] with potential for designating the analyte and the adapted internal standard — Maprotiline/maprotiline-d₃

CD Group[c]	Maprotiline			Maprotiline-d₃		
	Ion (m/z)[d]	Rel. int.	Analog's cont.	Ion (m/z)[d]	Rel. int.	Analog's cont.
None	<u>44</u>	100	—	<u>47</u>	100	—
Acetyl	100	10.5	2.87	103	11.5	1.81
	<u>291</u>	100	0.52	<u>294</u>	100	0.21
	292	23.5	0.66	295	23.7	0.11
	319	33.3	0.63	322	3.10	0.30
TCA	304	4.94	1.22	307	5.18	0.57
	<u>393</u>	100	—	<u>396</u>	100	—
TFA	<u>345</u>	100	0.14	<u>348</u>	100	0.21
	346	23.5	0.35	349	23.7	0.06
	373	1.25	0.21	376	1.18	0.30
PFP	304	0.82	0.29	307	0.80	0.74
	<u>395</u>	100	0.09	<u>398</u>	100	0.23
	396	24.1	0.32	399	24.3	0.07
	423	0.95	0.00	426	0.93	0.31
HFB	240	4.52	0.74	243	4.72	1.10
	254	3.61	0.48	257	3.56	0.23
	304	0.46	0.37	307	0.48	1.97
	<u>445</u>	100	0.06	<u>448</u>	100	0.26
	446	25.6	0.28	449	26.4	0.07
	473	0.77	0.00	476	0.77	0.32
4-CB	482	1.90	1.01	485	1.89	1.45
	<u>499</u>	100	0.75	<u>502</u>	100	0.47
	500	29.2	1.00	503	29.3	0.25
TMS	<u>116</u>	100	0.65	<u>119</u>	100	0.25
	277	3.26	1.76	280	3.08	0.48
	334	2.20	2.32	337	2.38	1.33
	349	10.2	0.28	352	10.8	1.36
t-BDMS	102	50.8	0.87	105	48.2	0.27
	158	43.1	0.71	161	41.9	0.46
	306	17.7	0.32	309	17.7	0.98
	<u>334</u>	100	0.07	<u>337</u>	100	1.24
	335	30.4	3.42	338	31.0	0.42
	391	8.13	0.28	394	8.58	1.61

[a–d] See the corresponding footnotes in Table I-1a.

Table VI — Antidepressants

**Summary of Drugs, Isotopic Analogs, and Chemical Derivatization Groups Included in
Table VII (Others)**

Compound	Isotopic analog	Chemical derivatization group	Table #
Diphenhydramine	d_3	None	VII-1
Cotinine	d_3	None	VII-2
Nicotine	d_4	None	VII-3
5-α-Estran-3α-ol-17-one	d_3	None, acetyl, TMS	VII-4
5-β-Estran-3α-ol-17-one	d_3	None, acetyl, TMS	VII-5
Stanozolol	d_3	None, acetyl, [TMS]$_2$, t-BDMS	VII-6
3-Hydroxystanozolol	d_3	[TMS]$_2$, [t-BDMS]$_2$	VII-7
Promethazine	d_3	None	VII-8
Chlorpromazine	d_3	None	VII-9
Acetaminophen	d_4	None, [acetyl]$_2$, TCA, TFA, PFP, HFB, 4-CB, TMS, [TMS]$_2$, t-BDMS, [t-BDMS]$_2$	VII-10
Clonidine	d_4	None, acetyl, [acetyl]$_2$, TMS, [TMS]$_2$, [t-BDMS]$_2$	VII-11
Chloramphenicol	d_5	None, [acetyl]$_2$, TMS, [TMS]$_2$	VII-12
Melatonin	d_7	None, acetyl, TFA, PFP, HFB, TMS	VII-13

Table VII — Others

Appendix Two — Table VII
Cross-Contributions Between Ions Designating the Drugs and Their Isotopically Labeled Analogs
in Various Derivatization Forms — Others

Table VII — Others

Table VII-1. Relative intensity and cross-contribution data[a] of ions[b] with potential for designating the analyte and the adapted internal standard — Diphenhydramine/diphenhydramine-d_3

CD Group[c]	Diphenhydramine			Diphenhydramine-d_3		
	Ion (m/z)[d]	Rel. int.	Analog's cont.	Ion (m/z)[d]	Rel. int.	Analog's cont.
None	<u>58</u>	100	0.93	<u>61</u>	100	0.09

[a–d] See the corresponding footnotes in Table I-1a.

Table VII-2. Relative intensity and cross-contribution data[a] of ions[b] with potential for designating the analyte and the adapted internal standard — Cotinine/cotinine-d_3

CD Group[c]	Cotinine			Cotinine-d_3		
	Ion (m/z)[d]	Rel. int.	Analog's cont.	Ion (m/z)[d]	Rel. int.	Analog's cont.
None	<u>98</u>	100	1.82	<u>101</u>	100	0.25
	175	10.0	2.04	178	11.0	3.00
	176	37.0	1.89	179	41.0	0.25

[a–d] See the corresponding footnotes in Table I-1a.

Table VII-3. Relative intensity and cross-contribution data[a] of ions[b] with potential for designating the analyte and the adapted internal standard — Nicotine/nicotine-d_4

CD Group[c]	Nicotine			Nicotine-d_4		
	Ion (m/z)[d]	Rel. int.	Analog's cont.	Ion (m/z)[d]	Rel. int.	Analog's cont.
None	92	7.00	17.0	96	6.00	0.13
	<u>133</u>	36.0	—	<u>136</u>	22.0	—
	161	21.0	2.12	165	19.0	0.02

[a–d] See the corresponding footnotes in Table I-1a.

Table VII-4. Relative intensity and cross-contribution data[a] of ions[b] with potential for designating the analyte and the adapted internal standard — 5-α–Estran–3α–ol-17-one/5-α–estran–3α–ol-17-one-d_3

CD Group[c]	5-α-Estran-3α-ol-17-one			5-α-Estran-3α-ol-17-one-d_3		
	Ion (m/z)[d]	Rel. int.	Analog's cont.	Ion (m/z)[d]	Rel. int.	Analog's cont.
None	<u>276</u>	100	2.47	<u>279</u>	100	0.21
Acetyl	<u>258</u>	100	1.05	<u>261</u>	100	0.53
	318	7.60	1.19	321	18.3	0.53
TMS	<u>129</u>	34.7	—	<u>130</u>	43.3	—
	333	100	0.45	336	87.4	1.91
	348	25.6	0.83	351	11.4	4.03

[a–d] See the corresponding footnotes in Table I-1a.

Table VII-5. Relative intensity and cross-contribution data[a] of ions[b] with potential for designating the analyte and the adapted internal standard — 5-β–Estran–3α–ol–17-one/5-β–estran–3α–ol–17-one-d3

CD Group[c]	5-β-Estran-3α-ol-17-one			5-β-Estran-3α-ol-17-one-d3		
	Ion (m/z)[d]	Rel. int.	Analog's cont.	Ion (m/z)[d]	Rel. int.	Analog's cont.
None	<u>276</u>	100	—	<u>279</u>	95.9	—
Acetyl	<u>258</u>	100	1.77	<u>261</u>	67.5	0.56
	318	33.9	1.18	321	24.9	0.24
TMS	<u>216</u>	86.8	—	<u>217</u>	100	—
	258	100	3.73	261	76.4	0.43
	333	47.3	1.06	336	34.5	2.10
	348	8.08	2.05	351	6.12	1.99

[a–d] See the corresponding footnotes in Table I-1a.

Table VII-6. Relative intensity and cross-contribution data[a] of ions[b] with potential for designating the analyte and the adapted internal standard — Stanozolol/stanozolol-d3

CD Group[c]	Stanozolol			Stanozolol-d3		
	Ion (m/z)[d]	Rel. int.	Analog's cont.	Ion (m/z)[d]	Rel. int.	Analog's cont.
None	<u>328</u>	61.5	—	<u>331</u>	59.2	—
Acetyl	<u>327</u>	21.2	—	<u>330</u>	20.7	—
[TMS]2	<u>143</u>	100	3.49	<u>146</u>	100	2.02
	473	7.60	4.98	476	7.62	3.55
t-BDMS	358	12.0	3.49	361	11.7	1.44
	<u>386</u>	100	1.97	<u>389</u>	100	1.33
	387	30.4	4.34	390	31.6	0.56
	442	7.47	1.97	445	7.19	2.28

[a–d] See the corresponding footnotes in Table I-1a.

Table VII-7. Relative intensity and cross-contribution data[a] of ions[b] with potential for designating the analyte and the adapted internal standard — 3-Hydroxystanozolol/3-hydroxystanozolol-d3

CD Group[c]	3-Hydroxystanozolol			3-Hydroxystanozolol-d3		
	Ion (m/z)[d]	Rel. int.	Analog's cont.	Ion (m/z)[d]	Rel. int.	Analog's cont.
[TMS]2	<u>473</u>	96.2	—	<u>476</u>	100	—
	488	36.8	3.52	491	39.4	3.94
	489	15.1	4.63	492	15.7	1.58
[t-BDMS]2	459	26.0	2.55	462	24.6	4.23
	<u>515</u>	100	—	<u>518</u>	100	—
	516	71.4	1.31	519	71.5	2.63
	517	28.5	3.10	520	28.4	0.77

[a–d] See the corresponding footnotes in Table I-1a.

Table VII — Others

Table VII-8. Relative intensity and cross-contribution data[a] of ions[b] with potential for designating the analyte and the adapted internal standard — Promethazine/promethazine-d$_3$

CD Group[c]	Promethazine			Promethazine-d$_3$		
	Ion (m/z)[d]	Rel. int.	Analog's cont.	Ion (m/z)[d]	Rel. int.	Analog's cont.
None	<u>72</u>	100	1.18	<u>75</u>	100	0.58
	284	6.91	1.51	287	6.49	1.21

[a-d] See the corresponding footnotes in Table I-1a.

Table VII-9. Relative intensity and cross-contribution data[a] of ions[b] with potential for designating the analyte and the adapted internal standard — Chlorpromazine/chlorpromazine-d$_3$

CD Group[c]	Chlorpromazine			Chlorpromazine-d$_3$		
	Ion (m/z)[d]	Rel. int.	Analog's cont.	Ion (m/z)[d]	Rel. int.	Analog's cont.
None	<u>58</u>	100	2.26	<u>61</u>	100	0.36

[a-d] See the corresponding footnotes in Table I-1a.

Table VII-10. Relative intensity and cross-contribution data[a] of ions[b] with potential for designating the analyte and the adapted internal standard — Acetaminophen/acetaminophen-d$_4$

CD Group[c]	Acetaminophen			Acetaminophen-d$_4$		
	Ion (m/z)[d]	Rel. int.	Analog's cont.	Ion (m/z)[d]	Rel. int.	Analog's cont.
None	<u>109</u>	100	—	<u>112</u>	100	—
[Acetyl]$_2$	<u>109</u>	100	0.40	<u>113</u>	100	0.02
	151	53.5	0.15	155	56.0	0.01
	193	36.4	0.09	197	38.2	0.02
	235	3.09	0.23	239	3.33	0.05
TCA	<u>108</u>	100	0.32	<u>112</u>	100	1.48
	134	18.0	0.83	138	19.1	0.89
	255	36.5	0.40	259	39.2	3.49
	297	15.6	0.46	301	17.2	3.59
TFA	<u>108</u>	100	0.76	<u>112</u>	100	0.04
	205	63.9	0.56	209	63.7	0.07
	206	6.14	1.10	210	6.23	0.10
	247	33.7	0.54	251	34.9	0.04
PFP	<u>108</u>	100	0.46	<u>112</u>	100	0.20
	208	3.72	0.65	212	3.60	0.37
	236	2.86	0.64	240	2.96	4.24
	255	65.7	0.27	259	62.9	0.14
	297	36.4	0.29	301	35.6	0.11
HFB	80	9.25	4.39	84	8.96	0.56
	<u>108</u>	100	0.40	<u>112</u>	100	0.14
	134	5.89	4.63	138	5.96	0.46
	286	3.27	0.26	290	3.18	0.35
	305	50.4	0.21	309	49.0	0.07
	347	27.9	0.20	351	27.6	0.07
4-CB	<u>108</u>	100	0.94	<u>112</u>	100	0.04
	243	3.43	1.04	247	3.28	0.04
	359	34.1	1.02	363	33.9	0.01
	401	19.1	0.75	405	19.3	0.01

Table VII-10. (Continued)

CD Group[c]	Acetaminophen			Acetaminophen-d$_4$		
	Ion (m/z)[d]	Rel. int.	Analog's cont.	Ion (m/z)[d]	Rel. int.	Analog's cont.
TMS	166	84.3	1.39	170	78.7	0.33
	181	100	0.47	185	100	0.34
	208	23.9	0.40	212	23.7	0.26
	223	75.9	0.39	227	73.9	0.20
[TMS]$_2$	181	26.2	2.86	185	27.1	0.18
	206	100	0.39	210	100	0.43
	223	17.7	1.55	227	18.9	0.32
	280	87.3	0.23	284	89.0	0.40
	295	64.5	0.25	299	65.3	0.57
t-BDMS	166	11.4	2.49	170	11.6	0.38
	208	100	0.12	212	100	0.12
	209	22.0	0.28	213	21.6	0.05
	250	2.46	0.55	254	2.63	0.03
	265	31.0	0.06	269	31.6	0.03
[t-BDMS]$_2$	223	14.0	0.71	227	14.2	0.42
	248	44.2	0.24	252	43.1	4.32
	308	17.9	0.37	312	18.2	0.34
	322	100	0.13	326	100	0.47
	323	35.1	0.75	327	34.9	0.30
	379	5.93	0.14	383	6.06	1.85

[a–d] See the corresponding footnotes in Table I-1a.

Table VII-11. Relative intensity and cross-contribution data[a] of ions[b] with potential for designating the analyte and the adapted internal standard — Clonidine/clonidine-d$_4$

CD Group[c]	Clonidine			Clonidine-d$_4$		
	Ion (m/z)[d]	Rel. int.	Analog's cont.	Ion (m/z)[d]	Rel. int.	Analog's cont.
None	194	32.8	2.90	198	32.2	0.60
	229	100	1.45	233	100	10.9
Acetyl	194	95.9	2.83	198	99.9	0.37
	208	9.27	3.33	212	10.6	1.87
	236	100	2.89	240	100	0.48
	238	32.7	2.82	242	32.3	0.23
[Acetyl]$_2$	194	87.4	0.79	198	86.7	0.25
	236	100	0.82	240	100	0.34
	238	33.7	0.72	242	33.2	0.46
	278	71.8	0.28	282	71.7	0.56
	280	24.3	0.23	284	23.9	0.03
TMS	142	38.8	2.74	146	44.7	4.70
	266	100	1.35	270	100	2.92
	268	36.9	3.46	272	37.6	1.42
[TMS]$_2$	214	36.0	1.34	218	40.2	2.83
	322	17.3	0.46	326	14.4	4.43
	338	100	0.89	342	100	3.74
	340	42.5	2.73	344	42.7	0.44
[t-BDMS]$_2$	252	100	0.44	256	100	1.81
	254	39.8	4.23	258	39.9	0.24

[a–d] See the corresponding footnotes in Table I-1a.

Table VII — Others

Table VII-12. Relative intensity and cross-contribution data[a] of ions[b] with potential for designating the analyte and the adapted internal standard — Chloramphenicol/chloramphenicol-d$_5$

CD Group[c]	Chloramphenicol			Chloramphenicol-d$_5$		
	Ion (m/z)[d]	Rel. int.	Analog's cont.	Ion (m/z)[d]	Rel. int.	Analog's cont.
None	115	26.2	—	120	17.9	—
	162	23.9	3.25	167	22.4	0.11
[Acetyl]$_2$	153	100	—	158	90.2	—
TMS	224	99.6	0.79	229	100	0.19
	225	17.2	3.37	230	17.1	1.28
	235	11.9	1.49	240	12.4	0.88
	252	14.1	1.85	257	15.7	0.36
	321	14.3	0.67	326	15.6	0.71
	351	8.39	0.43	356	9.79	1.13
[TMS]$_2$	297	79.8	0.21	302	84.1	0.12
	298	20.3	1.62	303	21.0	0.42

[a–d] See the corresponding footnotes in Table I-1a.

Table VII-13. Relative intensity and cross-contribution data[a] of ions[b] with potential for designating the analyte and the adapted internal standard — Melatonin/melatonin-d$_7$

CD Group[c]	Melatonin			Melatonin-d$_7$		
	Ion (m/z)[d]	Rel. int.	Analog's cont.	Ion (m/z)[d]	Rel. int.	Analog's cont.
None	160	100	4.17	162	100	0.84
	173	97.9	3.19	176	79.3	0.41
	174	13.1	4.54	177	26.5	0.34
	232	20.8	2.25	239	18.5	0.25
Acetyl	117	14.4	4.53	119	14.5	0.84
	145	18.7	3.42	147	17.6	3.11
	160	98.3	0.81	162	100	0.62
	173	100	0.26	176	92.4	0.12
	174	14.5	1.24	177	16.6	0.13
	274	15.6	0.88	281	15.1	2.90
TFA	144	11.3	1.93	146	13.6	0.29
	159	19.9	1.60	161	21.5	1.49
	256	14.8	2.03	258	17.1	1.37
	269	100	0.63	272	100	0.07
	270	15.0	2.15	273	22.8	0.02
	328	9.55	1.00	335	9.49	0.00
PFP	159	24.1	2.95	161	23.6	3.60
	306	15.4	2.26	308	17.0	2.39
	319	100	0.52	322	100	0.54
	320	15.9	1.58	323	23.5	0.43
HFB	159	25.8	2.64	161	27.0	1.97
	356	13.5	1.41	358	14.4	1.64
	359	100	0.43	372	100	0.15
	428	8.17	0.76	435	5.81	0.68
TMS	232	100	—	234	100	—
	245	74.1	0.68	248	73.8	0.94
	246	16.6	2.31	249	19.0	0.42
	304	18.5	0.70	311	12.6	0.20

[a–d] See the corresponding footnotes in Table I-1a.

Index

N

O

P

Q

R

S